BUSINESS MATHEMATICS AND STATISTICS

BUSINESS MATHEMATICS AND STATISTICS

G SRINIVASA
MSc.
Former Head
Department of Mathematics
T John College
Bangalore, Karnataka

DEEPA GEORGE
MSc., MPhil.
Former Lecturer
Department of Statistics
Christ College
Bangalore, Karnataka

Publishing Globally

NEW AGE INTERNATIONAL (P) LIMITED, PUBLISHERS
LONDON • NEW DELHI • NAIROBI
Bangalore • Chennai • Cochin • Guwahati • Hyderabad • Kolkata • Lucknow • Mumbai
Visit us at www.newagepublishers.com

Published by New Age International (P) Ltd., Publishers
First Edition: 2002
Reprint: 2020

GLOBAL OFFICES

- **New Delhi** **NEW AGE INTERNATIONAL (P) LIMITED, PUBLISHERS**
  7/30 A, Daryaganj, New Delhi-110002, (INDIA)
  Tel.: (011) 23253771, 23253472, **Telefax:** 23267437, 43551305
  E-mail: contactus@newagepublishers.com • Visit us at www.newagepublishers.com
- **London** **NEW AGE INTERNATIONAL (UK) LTD.**
  27 Old Gloucester Street, London, WC1N 3AX, UK
  E-mail: info@newacademicscience.co.uk • Visit us at www.newacademicscience.co.uk
- **Nairobi** **NEW AGE GOLDEN (EAST AFRICA) LTD.**
  Ground Floor, Westlands Arcade, Chiromo Road (Next to Naivas Supermarket)
  Westlands, Nairobi, KENYA, **Tel.:** 00-254-713848772, 00-254-725700286
  E-mail: kenya@newagepublishers.com

BRANCHES

- **Bangalore** 37/10, 8th Cross (Near Hanuman Temple), Azad Nagar, Chamarajpet, Bangalore- 560 018
  Tel.: (080) 26756823, **Telefax:** 26756820, **E-mail: bangalore@newagepublishers.com**
- **Chennai** 26, Damodaran Street, T. Nagar, Chennai-600 017, **Tel.:** (044) 24353401
  Telefax: 24351463, **E-mail: chennai@newagepublishers.com**
- **Cochin** CC-39/1016, Carrier Station Road, Ernakulam South, Cochin-682 016
  Tel.: (0484) 2377303, **Telefax:** 4051304, **E-mail: cochin@newagepublishers.com**
- **Guwahati** Hemsen Complex, Mohd. Shah Road, Paltan Bazar, Near Starline Hotel
  Guwahati-781 008, **Tel.:** (0361) 2513881, **Telefax:** 2543669
  E-mail: guwahati@newagepublishers.com
- **Hyderabad** 105, 1st Floor, Madhiray Kaveri Tower, 3-2-19, Azam Jahi Road, Near Kumar Theater
  Nimboliadda Kachiguda, Hyderabad-500 027, **Tel.:** (040) 24652456, **Telefax:** 24652457
  E-mail: hyderabad@newagepublishers.com
- **Kolkata** RDB Chambers (Formerly Lotus Cinema) 106A, 1st Floor, S N Banerjee Road
  Kolkata-700 014, **Tel.:** (033) 22273773, **Telefax:** 22275247
  E-mail: kolkata@newagepublishers.com
- **Lucknow** 16-A, Jopling Road, Lucknow-226 001, **Tel.:** (0522) 2209578, 4045297, **Telefax:** 22040
  E-mail: lucknow@newagepublishers.com
- **Mumbai** 142C, Victor House, Ground Floor, N.M. Joshi Marg, Lower Parel, Mumbai-400 013
  Tel.: (022) 24927869, **Telefax:** 24915415, **E-mail: mumbai@newagepublishers.com**
- **New Delhi** 22, Golden House, Daryaganj, New Delhi-110 002, **Tel.:** (011) 23262368, 23262370
  Telefax: 43551305, **E-mail: sales@newagepublishers.com**

ISBN: 978-81-224-1433-2
C-19-09-12234

Printed in India at Ajit Printing Press, Delhi.

NEW AGE INTERNATIONAL (P) LIMITED, PUBLISHERS
7/30 A, Daryaganj, New Delhi-110002
Visit us at **www.newagepublishers.com**
(CIN: U74899DL1966PTC004618)

To

our

beloved parents

Preface

This book has been written as a comprehensive text to satisfy the requirement of the new scheme in "Business Mathematics and Statistics—Paper IV" for the first year B.B.M. students of Bangalore University.

The basic concepts and fundamentals have been explained in a lucid manner and a wide variety of completely solved problems have been included, many problems are included along with their solutions. The other important features of this text book are a list of formulae and previous years' examination question papers placed in appropriate sections.

We are sure that the students and faculty of Business Management will find this book "good learning material".

We are extremely grateful to the student body who has given us valuable suggestions, Wg. Cdr. D.A.K. Yohan (Retd.), Principal, T. John College, Mr. Philip Machado, Vice-principal, also of T. John College, Bangalore, Mr. P.S. Shivakumar, Head of Department of Statistics, Staff and Management of Christ College, and well wishers for their support and encouragement. We would also like to express our gratitude to our families.

We acknowledge the efforts and co-operation of our publishers, M/s New Age International Publishers and Mr. V.R. Babu, Executive, Bangalore Division in bringing out this book.

Any suggestions towards the improvement of this book will be highly appreciated and gratefully acknowledged.

G. Srinivasa
Head of Department, Mathematics
T. John College

Deepa George
Lecturer, Department of Statistics
Christ College

Contents

Part I

Business Mathematics

Internal Marks: 5
Examination Marks: 45
Total Marks: 50

Isaac Barrow
(1630–1677)

"Mathematics–the unshaken foundation of sciences, and the plentiful fountain of advantage to human affairs".

–Isaac Barrow

1

Permutations and Combinations

Contents

INTRODUCTION

Combinatorial analysis involves determining the number of possibilities of some event without enumerating all the possibilities. In order to develop the general procedure for obtaining possibilities we have to introduce the concepts called permutation and combinations.

Two Fundamental Rules

1. The Sum Rule

Suppose a work A can occur in m ways and B can occur in n ways and both cannot occur simultaneously. Then A or B (at least one of them) can occur in $(m + n)$ ways. This rule is also applicable for two or more exclusive events.

2. The Product Rule

Suppose there are two works A and B. Let A can occur in m ways and B in n ways. Suppose that the ways for A and B are not related, in the sense that B occur in n ways regardless the outcome of A. Then both A and B occur in mn ways.

Example: Let there be two questions A and B which can be solved by two methods and 3 methods respectively. Then A or B can be solved in $2 + 3 = 5$ ways and both A and B in $2 \times 3 = 6$ ways.

PERMUTATIONS AND COMBINATIONS

Definition: The number of ways of arrangements of 'n' different things taken 'r' at a time $(r \leq n)$ is called the number of permutations of 'n' different things taken 'r' at a time and it is written as $^{n}P_r$ or $_{n}P_r$.

or

An arrangement without replacement is called a permutation.

Example: Arrangements of cards, number problems without repetitions.

Definition: The number of ways of selection of 'n' different things taken 'r' at a time $(r \leq n)$ is called the number of combinations of 'n' different things taken 'r' at a time and it is written as $^{n}C_r$ or $_{n}C_r$.

or

A selection of r-objects without replacement is called an r-combination (or combination).

Factorial Notation

The product of the first 'n' natural numbers i.e., 1.2.3.4... $(n - 1)$ n is represented by the symbol $\lfloor n$ or $n!$ and it is called 'factorial n'.

Note:

1. To find the number of arrangements of 'n' objects taken all at a time, of p things are alike and the rest are all different. i.e., no. of arrangements $= \dfrac{\lfloor n}{\lfloor p}$.
2. To find the number of arrangements of 'n' things taken all at a time, whose 'p' things are alike of one kind, 'q' things are alike of another kind and the rest are different. i.e., no. of arrangements $= \dfrac{\lfloor n}{\lfloor p \cdot \lfloor q \cdot \lfloor r}$.

List of Formulae:

1. $\lfloor o$ or $o! = 1$
2. $^{n}P_r or\ _{n}P_r = \dfrac{\lfloor n}{\lfloor n - r}$ or $\dfrac{n!}{(n-r)!}$
3. $^{n}P_n = \lfloor n$
4. $\lfloor n = n(n-1)(n-2)(n-3)...1$

5. $^nP_r = n(n-1)(n-2)(n-3)\ldots[n-(r-1)]$

6. ^nC_r or $_nC_r = \dfrac{^nP_r}{r!}$ or $\dfrac{_nP_r}{r!}$

7. $^nC_r = \dfrac{n(n-1)(n-2)(n-3)\ldots[n-(r-1)]}{r!}$

8. $^nC_r = \dfrac{n!}{r!\cdot(n-r)!},\ r = 0, 1, 2, \ldots n.$

9. $^nC_0 = 1$

10. $^nC_n = 1$

11. $^nC_{n-r} = {^nC_r},\ r = 0, 1, 2, \ldots, n.$

12. $^nC_r + {^nC_{r-1}} = {^{n+1}C_r}$

13. $^nC_{n-r} = \dfrac{n!}{(n-r)!\cdot r!},\ r = 0, 1, 2, \ldots, n.$

Solved Examples

Example1. Find the values of

(a) $^{10}P_3$, (b) $^{10}P_6$, (c) $^5P_3$, (d) $^9P_6$, (e) $^{10}P_2$, (f) $^{16}P_5$

Solution:

(a) $^{10}P_3 = 10 \times 9 \times 8 = 720$

(b) $^{10}P_6 = 10 \times 9 \times 8 \times 7 \times 6 \times 5 = 1,51,200$

(c) $^5P_3 = 5 \times 4 \times 3 = 60$

(d) $^9P_6 = 9 \times 8 \times 7 \times 6 \times 5 \times 4 = 60,480$

(e) $^{10}P_2 = 10 \times 9 = 90$

(f) $^{16}P_5 = 16 \times 15 \times 14 \times 13 \times 12 = 5,24,160$

Example 2. Using the formula $^nP_r = \dfrac{\lfloor n}{\lfloor n-r}$, find the values of (a) $^{12}P_2$, (b) $^5P_4$.

Solution:

(a) $$^{12}P_2 = \frac{\lfloor 12}{\lfloor 12-2} = \frac{\lfloor 12}{\lfloor 10} = \frac{12 \times 11 \times \lfloor 10}{\lfloor 10} = 12 \times 11$$

$$\therefore\ ^{12}P_2 = 132$$

(b) $$^5P_4 = \frac{\lfloor 5}{\lfloor 5-4} = \frac{\lfloor 5}{\lfloor 1} = \lfloor 5 = 5 \times 4 \times 3 \times 2 \times 1$$

$$\therefore\ ^5P_4 = 120$$

Example 3. Find the values of

(a) $^8C_3$, (b) $^{12}C_6$, (c) $^{15}C_8$, (d) $^9C_6$, (e) $^8C_5$, (f) $^{14}C_{10}$, (g) $^9C_2$,

(h) $^{10}C_2$, (i) $^{12}C_8$

Solution:

(a) $$^8C_3 = \frac{8 \times 7 \times 6}{3 \times 2 \times 1} = 8 \times 7 = 56$$

(b) $$^{12}C_6 = \frac{12 \times 11 \times 10 \times 9 \times 8 \times 7}{6 \times 5 \times 4 \times 3 \times 2 \times 1} = 11 \times 3 \times 4 \times 7$$

$$\therefore\ ^{12}C_6 = 924$$

(c) $$^{15}C_8 = \frac{15 \times 14 \times 13 \times 12 \times 11 \times 10 \times 9 \times 8}{8 \times 7 \times 6 \times 5 \times 4 \times 3 \times 2 \times 1}$$

$$\therefore\ ^{15}C_8 = 15 \times 13 \times 11 \times 3 = 6{,}435$$

(d) $$^9C_6 = \frac{9 \times 8 \times 7 \times 6 \times 5 \times 4}{6 \times 5 \times 4 \times 3 \times 2 \times 1} = 3 \times 4 \times 7 = 84$$

(e) $$^8C_5 = \frac{8 \times 7 \times 6 \times 5 \times 4}{5 \times 4 \times 3 \times 2 \times 1} = 8 \times 7 = 56$$

(f) $$^{14}C_{10} = \frac{14 \times 13 \times 12 \times 11 \times 10 \times 9 \times 8 \times 7 \times 6 \times 5}{10 \times 9 \times 8 \times 7 \times 6 \times 5 \times 4 \times 3 \times 2 \times 1} = 7 \times 13 \times 11$$

$$\therefore\ ^{14}C_{10} = 1001.$$

(g) $$^9C_2 = \frac{9 \times 8}{2 \times 1} = 9 \times 4 = 36$$

(h) $\quad {}^{10}C_2 = \dfrac{10 \times 9}{2 \times 1} = 5 \times 9 = 45$

(i) $\quad {}^{12}C_8 = \dfrac{12 \times 11 \times 10 \times 9 \times 8 \times 7 \times 6 \times 5}{8 \times 7 \times 6 \times 5 \times 4 \times 3 \times 2 \times 1}$

$= 11 \times 5 \times 9 = 3025$

Example 4. Using the formula ${}^nC_r = \dfrac{\lfloor n}{\lfloor r \cdot \lfloor n - r}$, find the values of

(a) ${}^5C_3$, (b) ${}^{11}C_4$

Solution:

(a) $\quad {}^5C_3 = \dfrac{\lfloor 5}{\lfloor 3 \cdot \lfloor 5 - 3} = \dfrac{\lfloor 5}{\lfloor 3 \cdot \lfloor 2}$

$= \dfrac{5 \times 4 \times 3 \times 2 \times 1}{3 \times 2 \times 1 \times 2 \times 1}$

$\therefore {}^5C_3 = 10$

(b) $\quad {}^{11}C_4 = \dfrac{\lfloor 11}{\lfloor 4 \cdot \lfloor 11 - 4} = \dfrac{\lfloor 11}{\lfloor 4 \cdot \lfloor 9}$

$= \dfrac{11 \times 10 \times 9 \times 8 \times 7 \times 6 \times 5 \times 4 \times 3 \times 2}{4 \times 3 \times 2 \times 1 \times 9 \times 8 \times 7 \times 6 \times 5 \times 4 \times 3 \times 2}$

$\therefore {}^{11}C_4 = \dfrac{55}{12}$

Example 5. Find 'n' if (a) ${}^nP_2 = 90$, (b) ${}^np_4 = 360$

Solution:

(a) ${}^nP_2 = 90$

$n(n-1) = 90$

$n(n-1) = 10 \times 9 \qquad \therefore n = 10$

3	90
3	30
	10

(b) ${}^nP_4 = 360$

$n(n-1)(n-2)(n-3) = 6 \times 5 \times 4 \times 3$

$\therefore n = 6$

3	360
3	120
2	40
2	20
2	10
	5

Example 6. Find 'r' if (a) ${}^{6}P_r = 360$, (b) ${}^{13}P_r = 156$

Solution:

(a) ${}^{6}P_r = 360$

$${}^{6}P_r = 6 \times 5 \times 4 \times 3$$

$$\therefore r = 4$$

6	360
6	60
5	10
	2

(b) ${}^{13}P_r = 156$

$${}^{13}P_r = 13 \times 12$$

$$\therefore r = 2$$

13	156
	12

Example 7. Find 'n' if (a) ${}^{n}C_2 = 10$, (b) ${}^{n}C_3 = 20$

Solution:

(a) ${}^{n}C_2 = 10$

$$\frac{n(n-1)}{2 \times 1} = 10$$

$$\frac{n(n-1)}{2} = 10$$

$$n(n-1) = 20$$

$$n(n-1) = 5 \times 4$$

$$\therefore n = 5$$

(b) ${}^{n}C_3 = 20$

$$\frac{n(n-1)(n-2)}{3 \times 2 \times 1} = 20$$

$$n(n-1)(n-2) = 120$$

$$n(n-1)(n-2) = 6 \times 5 \times 4$$

$$\therefore n = 6$$

Example 8. Find the value of 'n' if

(a) ${}^{(n-1)}P_3 : {}^{(n+1)}P_3 = 5 : 12$, (b) $2 \cdot {}^{n}P_3 = 100 \cdot {}^{n}P_2$

Solution:

(a) $$\frac{(n-1)(n-2)(n-3)}{(n+1)(n+1-1)(n+1-1-1)} = \frac{5}{12} \qquad \left(\because \text{In general } a : b = \frac{a}{b}\right)$$

$$\frac{(n-1)(n-2)(n-3)}{(n+1)(n)(n-1)} = \frac{5}{12}$$

$$\frac{(n-2)(n-3)}{n\,(n+1)} = \frac{5}{12}$$

$12\,(n-2)\,(n-3) = 5n\,(n+1)$

$12\,(n^2 - 5n + 6) = 5n^2 + 5n$

$12n^2 - 60n + 72 - 5n^2 - 5n = 0$

$7n^2 - 65n + 72 = 0$

$7n^2 - 56n - 9n + 72 = 0$ 504 ↙ ↘ −56 −9

$7n\,(n-8) - 9\,(n-8) = 0$

$(n-8)\,(7n-9) = 0$

$n - 8 = 0$ or $7n - 9 = 0$

$n = 8$ or $7n = 9$

$\therefore\ n = 8$ or $\therefore\ n = \frac{9}{7}$

$\therefore\ n = 8$ (Since n cannot be a fraction)

(b) $2 \cdot {}^nP_3 = 100 \cdot {}^nP_2$

$2\,[n\,(n-1)\,(n-2)] = 100\,[n\,(n-1)]$

$2\,[(n^2 - n)\,(n-2)] = 100\,[(n^2 - n)]$

$2\,[n^3 - n^2 - 2n^2 + 2n] = 100n^2 - 100n$

$2\,(n^3 - 3n^2 + 2n) - 100n^2 + 100n = 0$

$2n^3 - 6n^2 + 4n - 100n^2 + 100n = 0$

$2n^3 - 106n^2 + 104n = 0$

$2n\,(n^2 - 53n + 52) = 0$

$2n = 0$ or $n^2 - 53n + 52 = 0$

$n^2 - 52n - n + 52 = 0$

$\therefore\ n = 0$ $n\,(n-52) - (n-52) = 0$ 52 ↙ ↘ −52 −1

$(n-52)\,(n-1) = 0$

$n = 52$ or $n = 1$

$\therefore\ n = 52\ (\because\ n \geq r)$

Example 9. Find the values of 'r' if (a) ${}^{10}P_{r+1} : {}^{11}P_r = 30 : 11$, (b) ${}^5P_r = 2 \cdot {}^6P_{r-1}$

Solution:

(a) $$\frac{{}^{10}P_{r+1}}{{}^{11}P_r} = \frac{30}{11}$$

$$\frac{\dfrac{\lfloor 10}{\lfloor 10-(r+1)}}{\dfrac{\lfloor 11}{\lfloor 11-r}} = \frac{30}{11}$$

$$\frac{\lfloor 10}{\lfloor 9-r} \times \frac{\lfloor 11-r}{\lfloor 11} = \frac{30}{11}$$

$$\frac{\lfloor 10 \times \lfloor 11-r}{11 \times \lfloor 10 \times \lfloor 9-r} = \frac{30}{11}$$

$$\frac{\lfloor 11-r}{11\lfloor 9-r} = \frac{30}{11}$$

$$\frac{(11-r)(10-r)(9-r)}{11\lfloor 9-r} = \frac{30}{11}$$

$$\frac{(11-r)(10-r)}{11} = \frac{30}{11}$$

$(11 - r)(10 - r) = 30$

$110 - 10r - 11r + r^2 = 30$

$r^2 - 21r + 110 - 30 = 0$

$r^2 - 21r + 80 = 0$

$r^2 - 16r - 5r + 80 = 0$ 80 ↙ ↘ −16 −5

$r(r - 16) - 5(r - 16) = 0$

$(r - 16)(r - 5) = 0$

$r = 16$ or $r = 5$

$\therefore r = 5 \qquad (\because r \le n)$

(b) $^5P_r = 2 \cdot {}^6P_{r-1}$

$$\frac{\lfloor 5}{\lfloor 5-r} = 2\left[\frac{\lfloor 6}{\lfloor 6-(r-1)}\right]$$

$$\frac{\lfloor 5}{\lfloor 5-r} = 2\left[\frac{\lfloor 6}{\lfloor 7-r}\right]$$

$$\frac{\lfloor 7-r}{\lfloor 5-r} = 2\left[\frac{\lfloor 6}{\lfloor 5}\right]$$

$$\frac{(7-r)(6-r)\lfloor 5-r}{\lfloor 5-r} = 2\left[\frac{6\lfloor 5}{\lfloor 5}\right]$$

$(7-r)(6-r)=12$
$42-6r-7r+r^2=12$
$r^2-13r+42-12=0$
$r^2-13r+30=0$ 30
$r^2-10r-3r+30=0$
$r(r-10)-3(r-10)=0$ −10 −3
$(r-3)(r-10)=0$
$\therefore r=3$ or $\therefore r=10$

$\therefore r=3\ (\because r\le n)$

Example 10. Find the values of n if

(a) ${}^nC_{12}={}^nC_8$, (b) ${}^nC_3:{}^{2n-1}C_2=8:15$, (c) ${}^{n+2}C_3=4\cdot{}^nC_2$.

Solution:

(a) ${}^nC_{12}={}^nC_8$

$$\frac{\lfloor n}{\lfloor 12\cdot\lfloor n-12}=\frac{\lfloor n}{\lfloor 8\cdot\lfloor n-8}$$

$\lfloor 8\cdot\lfloor n-8=\lfloor 12\cdot\lfloor n-12$

$\lfloor 8\cdot\lfloor n-8=12\times11\times10\times9\times\lfloor 8\cdot\lfloor n-12$

$\lfloor n-8=12\times11\times10\times9\times\lfloor n-12$

$(n-8)(n-9)(n-10)(n-11)\lfloor n-12=12\times11\times10\times9\lfloor n-12$
$(n-8)(n-9)(n-10)(n-11)=12\times11\times10\times9$
$(n^2-19n+88)(n^2-19n+90)=11880$
$(x+88)(x+90)=11880$ (where $x=n^2-9n$)
$x^2+178x+7920=11880$
$x^2+178x-3960=0$
$(x+198)(x-20)=0$
$x=-198$ or $x=20$
if $x=-198$,
$n^2-19n=-198$
$n^2-19n+198=0$

Here n is imaginary.
and if $x=20$,
$n^2-19n=20$
$n^2-19n-20=0$ −20
$n^2-20n+n-20=0$
$n(n-20)+(n-20)=0$ −20 1
$(n-20)(n+1)=0$
$\therefore n=20$ or $n=-1$

$\therefore\ n = 20$ (Since n cannot be negative).

(b) ${}^{n}C_3 : {}^{2n-1}C_2 = 8 : 15$

$$\frac{{}^{n}C_3}{{}^{2n-1}C_2} = \frac{8}{15}$$

$$\frac{\dfrac{\lfloor n}{\lfloor 3 \cdot \lfloor n-3}}{\dfrac{\lfloor 2n-1}{\lfloor 2 \cdot \lfloor 2n-1-2}} = \frac{8}{15}$$

$$\frac{\lfloor n}{\lfloor 3 \cdot \lfloor n-3} \times \frac{\lfloor 2 \cdot \lfloor 2n-3}{\lfloor 2n-1} = \frac{8}{15}$$

$$\frac{n(n-1)(n-2)\lfloor n-3}{3 \times \lfloor 2 \cdot \lfloor n-3} \times \frac{\lfloor 2 \lfloor 2n-3}{(2n-1)(2n-2)\lfloor 2n-3} = \frac{8}{15}$$

$$\frac{n(n-1)(n-2)}{3(2n-1)(2n-2)} = \frac{8}{15}$$

$$\frac{n(n-1)(n-2)}{2 \times 3(2n-1)(n-1)} = \frac{8}{15}$$

$15\,(n^2 - 2n) = 48\,(2n - 1)$

$15n^2 - 30n = 96n - 48$

$15n^2 - 30n - 96n + 48 = 0$

$15n^2 - 126n + 48 = 0$

$\div\, 3\ ,\ 5n^2 - 42n + 16 = 0$

$5n^2 - 40n - 2n + 16 = 0$

$5n\,(n - 8) - 2\,(n - 8) = 0$

$(n - 8)\,(5n - 2) = 0$

$\therefore\ n = 8$ or $5n = 2$

$\therefore\ n = \dfrac{2}{5}$

$\therefore\ n = 8$ (Since n cannot be a fraction)

80 → −40, −2

2	80
2	40
2	20
2	10
	5

(c) ${}^{n+2}C_3 = 4 \cdot {}^{n}C_2$

$$\frac{\lfloor n+2}{\lfloor 3 \cdot \lfloor n+2-3} = 4\left[\frac{\lfloor n}{\lfloor 2 \cdot \lfloor n-2}\right]$$

$$\frac{\lfloor n+2}{\lfloor 3 \cdot \lfloor n-1} = 4\left[\frac{\lfloor n}{2\lfloor n-2}\right]$$

$$\frac{(n+2)(n+1)\,n\lfloor n-1}{3\lfloor 2 \cdot \lfloor n-1} = 4\left[\frac{\lfloor n}{2\lfloor n-2}\right]$$

$$\frac{n(n+2)(n+1)}{3} = 4\,\frac{n(n-1)\lfloor n-2}{\lfloor n-2}$$

$$\frac{(n+2)(n+1)}{3} = 4(n-1)$$

$n^2 + 2n + n + 2 = 12n - 12$

$n^2 + 3n + 2 - 12n + 12 = 0$

$n^2 - 9n + 14 = 0$ 14

$n^2 - 7n - 2n + 14 = 0$ $\swarrow \searrow$

$n(n-7) - 2(n-7) = 0$ -7 -2

$(n-2)(n-7) = 0$

$\therefore\ n = 2$ or $n = 7$

Example 11. Find the value of 'r' if ${}^{15}C_{r+3} = {}^{15}C_{2r-3}$

Solution:

${}^{15}C_{r+3} = {}^{15}C_{2r-3}$

$\Rightarrow\ r + 3 = 2r - 3$ or $(r+3) + (2r-3) = 15$

$2r - r = 3 + 3$ or $3r + 0 = 15$

$\therefore\ r = 6$ or $3r = 15$

$\therefore\ r = 5$

$\therefore\ r = 5$ or $r = 6$

Example 12. Find n and r in each of the following cases:

(a) ${}^{n}C_r = 56,\ {}^{n}P_r = 336$

(b) ${}^{n}C_{r+3} = {}^{n}C_{2r-3},\ {}^{n}C_2 = 105$

(c) ${}^{n}P_r = 240,\ {}^{n}C_r = 120$

Solution:

(a) $${}^{n}C_r = \frac{{}^{n}P_r}{\lfloor r}$$

$$56 = \frac{336}{\lfloor r}$$

$$\lfloor r = \frac{336}{56} = 6 = 3 \times 2 \times 1 = \lfloor 3$$

$\therefore r = 3$

$\therefore {}^nP_r = 336$

Put $r = 3$, ${}^nP_3 = 336$

$n(n-1)(n-2) = 336$

$n(n-1)(n-2) = 8 \times 7 \times 6$

$\therefore n = 8$

$\therefore n = 8$ and $r = 3$.

(b) ${}^nC_2 = 105$

$$\frac{n(n-1)}{2 \times 1} = 105$$

$n(n-1) = 210$

$n(n-1) = 15 \times 14$

$\therefore n = 15$

$\therefore {}^{15}C_{r+3} = {}^{15}C_{2r-3}$

$r + 3 = 2r - 3 \Rightarrow r = 6$ or

$(r+3) + (2r-3) = 15$

$3r = 15$

$\therefore r = 5$

$\therefore n = 15$ and $r = 5$ or 6

(c) $${}^nC_r = \frac{{}^nP_r}{\lfloor r}$$

$$120 = \frac{240}{\lfloor r}$$

$$\lfloor r = \frac{240}{120} = 2$$

$\lfloor r = 2 \times 1 = \lfloor 2$

$\therefore r = 2$

$\therefore {}^nP_r = 240$

${}^nP_2 = 240$

$n(n-1) = 16 \times 15$

$\therefore\ n = 16$

$\therefore\ n = 16$ and $r = 2$

Example 13. Show that ${}^{10}P_3 = {}^{9}P_3 + 3 \cdot {}^{9}P_2 \cdot$

Solution:

$$\text{L.H.S} = {}^{10}P_3 = 10 \times 9 \times 8 = 720$$

$$\begin{aligned} \text{R.H.S} &= {}^{9}P_3 + 3 \cdot {}^{9}P_2 \\ &= 9 \times 8 \times 7 + 3\,[9 \times 8] \\ &= 504 + 3\,[72] \\ &= 504 + 216 \\ \text{R.H.S} &= 720 \end{aligned}$$

$\therefore$ L.H.S = R.H.S

Example 14. Find the number of permutations of the letters of the word INSTITUTION (i) How many of these begin with *I*? (ii) How many of these begin with *I* and end with *N*? (iii) In how many the 3 *T*'s are together?

Solution: The word contains 11 letters of which 3 are *I*'s, 2 *N*'s, 3 *T*'s, 1 *S*, 1 *U* 1 *O*.

$$\text{The number of permutations} = \frac{\lfloor 11}{\lfloor 3 \cdot \lfloor 2 \cdot \lfloor 3} = 5{,}54{,}400.$$

(i) Put *I* in the first place and arrange the rest. Now there are 10 letters of which 3-*T*, 2-*I*, 2-*N* and the rest different.

$$\therefore \text{No. of permutations which begin with } I \text{ is} = \frac{\lfloor 10}{\lfloor 3 \cdot \lfloor 2 \cdot \lfloor 2} = 1{,}51{,}200.$$

(ii) Put *I* in the first place and *N* in the last place and arrange the rest. Now there are 9 letters of which 3-*T*, 2-*I*, 1-*N*, 1-*S*, 1-*U*, 1-0.

$\therefore$ The number of permutations which begin with *I* and end with *N* is

$$= \frac{\lfloor 9}{\lfloor 3 \lfloor 2} = 30{,}240.$$

(iii) Regard 3 *T*'s as one letter say *X*. then we have 9 letters of which 1-*X*, 3-*I*, 2-*N*, 1-*S*, 1-*U*, 1-0.

$$\therefore \text{The number of permutations in which 3 } T\text{'s are together} = \frac{\lfloor 9}{\lfloor 3 \lfloor 2} = 30{,}240.$$

Example 15. In how many ways the letters of the word MISSISSIPPI be arranged so that 4 *S*'s are not together?

Solution: The given word contains 11 letters of which 4-*S*, 4-*I*, 2-*P*, 1-*M*.

$$\text{The number of arrangements of these letters} = \frac{\lfloor 11}{\lfloor 4 \lfloor 4 \lfloor 2} = 34{,}650$$

Regard 4-*S* as a single letter say *X*, then there are 8 letters of which 1-*X*, 4-*I*, 2-*P*, 1-*M*.

$\therefore$ The number of arrangements in which 4-*S* are together $= \dfrac{\lfloor 8}{\lfloor 4 \lfloor 2} = 840.$

Thus the number of arrangements in which 4-*S* are not together is

$= 34650 - 840$

$= 33810.$

Example 16. How many different words can be formed with the letters of 'HARYANA'? In how many of these *H* and *N* are together? How many of these words begin with *H* and end with *N*?

Solution: Total number of letters in HARYANA is 7 with 3 *A*'s.

So total number of different words

$$= \frac{\lfloor 7}{\lfloor 3} = 840$$

Counting *HN* as one letter, we are left with 6 letters having 3 *A*'s. *H* and *N* can be put together in two different manners namely *HN* and *NH*.

Regd. No. of words, in which *H* and *N* are together

$$= 2 \times \frac{\lfloor 6}{\lfloor 3} = 240$$

Finally the number of words beginning with *H* and ending with *N* is same as the number of arrangements of ARYAA, which is equal to

$$= \frac{\lfloor 5}{\lfloor 3} = 20$$

Example17. How many different ways can be the letters of the word VOWEL be arranged?

(i) How many of these arrangements begin with w?

(ii) How many of these arrangements, in which the letters *O*, *E* occupy even places?

(iii) How many of these arrangements which begin with *V* and end with *L*.?

(iv) In how many arrangements *A* and *E* are together?

Solution: The word VOWEL, has 5 different letters. These letters can be arranged in 5! Ways.

$\therefore$ 5 different letters can be arranged in 5! = 120 ways.

(i) We shall fix the first letter by *w*. The remaining 4 places can be filled up with remaining 4 letters in 4! Ways. Thus the required number is 4! = 24.

(ii) There are two even places - i.e., 2nd and the 4th places. These two places can be filled by *O* and *E*, in 2 way.

The remaining 3 places can be filled up by the remaining 3 letters by 3! Ways.

$\therefore$ The required number $= 2 \times 3! = 2 \times 6 = 12$

(iii) The first and the last places are fixed with *V* and *L*. Now the remaining three places can be filled with the remaining 3 letters by 3! Ways. Thus the required number of arrangements is 6.

(iv) Let us regard *A* and *E* together as one letter. The 4 remaining letters together with one unit of *A* and *E*, can be arranged in 4! Ways. The letters *A* and *E* wherever they are can be arranged in 2 ways. Thus the required number of ways is $4! \times 2 = 24 \times 2 = 48$.

Example 18. How many different permutations are there if all the letters of the word ALABAMA? Of these permutations how many contain the word LAMB?

Solution: There are 7 letters = A, A, A, A, L, B, M

$$\therefore \text{ Number of arrangements} = \frac{\lfloor 7}{\lfloor 4} = 210$$

Consider LAMB as one unit

Now the units are LAMB, A, A, A.

$$\therefore \text{ Total number of arrangements} = \frac{\lfloor 4}{\lfloor 3} = 4$$

Example 19. The letters of the word EQUATION are arranged taking all at a time so that the consonants occupy odd places. What is the total number of arrangements?

Solution: There are 8 letters in the given word of which 3 are consonants (Q, T, N)

There are 4 odd places (1) 2 (3) 4 (5) 6 (7) 8 and 3 consonants can be put in 4 places in ${}^4P_3$ ways. After this the remaining 5 letters can be arranged in $\lfloor 5$ ways.

$$\therefore \text{ No. of arrangements} = {}^4P_3 \times \lfloor 5 = 4 \times 3 \times 2 \times 120 = 2880$$

Example 20. Find the number of permutations of the letters of the word ENGINEERING. How many of these begin with *E*? How many begin with *E* and end with *G*? In how many arrangements tne 3 *E*'s are together?

Solution: There are 11 letters of which 3-*E*, 3-*N*, 2-*G*, 2-*I*, 1-*R*

$$\text{The number of permutations} = \frac{\lfloor 11}{\lfloor 3 \cdot \lfloor 3 \cdot \lfloor 2 \cdot \lfloor 2} = 2{,}77{,}200$$

Put *E* in the first place and arrange the rest. Now there are 10 letters of which 2-*E*, 3-*N*, 1-*G*, 2-*I*, 1-*R*.

$$\therefore \text{ Number of arrangements begining with } E \text{ is} = \frac{\lfloor 10}{\lfloor 2 \cdot \lfloor 3 \cdot \lfloor 2 \cdot \lfloor 2} = 75{,}600$$

Put *E* in the first place and *G* in the last place and arrange the rest. Now there are 9 letters of which 2-*E*, 3-*N*, 1-*G*, 2-*I*, 1-*R*.

$\therefore$ Number of arrangements which begin with *E* and end with *G* is

$$= \frac{\lfloor 9}{\lfloor 2 \cdot \lfloor 3 \cdot \lfloor 2} = 15{,}120$$

Regard 3 E's as single letter *X*. Then there are 9 letters of which 1-*X*, 3-*N*, 2-*G*, 2-*I*, 1-*R*.

$$\therefore \text{ Number of arrangements in which 3 } E\text{'s are together} = \frac{\lfloor 9}{\lfloor 3 \cdot \lfloor 2 \cdot \lfloor 2} = 15{,}120$$

Example 21. Find the number of permutations of the letters of the word ASSASSINATION. How many of these (a) have 3 *A*'s together (b) begin with 2 *A*'s.

Solution: The word contains 13 letters of which 3-*A*, 4-*S*, 2-*N*, 2-*I*, 1-*O* and 1-*T*.

$$\text{The number of permutations} = \frac{\lfloor 13}{\lfloor 3 \cdot \lfloor 4 \cdot \lfloor 2 \cdot \lfloor 2} = 21{,}62{,}160$$

(a) Regard 3 *A*'s as one unit *X*. Then there are 11 letters of which 1-*X*, 4-*S*, 2-*I*, 2-*N*, 1-*O* and 1-*T*.

$$\text{Number of arrangements having 3 } E\text{'s together} = \frac{\lfloor 11}{\lfloor 4 \cdot \lfloor 2 \cdot \lfloor 2} = 4{,}15{,}800$$

(b) Put 2 *A*'s in the first 2 places and arrange the rest.

$$\text{The number of arrangements beginning with 2 } A\text{'s is} = \frac{\lfloor 11}{\lfloor 4 \cdot \lfloor 2 \cdot \lfloor 2} = 4{,}15{,}800$$

Example 22. The letters of the word NATURE are permuted and the words so formed are arranged as in a dictionary. Find the rank of the word NATURE.

Solution: The alphabetical order of the letters are A E N R T U

In this dictionary.

Number of words beginning with A is $\lfloor 5$	=	120
Number of words beginning with E is $\lfloor 5$	=	120
Number of words beginning with NAE is $\lfloor 3$	=	6
Number of words beginning with NAR is $\lfloor 3$	=	6
Number of words beginning with NATE is $\lfloor 2$	=	2
Number of words beginning with NATR is $\lfloor 2$	=	2
Number of words beginning with NATUE is $\lfloor 1$	=	2
Number of words beginning with NATURE is	=	1
		259

∴ The rank of the word Nature is 259.

Example 23. How many words can be formed from the letters of VALEDICTORY provided the letters A, E, I, O, Y are not to be separated at all?

Solution: Put together A, E, I, O, Y and count them as one letter. Thus we have V, L, D, C, T, R, (AEIOY). These seven, letters can be arranged in 7! Ways. But the group (AEIOY) consisting of five letters can itself be arranged in 5! Ways.

Hence the required number = 5! × 7! = 6,04,800.

Example 24. Find the number of combinations if the letters of the word EXAMINATION taken one at a time.

Solution: There are 11 letters in EXAMINATION consisting of (A, A) (I, I), (N, N), E, X, T, O, M. Following cases arise:

(i) All the four letters are different.

(ii) Two are different two are alike.

(iii) Two pairs of alike letters.

Case I. There are 8 different letters; we are to choose 4. Number of combinations formed in this manner = ${}^8C_4 = 70$.

Case II. One pair of alike letters can be chosen in three ways and the remaining two different letters are to be chosen from 7.

Number of such combinations = $3 \times {}^7C_2 = 3 \times 21 = 63$

Case III. There are three pairs of alike letters:

So the number of combinations having two pairs of alike letters = ${}^3C_2 = 3$.

Consequently total number of combinations

$= 70 + 63 + 3$

$= 136$.

Example 25. A shelf contains 7 Kannada books. 4 Hindi books and 5 English books. In how many ways can they be arranged so that

(a) The Hindi books are together.

(b) Kannada books are together and Hindi books are together.

(c) No two English books are together.

(d) Books of the same language are together.

Solution:

(a) Regard 4 Hindi books as one unit. Then the number of arrangements of 13 books (7 Kannada, 5 English and 1 unit of Hindi books) can be arranged in $\lfloor 13$ ways. Further 4 Hindi books can be arranged among themselves in $\lfloor 4$ ways.

$\therefore$ Total number of arrangements $= \lfloor 13 \times \lfloor 4$.

(b) Regard Hindi books and Kannada books as separate units.

Then the total number of arrangements $= \lfloor 7 \times \lfloor 7 \times \lfloor 4$

(c) The number of arrangements of Kannada and Hindi books is $\lfloor 11$. In each of these arrangements there are 12 possible positions for 5 English books to occupy so that no two English books are together. The No. of arrangements of 5 English books is ${}^{12}P_5$ ways.

$\therefore$ Total No. of arrangements $= {}^{12}P_5 \times \lfloor 11$

(d) Regard the books of same language as a unit. Then 3 units of books may be arranged in $\lfloor 3$ ways. In each of these arrangements the 7 Kannada books, 4 Hindi books and 5 English books may be arranged in $\lfloor 7 \times \lfloor 4 \times \lfloor 5$ ways.

$\therefore$ Total number of arrangements $= \lfloor 3 \times \lfloor 7 \times \lfloor 4 \times \lfloor 5$

Example 26. In how many ways can 7 gentlemen and 7 ladies be seated at a round table so that no 2 ladies shall sit next to each other.

Solution: The seven gentlemen can be seated round a table in $\lfloor (7-1) = \lfloor 6$ ways. Consider any one of those arrangements. There are 7 possible places for the 7 ladies to sit. So the ladies can be seated in $\lfloor 7$ ways. Note the arrangements of ladies is linear.

$\therefore$ Total number of arrangements $= \lfloor 6 \times \lfloor 7$

Example 27. In how many ways (a) 12 persons can be seated round a table and (b) 12 beads of different colour be strung to form a necklace?

***Solution*:** (a) 12 persons can be seated round a table in $\lfloor(12-1) = \lfloor 11$. Here the clockwise and anticlockwise arrangements should be regarded as different.

(b) 12 beads of different colour can be strung in $\frac{1}{2}\lfloor(12-1) = \frac{1}{2}\lfloor 11$

Example 28. Find the number of permutations in which 5 boys and 4 girls can be seated in a row, so that (a) no two girls shall sit together (b) all the girls sit together.

***Solution*:** (a) First let the 5 boys be seated and this can be done in ${}^5P_5$ or $\lfloor 5$ ways. Consider one of the arrangements of boys.

$$\uparrow B1 \uparrow B2 \uparrow B3 \uparrow B4 \uparrow B5 \uparrow$$

There are six possible places (indicated by $\uparrow$) for the four girls to occupy so that no two of them shall sit together . So the 4 girls can be seated in these 6 places in ${}^6P_4$ ways.

$\therefore$ The number of permutations $= \lfloor 5 \times {}^6P_4$

$= 120 \times 6 \times 5 \times 4 \times 3 = 120 \times 360 = 43,200.$

(b) As the four girls are to be seated together, we regard them as one unit. Then 5 boys and 1 unit of girls (6 members) can be seated in $\lfloor 6$ ways. In each of these arrangements, the four girls can be permuted among themselves in $\lfloor 4$ ways. So the number of permutations so that all the girls are seated together is $\lfloor 6 \times \lfloor 4 = 720 \times 24 = 17,280.$

Example 29. If the letters of the word 'FATE' are arranged as in a dictionary, what is the rank of the word?

***Solution*:** The alphabetical order of the letter is A, E, F, T. The number of arrangements beginning with A is got by putting A in the first place and arranging the other 3 letters. This can done in $\lfloor 3$ ways.

A			

$\therefore$ Number of arrangements which begin with A $= \lfloor 3$	=	6
Number of arrangements which begin with E $= \lfloor 3$	=	6
Next arrangement is – FAET	=	1
Next arrangement is – FATE	=	1
Total	=	14

The word FATE comes as 14th arrangements $\therefore$ The rank of the word is 14.

Example 30. If the letters of the word MOTHER be arranged as in the dictionary, what is the rank of the word?

***Solution*:** Alphabetical order of letters of the given word is EHMORT.

Number of arrangements beginning with E is $\lfloor 5$	= 120
Number of arrangements beginning with H is $\lfloor 5$	= 120
Number of arrangements beginning with ME is $\lfloor 4$	= 24
Number of arrangements beginning with MH is $\lfloor 4$	= 24
Number of arrangements beginning with MOE is $\lfloor 3$	= 6
Number of arrangements beginning with MOH is $\lfloor 3$	= 6
Number of arrangements beginning with MOR is $\lfloor 3$	= 6
Number of arrangements beginning with MOTE is $\lfloor 2$	= 2
Next arrangement is MOTHER is 1	= 1
Total	= 309

∴ Rank of the word MOTHER is 309.

Example 31. Find the number of arrangements of 5 boys and 5 girls in a row so that no two boys and no two girls sit together.

***Solution*:** Suppose the arrangements start with a boy; then the arrangements are of the form BG BG BG BG BG. The 5 boys can be arranged in the odd places in $\lfloor 5$ ways and the 5 girls can be arranged in the even places in $\lfloor 5$ ways.

∴ Total number of arrangements $= \lfloor 5 \times \lfloor 5 = 14{,}400$

If the arrangements start with a girl then the total number of arrangements $= \lfloor 5 \times \lfloor 5$

∴ Total number of arrangements

$= \lfloor 5 \times \lfloor 5 + \lfloor 5 \times \lfloor 5 = 120 \times 120 + 120 \times 120$

$= 28{,}800$

Example 32. Find the number of arrangements in which 6 boys and 4 girls can be arranged in a line so that all the girls sit together and all the boys sit together.

***Solution*:** Consider the 6 boys as one unit and 4 girls as another unit

BBBBBB	GGGG

The 2 units can be arranged in $\lfloor 2$ ways. In each of these arrangements the 6 boys can be arranged among themselves in $\lfloor 6$ ways and 4 girls can be arranged among themselves in $\lfloor 4$ ways.

∴ The total number of arrangements $= 2 \times \lfloor 6 \times \lfloor 4$

$= 2 \times 720 \times 24 = 34566$

Example 33. A family of 4 brothers and 3 sisters are to be arranged for a photograph in one row. In how many ways can they be seated if all the sisters sit together?

Solution: Consider the 3 sisters as one unit and 4 brothers as 4 different units. There are now 5 units. They can be arranged among themselves in $\lfloor 5$ ways. In each of these arrangements the 3 sisters can be arranged in $\lfloor 3$ ways.

$\therefore$ Total number of arrangements $= \lfloor 5 \lfloor 3 = 720$

Example 34. There are 6 books on Economics, 3 on Mathematics and 2 on Accountancy. In how many ways can they be arranged on a shelf if the books of the same subject are always to be together?

Solution: Consider the 6 Economics books as one unit, 3 Mathematics books as one unit and 2 Accountancy books as one unit. These 3 units can be arranged in $\lfloor 3$ ways. In each of these arrangements, the 6 Economics books can be arranged in $\lfloor 6$ ways, 3 Mathematics books in $\lfloor 3$ ways and 2 Accountancy books in $\lfloor 2$ ways.

$\therefore$ Total number of arrangements $= \lfloor 3 \times \lfloor 6 \times \lfloor 3 \times \lfloor 2$

$= 51,840$

Example 35. In how many ways 3 boys and 5 girls can be arranged in a row so that

(a) No two boys are together.

(b) All the girls are together.

Solution: (a) Let us denote the boys by B_1, B_2 and B_3 and the girls are denoted by G_1, G_2, G_3, G_4 and G_5.

Since no two boys are to be together, first we shall arrange the girls in a row. This can be done in 5! ways.

Now the two boys can only be placed in the places matched x in the following arrangement.

$x\ G_1,\ x\ G_2,\ x\ G_3,\ x\ G_4,\ x$

That is the three boys can be placed in any one 6 places which are marked *x*. This can be done in $6P_3$ ways.

Hence the total number of arrangements is given by

$5! \times 6P_3 = 120 \times 6 \times 5 \times 4 = 14,400.$

(b) We require all the 5 girls to be together. We shall consider 5 girls as one unit.

Now 3 boys and 1 unit of 5 girls can be arranged in 4 ! ways.

Again the 5 girls wherever they are can be arranged in 5! ways.

Thus the required total number of ways $= 4! \times 5! = 24 \times 120 = 2,880.$

Example 36. In how many ways may a party of 4 ladies and 4 gentlemen be seated at a round table, the ladies and gentlemen being seated alternately?

Solution: Let the 4 ladies be seated around a table first. This can be done in $\lfloor 4 - 1 = \lfloor 3$ ways. Consider one of the arrangement of ladies. There are 4 possible places for the 4 gentlemen to sit. As the arrangements of gentlemen is linear [$\because$ if we shift G_1 to G_2, G_2 to G_3, G_3 to G_4 and G_4 to G_1 we get a different arrangements as the relative positions are no longer the same], the 4 gentlemen can be seated in $\lfloor 4$ ways.

$\therefore$ Total number of arrangements $= \lfloor 3 \times \lfloor 4 = 6 \times 24 = 144$

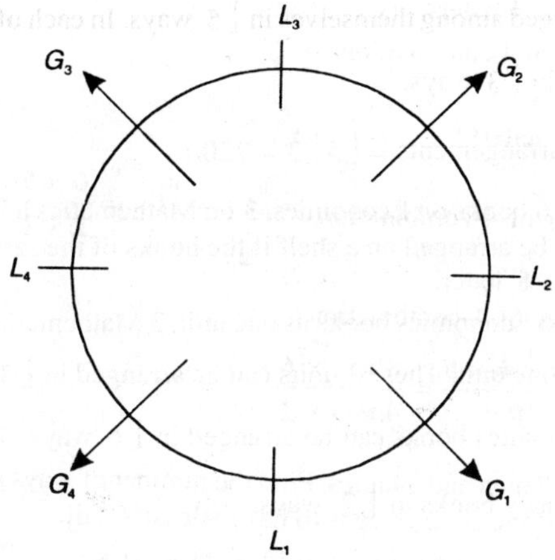

Example 37. In how many ways can 5 gentlemen and 3 ladies can be arranged along a round table so that no two ladies are together?

Solution: 5 gentlemen can be arranged along a circle in $\lfloor 4$ ways. There are 5 places for the ladies and the three ladies can be arranged in these places in $5P_3$ ways.

$\therefore$ Total number of arrangements $= \lfloor 4 \times 5P_3 = 24 \times 60$

$= 1{,}440$

Example 38. In how many ways can 3 boys and 5 girls be arranged in a row so that all the 3 boys are together. ***(C. A., May, 1977)***

Solution: We count the 3 boys as one boy so that number of persons involved is now 6. They can be arranged in 6! ways. But these 3 boys themselves can be arranged in 3! ways. Hence required number of arrangements is

$6! \times 3! = 720 \times 6 = 4320.$

Example 39. In how many ways 5 passengers can sit in a compartment having 16 vacant seats?

Solution: Required number of ways $= {}^{16}P_5$

$= \dfrac{16\,!}{(16-5)\,!}$

$= \dfrac{16\,!}{11\,!}$

$= \dfrac{16 \cdot 15 \cdot 14 \cdot 13 \cdot 12 \cdot \lfloor 11}{\lfloor 11}$

$= 16 \cdot 15 \cdot 14 \cdot 13 \cdot 12$

$= 5{,}24{,}160.$

Example 40, Find the total number of 4-combinations of the letters of the following words.

(a) WORD (b) SEARCH (c) PROBLEMS

Solution: (a) WORD has 4 letters.

$\therefore$ The total number of 4 combinations = $^4C_4 = 1$

(b) SEARCH has 6 letters

$\therefore$ The total number of 4-combinations = $^6C_4 = {}^6C_2 = \frac{6 \times 5}{2 \times 1} = 15$

(c) PROBLEMS has 8 letters.

$\therefore$ The total number of 4-combinations

$$= {}^8C_4 = \frac{8 \times 7 \times 6 \times 5}{\underline{|4}} = \frac{8 \times 7 \times 6 \times 5}{4 \times 3 \times 2 \times 1} = 70$$

Example 41. There are 7 men and 3 ladies. Find the number of ways in which a committee of 6 persons can be formed if the committee is to have atleast 2 ladies.

Solution:

No. persons cases	*Men* 7	*Ladies* 3	*No. of ways of selection*
1.	4	2	$^7C_4 \times {}^3C_2 = 105$
2.	3	3	$^7C_3 \times {}^3C_3 = 35$
	Total no. of ways		= 140

Example 42. In an election a voter may vote for any number of candidates not greater than the number to be chosen. There are 7 candidates and 4 members are chosen. In how many ways can a person vote?

Solution: 4 members are to be chosen out of 7 candidates. A person can vote to a maximum of 4 persons.

$\therefore$ A person can vote for 1 person or 2 persons or 3 persons or 4 persons.

$\therefore$ The number of possible ways.

$$= {}^7C_1 + {}^7C_2 + {}^7C_3 + {}^7C_4$$
$$= 7 + 21 + 35 + 35 = 98$$

Example 43. There are 12 points in a plane of which 4 are collinear. Find the number of (a) Straight lines (b) Triangles which can be formed from these points.

Solution: (a) We need two points for a line. If none of the 3 given points are collinear, then we would get $^{12}C_2$ lines. But 4 points are given to be collinear. So we would not get $^4C_2$ lines from these points. Instead we get only one line containing all the 4 points.

$\therefore$ Number of straight lines = $^{12}C_2 - {}^4C_2 + 1$

$$= \frac{12 \times 11}{2 \times 1} - \frac{4 \times 3}{2 \times 1} + 1 = 66 - 6 + 1 = 61$$

(b) We need three non-collinear points for a triangle. If none of the 12 given points are collinear, we would get $^{12}C_3$ triangles. As 4 points are collinear we could not get $^4C_3$ triangles from these points.

$$\therefore \text{ Number of triangles } = {}^{12}C_3 - {}^{4}C_3 = \frac{12 \times 11 \times 10}{3 \times 2 \times 1} - 4$$

$= 220 - 4 = 216$

Example 44. A committee of 3 experts is to be selected out of a panel of 7 persons; three of them are Lawyers, three of them are Chartered Accountants and one is both a Chartered Accountant and a Lawyer. In how many ways can the committee be selected if it must have atleast a Chartered Accountant and a lawyer?

Solution: Number of ways of selecting 3 experts such that the committee contains atleast a CA and a lawyer = Number of ways of selecting 3 persons out of the 7 persons – Number of ways of selecting all 3 lawyers – Number of ways of selecting all 3*C.A*'s.

$= {}^{7}C_3 - {}^{3}C_3 - {}^{3}C_3.$

$$= \frac{7 \times 6 \times 5}{1 \times 2 \times 3} - 1 - 1 = 35 - 2 = 33$$

Example 45. How many diagonals are there in a polygon of 10 sides?

Solution: A polygon of 10 sides has 10 vertices.

By selecting and joining two vertices in all possible ways we get all sides and diagonals.

$\therefore$ No. of sides + No. of diagonals = ${}^{10}C_2$

10 + No. of diagonals = ${}^{10}C_2$

No. of diagonals = ${}^{10}C_2 - 10$

So the number of diagonals = ${}^{10}C_2 - 10$

$$= \frac{10 \times 9}{2 \times 1} - 10 = 35$$

Note: A general formula can be derived for the number of diagonals found in an *n*-gon. It will be just ${}^{n}C_2 - n$.

$$\therefore \frac{n(n-1)}{2 \times 1} - n = \frac{n^2 - n - 2n}{2} = \frac{n^2 - 3n}{2} = \frac{n(n-3)}{2}$$

$\therefore$ The number of all possible diagonals in a polygon of *n*-sides is given by $= \dfrac{n(n-3)}{2}$

Example 46: There are 6 points no three of which are collinear. How many (a) Straight lines (b) triangles can be got by joining the points?

Solution: (a) There are 6 points, no three of which are collinear. A straight line is got by joining 2 points.

$$\therefore \text{ Total number of straight lines } = {}^{6}C_2 = \frac{6 \times 5}{2 \times 1} = 15$$

$\therefore$ Total number of straight lines = 15

(b) There are 6 points, no three of which are collinear. A triangle is got by joining 3 points in pairs.

$$\therefore \text{ Total number of triangles } = {}^{6}C_3 = \frac{6 \times 5 \times 4}{3 \times 2 \times 1} = 20$$

Example 47. A school has 8 teachers, one of whom is the head master (a) How many committees of five can be formed? (b) How many of these has the head master?

Solution: (a) There are 8 teachers. A committee of 5 is to be formed.

$$\therefore \text{ Number of committees } = {}^8C_5 = {}^8C_3 = \frac{8 \times 7 \times 6}{3 \times 2 \times 1} = 56$$

(b) In finding the number of committees, having the head master, the choice is limited to 7 teachers of whom 4 are to be selected.

Hence the number of committees having the head master

$$= {}^7C_4 = {}^7C_3 = \frac{7 \times 6 \times 5}{3 \times 2 \times 1} = 35$$

Example 48. In an examination paper, there are 7 questions in part A out of which any 4 are to be attempted and there are 6 questions in part *B* out of which 3 are to be attempted. In how many different ways can a candidate answer part *A* and part *B* in full?

Solution:

	Part A	*Part B*
No. of questions	7	6
No. to be attempted	4	3

Number of ways of choosing part A questions = ${}^7C_4$

Number of ways of choosing part B questions = ${}^6C_3$

$\therefore$ Total number of ways = ${}^7C_4 \times {}^6C_3$

$= {}^7C_3 \times {}^6C_3$

$$= \frac{7 \times 6 \times 5}{3 \times 2 \times 1} \times \frac{6 \times 5 \times 4}{3 \times 2 \times 1}$$

$= 35 \times 20$

$= 700$

Example 49. An examination paper consists of 12 questions divided into two parts *A* and *B*. Part *A* contains 7 questions and part *B* contains 5 questions. A candidate is required to answer 8 questions selecting atleast 3 questions from each part. In how many ways can he select the questions?

Solution: Number of questions

Cases	*Part A* 7	*Part B* 5	*No. of ways of selection*
1.	5	3	$7\,C_5 \times 5\,C_3 = 210$
2.	4	4	$7\,C_4 \times 5\,C_4 = 175$
3.	3	5	$7\,C_3 \times 5\,C_5 = 35$
		Total no. of ways	= 420

Example 50. A cricket team of 11 players is to be formed from 20 players including 6 bowlers and 3 wicket keepers. In how many ways can a team be formed so that the team contains exactly 2 wicket keepers and atleast 4 bowlers?

***Solution*:**

No. of Players	*Bowler* 6	*Wicket keepers* 3	*Others* 11	*No. of ways*		
1.	4	2	5	${}^6C_4 \times {}^3C_2 \times {}^{11}C_5$	=	16170
2.	5	2	4	${}^6C_5 \times {}^3C_2 \times {}^{11}C_4$	=	5940
3.	6	2	3	${}^6C_6 \times {}^3C_2 \times {}^{11}C_3$	=	165
				Total no. of ways	=	22275

Example 51. How many triangles can be made by joining the vertices of decagon? How many diagonals will it have?

***Solution*:** A triangle is obtained by joining any three vertices of a decagon. Thus number of triangles obtained from a decagon

$$= {}^{10}C_3 = \frac{10 \times 9 \times 8}{3 \times 2 \times 1} = 120$$

Now a diagonal is obtained by joining any two vertices which are not adjacent.

Total number of ways in which two vertices of a decagon can be joined

$$= {}^{10}C_2 = \frac{10 \times 9}{2} = 45$$

Number of lines obtained by joining two adjacent vertices

= Number of sides of decagon = 10.

Hence number of diagonals = 45 – 10 = 35.

Example 52. A committee consisting of 5 members is to be formed out of 6 men and 4 women. How many committees can be formed so that at least one woman is always there in the committee?

***Solution*:** Total number of committees = ${}^{10}C_5 = 252$

The number of committees in which no woman is there = ${}^6C_5 = {}^6C_1 = 6$

Hence the number of committees in which at least one woman is included

= 252 – 6 = 246.

Example 53. In an examination a candidate is required to answer 6 out of 10 questions which are divided into two groups each containing 5 questions and not permitted to attempt more than 4 questions from each group. In how many ways, can he make up his choice?

***Solution*:** The candidate can have following three choices:

(a) 2 questions from group *A* and 4 from group *B*.

(b) 3 questions from group *A* and 3 from group *B*.

(c) 4 questions from group *A* and 2 from group *B*:

Now first choice can be made up in ${}^5C_2 \times {}^5C_4$, i.e. 50 ways. Second choice can be made up in ${}^5C_3 \times {}^5C_3$ ways i.e. 100 ways and third choice can be made up in ${}^5C_4 \times {}^5C_2$ ways i.e. 50 ways.

So, total number of ways, he can make up his choice

$= 50 + 100 + 50 = 200.$

Example 54. From 6 boys and 4 girls, 5 are to be selected to admission for a particular house. In how many ways can this be done if there must be exactly 2 girls.

Solution: Two girls can be selected out of 4 girls in $^4C_2 = \dfrac{4 \times 3}{2 \times 1} = 6$ ways.

The remaining three are to be boys. They are to be selected out of 6 boys.

The number of ways in which 3 boys can be chosen out of 6 boys is

$$^6C_3 = \frac{6 \times 5 \times 4}{3 \times 2 \times 1} = 20.$$

$\therefore$ Required number of ways $= 6 \times 20 = 120.$

Example 55. Out of 10 consonants and 4 vowels how many words can be formed each containing 6 consonants and 3 vowels.

Solution: 6 consonants can be selected out of 10 in $^{10}C_6$ ways while 3 vowels can be selected out of 4 $^4C_3$ ways.

Total number of letters in each choice of 6 consonants and 3 vowels is 9.

So each choice can be arranged in 9! Different ways.

Hence total number of words

$$= {}^{10}C_6 \times {}^4C_3 \times 9!$$

$$= \frac{10 \times 9 \times 8 \times 7}{4 \times 3 \times 2 \times 1} \times 4 \times 9!$$

$$= 30,48,19,200.$$

Example 56. There are 6 boys and 3 girls in a class. An entertainment committee of 5 is to be selected such that there are 3 boys and 2 girls in the committee. In how many ways can the committee be selected? What is the number of ways, if there is atleast one girl in the committee.

Solution: The number of ways of selecting 3 boys out of 6 is $^6C_3$ and the number of ways of selecting 2 girls out of 3 is $^3C_2$.

Thus the required number $= {}^6C_3 \times {}^3C_2$

$$= \frac{6 \times 5 \times 4}{1 \times 2 \times 3} \times \frac{3 \times 2}{1 \times 2} = 60$$

As there should be atleast one girl, the committee may consist of as follows:

(a) 4 boys 1 girl

(b) 3 boys 2 girls

(c) 2 boys 3 girls

$$\text{Number of selection for (a)} = {}^6C_4 \times {}^3C_1 = \frac{6 \times 5 \times 4 \times 3}{1 \times 2 \times 3 \times 4} \times 3 = 45$$

$$\text{Number of selection for (b)} = {}^6C_3 \times {}^3C_2 = \frac{6 \times 5 \times 4}{1 \times 2 \times 3} \times \frac{3 \times 2}{1 \times 2} = 60$$

$$\text{Number of selection for (c)} = {}^6C_2 \times {}^3C_3 = \frac{6 \times 5}{1 \times 2} \times 1 = 15$$

$\therefore$ Total number of selections = 45 + 60 + 15 = 120

Example 57. In how many ways can a foot ball team of 11 players selected from 15 players. How many of these will

(a) Include one particular player

(b) Exclude one particular player.

Solution: The number of ways of selecting 11 players out of 15 players is

$$^{15}C_{11} = {}^{15}C_{15-11} = {}^{15}C_4 = \frac{15 \times 14 \times 13 \times 12}{1 \times 2 \times 3 \times 4} = 1,365$$

(a) If one player included, then we have to select 10 players out of remaining 14 players. This can be done in $^{14}C_{10}$ ways.

$$\therefore \; ^{14}C_{10} = {}^{14}C_{14-10} = {}^{14}C_4 = \frac{14 \times 13 \times 12 \times 11}{1 \times 2 \times 3 \times 4} = 1001$$

(b) If one player is always excluded, we have to select 11 players out of 14 players. This can be done in $^{14}C_{11}$ ways.

$$\therefore \; ^{14}C_{11} = {}^{14}C_{14-11} = {}^{14}C_3 = \frac{14 \times 13 \times 12}{3 \times 2 \times 1} = 364$$

Example 58. There are 15 points in a plane of which 5 are Collinear. Find the number of (a) straight lines (b) triangles which can be formed these points.

Solution: (a) Since we are given 5 collinear points, these 5 points give only one line.

Now the number of line with 15 points is given by $^{15}C_2$.

$\therefore$ Number of lines = $^{15}C_2 - {}^5C_2 + 1$.

$$= \frac{15 \times 14}{1 \times 2} - \frac{5 \times 4}{1 \times 2} + 1 = 105 - 10 + 1 = 95$$

(b) We need three points to form a triangle. Thus the total number of triangles would be $^{15}C_3$, provided none of the 3 points are collinear.

But 5 points are collinear, we would not get 5 C_3 triangles from these points.

$$\therefore \text{Number of triangles} = {}^{15}C_3 - {}^5C_3 = \frac{15\ 14\ 13}{1 \times 2 \times 3} - \frac{5 \times 4 \times 3}{1 \times 2 \times 3} = 445$$

Example 59. A man has 7 relatives, 4 of them are ladies and 3 gentlemen. His wife has also 7 relatives, 3 of them are ladies and 4 gentlemen. In how many ways can they invite a dinner party of 3 ladies and 3 gentlemen so that there are 3 of the man's relatives and 3 of the wife's relatives?

Solution: The different possibilities of selection are:

	Man's relatives		*Wife's relatives*	
Total	*4 ladies*	*3 gentlemen*	*3 ladies*	*4 gentlemen*
(1)	3 L	0 G	0 L	3 G
(2)	2 L	1 G	1 L	2 G
(3)	1 L	2 G	2 L	1 G
(4)	0 L	3 G	3 L	0 G

Thus the different ways of selection are

(1)	${}^4C_3 \times {}^3C_0 \times {}^3C_0 \times {}^4C_3$	=	16
(2)	${}^4C_2 \times {}^3C_1 \times {}^3C_1 \times {}^4C_2$	=	324
(3)	${}^4C_1 \times {}^3C_2 \times {}^3C_2 \times {}^4C_1$	=	144
(4)	${}^4C_0 \times {}^3C_3 \times {}^3C_3 \times {}^4C_0$	=	1
	Total No. of ways	=	485

Example 60. A football team of 11 has to be selected out of 14 players satisfying the following conditions. Govind will not play if Rama is selected but Rama will play only if Krishna is also selected. In how many ways can the team be selected?

Solution: Suppose Rama is selected, then we have to select Krishna and reject Govind. Then we have to select 9 players out of remaining 11 players. This can be done in ${}^{11}C_9$ ways. So the no. of selections when Rama is selected = ${}^{11}C_9 = {}^{11}C_2 = 55$.

Suppose we do not select Rama, then Govind may be selected and also Krishna may be selected. So reject Rama and select 11 players out of remaining 13 players. This can be done in ${}^{13}C_{11} = {}^{13}C_2 = 78$ ways.

$\therefore$ Total no. of selections = 55 + 78 = 133.

Example 61. A committee of 8 members is to be chosen from 9 teachers and 4 students. In how many ways can this be done if (a) the committee contains exactly 3 students (b) there is to be a majority of teachers. (c) there are atleast 3 students (d) there are at most 7 teachers.

Solution: (a) The committee consists of 5 teachers and 3 students. 5 teachers can be selected out of 9 in ${}^9C_5$ ways and 3 students can be selected out of 4 in ${}^4C_3$ ways. So the number of selections when there are exactly 3 students on the committee ${}^9C_5 \times {}^4C_3 = {}^9C_4 \times {}^4C_1$

$$= \frac{9 \times 8 \times 7 \times 6}{4 \times 3 \times 2 \times 1} \times \frac{4}{1} = 504$$

(b) As there should be a majority of teachers, the committee consists of

(a) 5 teachers and 3 students
(b) 6 teachers and 2 students
(c) 7 teachers and 1 students

Number of selections for (a) = ${}^9C_5 \times {}^4C_3 = 504$

Number of selections for (b) = ${}^9C_6 \times {}^4C_2 = {}^9C_3 \times {}^4C_2$

$$= \frac{9 \times 8 \times 7}{3 \times 2 \times 1} \times \frac{4 \times 3}{2 \times 1} = 504$$

Number of selection for (c) = ${}^9C_7 \times {}^4C_1 = 144$

Hence the total number of selections when there is a majority of teachers = 504 + 504 + 144 = 1,152

(c) As there are to be atleast 3 students, the committee consists of (a) 5 teachers and 3 students (b) 4 teachers and 4 students

(a) Number of selections for (a) = ${}^9C_5 \times {}^4C_3 = 504$

(b) Number of selections for (b) = ${}^9C_4 \times {}^4C_4 = 126$

Total number of selections when the committee has at least 3 students = 504 + 126 = 630.

(d) As there are to be at most 7 teachers, the committee consists of

(a) 7 teachers and 1 student

(b) 6 teachers and 2 students

(c) 5 teachers and 3 students

(d) 4 teachers and 4 students

Number of selections for (a) = ${}^9C_7 \times {}^4C_1 = 144$

Number of selections for (b) = ${}^9C_6 \times {}^4C_2 = 504$

Number of selections for (c) = ${}^9C_5 \times {}^4C_3 = 504$

Number of selections (d) = ${}^9C_4 \times {}^4C_4 = 126$

Total number of selections when there are at most 7 teachers on the committee is 144 + 504 + 504 + 126 = 1278.

Exercises

I. Question carrying THREE marks each (Short Answer):

1. Find the values of (a) $\lfloor 4$, (b) $\lfloor 5$, (c) ${}^4P_4$, (d) ${}^8P_5$

[*Ans: (a) 24, (b) 120, (c) 24, (d) 6,720*]

2. Find the value of (a) ${}^6P_4$ and (b) ${}^5P_2$. **[*Ans: (a) 360, (b) 20*]**

3. Find the value of ${}^6P_4 + {}^4P_3 - {}^2P_1$. **[*Ans: 382*]**

4. Using the formula ${}^nP_r = \dfrac{\lfloor n}{\lfloor n-r}$ find the values of

(a) ${}^5P_2$, (b) ${}^8P_3$, (c) ${}^{10}P_6$, (d) ${}^9P_4$, (e) ${}^7P_3$, (f) ${}^8P_7$, (g) ${}^{10}P_5$.

[*Ans: (a) 20, (b) 336, (c) 1,51,200, (d) 3024, (e) 210, (f) 40,320, (g) 30,240.*]

5. Find the values of

(a) ${}^8C_2$, (b) ${}^{10}C_7$, (c) ${}^{14}C_{12}$, (d) ${}^{21}C_{20}$, (e) ${}^{15}C_9$, (f) ${}^5C_3$, (g) ${}^{10}C_3$, (h) ${}^{100}C_{100}$, (i) ${}^{1000}C_0$.

[*Ans: (a) 28, (b) 120, (c) 91, (d) 21, (e) 5,005, (f) 10, (g) 120, (h) 1, (i) 1.*]

6. Find the values of

(a) ${}^3C_2 + {}^4C_3$, (b) ${}^5C_1 - {}^3C_3 + {}^4C_2$, (c) $4 \times {}^3C_1 + 2 \times {}^5C_3 + {}^3C_0$,

(d) ${}^4C_0 + {}^4C_1 + {}^4C_2 + {}^4C_3 + {}^4C_4$.

[*Ans: (a) 7, (b) 10, (c) 33, (d) 16.*]

7. Using the formula ${}^nC_r = \dfrac{\lfloor n}{\lfloor r \cdot \lfloor n-r}$ find the values of

(a) ${}^5C_2$, (b) ${}^6C_4$, (c) ${}^{10}C_7$, (d) ${}^{12}C_8$, (e) ${}^8C_3$, (f) ${}^{15}C_{15}$.

[*Ans: (a) 10, (b) 15, (c) 120, (d) 495, (e) 56, (f) 1.*]

8. Find 'n' of (a) ${}^nP_2 = 90$, (b) ${}^nP_3 = 210$, (c) $\lfloor n = 120$, (d) $\lfloor n = 720$,

(e) ${}^nP_2 = 72$, (f) $\lfloor n = 5040$, (g) ${}^nP_4 = 360$.

[*Ans: (a) 10, (b) 7, (c) 5, (d) 6, (e) 9, (f) 7, (g) 6.*]

9. Find 'r' of (a) ${}^6P_r = 360$, (b) ${}^{13}P_r = 156$, (c) ${}^{11}P_r = 990$, (d) ${}^5P_r = 60$.

[*Ans: (a) 4, (b) 2, (c) 3, (d) 3.*]

10. Find 'n' of (a) ${}^nC_2 = 10$, (b) ${}^nC_2 = 15$, (c) ${}^nC_3 = 1$, (d) ${}^nC_3 = 4$, (e) ${}^nC_2 = 28$, (f) ${}^nC_3 = 20$, (g) ${}^nC_3 = 120$, (h) ${}^nC_2 = 105$.

[*Ans: (a) 5, (b) 6, (c) 3, (d) 4, (e)8, (f) 6, (g) 10, (h) 15.*]

11. Find the values of 'n' if:

(a) ${}^{2n}P_3 = 2 \cdot {}^nP_4$, (b) ${}^nP_4 = 42 \cdot {}^nP_2$, (c) $2 \cdot {}^nP_2 + 50 = {}^{2n}P_2$,

(d) ${}^nP_4 = 12 \cdot {}^nP_2$, (e) ${}^nP_4 = 10 \cdot {}^nP_3$, (f) ${}^nP_5 = 42 \cdot {}^nP_3$ $(n \geq 4)$,

(g) ${}^nP_3 = 10 \cdot {}^nP_2$.

[*Ans: (a) 8, (b) 9, (c) 5, (d) 6, (e) 13, (f) 10, (g) 12.*]

12. Find the values of 'r' if:

(a) ${}^{11}P_r = {}^{12}P_{r-1}$, (b) ${}^{10}P_r = 2 \cdot {}^9P_r$, (c) $4 \cdot {}^6P_r = {}^6P_{r+1}$.

[*Ans: (a) 9, (b) 5, (c) 2.*]

13. Find the values of 'n' if:

(a) ${}^{2n}C_3 : {}^nC_2 = 44 : 3$, (b) ${}^nC_{10} = {}^nC_{12}$, (c) ${}^nC_8 = {}^nC_{12}$,

(d) ${}^nC_{30} = {}^nC_5$, (e) ${}^nC_5 = {}^nC_{15}$, (f) ${}^{2n}C_3 = 24 \cdot {}^nC_4$.

[*Ans: (a) 6, (b) 22, (c) 20, (d) 35, (e) 20, (f) 5.*]

14. Find the values of 'r' if:

(a) ${}^{18}C_r = {}^{18}C_{r+2}$, (b) ${}^{2n}C_r = {}^{2n}C_{r+2}$

[*Ans: (a) 8, (b) n – 1.*]

15. Find n and r if ${}^nP_r = 3024$, ${}^nC_r = 126$. [*Ans: n = 9, r = 4.*]

16. If ${}^{18}C_r = {}^{18}C_{r+2}$ find ${}^rC_5$. [*Ans: 56.*]

17. If ${}^nC_{12} = {}^nC_8$ find ${}^nC_{17}$ and ${}^{22}C_n$. [*Ans: 1140 and 231.*]

18. Verify $2 \cdot {}^7C_4 = {}^8C_4$

19. If ${}^nC_{10} = {}^nC_{12}$ find ${}^nC_5$. [*Ans: 26,334.*]

20. If ${}^nC_8 = {}^nC_{12}$ find ${}^nC_{18}$. [*Ans: 190.*]

21. If ${}^nC_{20} = {}^nC_8$ find ${}^nC_{26}$. [*Ans: 398.*]

22. If ${}^nC_5 = {}^nC_{15}$ find ${}^nC_{20}$. [*Ans: 1.*]

23. If ${}^nP_4 = 24 \cdot {}^nC_5$ find n and ${}^nC_3$. [*Ans: n = 9, ${}^nC_3$ = 84.*]

24. If ${}^nP_r = 240$ and ${}^nC_r = 120$ find n and r. [*Ans: n = 16, r = 2.*]

25. If ${}^{2n}C_3 = 24 \times {}^nC_4$ find ${}^nC_2$. [*Ans: 10.*]

II. Questions carrying FIVE marks each (Essay Type Answers):

1. Find the number of permutations of the letters of the word TOMORROW. In how many of these are the 3 *O*'s not together? ***[Ans: 3,360 and 3,000 ways.]***

2. Find the number of permutations of the letters of the word MANAGEMENT. In how many of these do the *M*'s come together?

   [Ans: 2,26,800 and 45,360 ways.]

3. How many different arrangements can be made by using all the letters in the word MATHEMATICS? How many of them begin with C?

   [Ans: 49,89,600 and 4,53,600 ways.]

4. In how many ways can the letters of the word MANAGEMENT be arranged? How many of these begin with *M*? ***[Ans: 2,26,800 ways and 45,360 ways.]***

5. Find the number of permutations of the letters of the word "ASSIGNMENT". In how many of these will the 2*N*'s come together?

   [Ans: 907200 and 181440 ways.]

6. In how many ways can the letters of the word STRANGE be arranged so that,

   (a) The vowels are never separated.

   (b) The vowels never come together.

   [Ans: (a) 1449 ways (b) 3600 ways.]

7. How many different words can be made out of the letters of the word ALLAHABAD? In how many of these will the vowels occupy the even places?

   [Ans: 7560 and 60 ways.]

8. In how many ways can the letters of the word COMMERCE be arranged? How many of these begin with '*M*'? ***[Ans: 5,040 and 1,260 ways.]***

9. In how many ways can the letters of the word 'INTERMEDIATE' be arranged so that, the vowels always occupy even places?

   [Ans: 21,600 ways.]

10. In how many ways can the letters of the word STRANGE be arranged so that the vowels may appear in the odd places? ***[Ans: 1,440 ways.]***

11. A committee of 6 is to be formed from 7 Indians and 4 Pakistanis. In how many ways can this be done when the committee must contain at least 2 Pakistanis?

    [Ans: 371 ways.]

12. In how many different ways can 10 examination papers be arranged in a row, so that the best and worst papers may never come together? ***[Ans: 29,03,040 ways.]***

13. A man has 6 friends. In how many ways may he invite one or more of them to a dinner? ***[Ans: 63 ways.]***

14. A student is allowed to choose 5 questions out of 7 in Geometry and 3 questions out of 5 in Trigonometry in an examination. In how many ways can he choose them? ***[Ans: 210 ways.]***

15. Find the number of permutations of the letters of the word ASSESSMENT. In how many of these do the *S*'s come together?

    [Ans: 75,600 ways and 2,520 ways.]

16. How many diagonals are there in a polygon of 15 sides? **[*Ans: 90.*]**

17. Find the number of permutations of the word "ACCOUNTANTS". **[*Ans: 24,94,800 ways.*]**

18. In how many ways 5 passengers can sit in a compartment having 16 vacant seats? **[*Ans: 5,24,160 ways.*]**

19. How many different arrangements can be made by using all the letters in the word "KARNATAKA"? **[*Ans: 7,560 ways.*]**

20. Find the number of arrangements of the letters of the word MATHEMATICS. In how many of these vowels occur together? **[*Ans: 49,89,600 ways and 1,20,960 ways.*]**

21. Find the number of permutations of the letters of the word ENGINEERING. How many of these begin with *E*? How many begin with *E* and end with *G*? In how many arrangements the 3 *E*'s are together? **[*Ans: 2,77,200, 75,600, 15,120, 15,120.*]**

22. Find the number of permutations of the letters of the word ASSASSINATION. How many of these have 3 *A*'s together? **[*Ans: 1,08,10,800, 4,15,800 ways.*]**

23. Find the number of permutations of the letters of the word STATISTICS. **[*Ans: 50, 400 ways*]**

24. How many words can be formed from the letters of the word DAUGHTER, So that (a) the vowels always come together. (b) the vowels are never together. **[*Ans: (a) 4,320 ways, (b) 36,000 ways*]**

25. How many permutations can be made out of the letters of the word INDEPENDENCE? In how many of them, the vowels occur together? **[*Ans: 16,63,200 ways and 16,800 ways*]**

26. In how many ways can 8 examinations papers be arranged in a row so that the best and the worst papers are never together? **[*Ans: 30,240 ways*]**

2

Vectors

Josiah Willard Gibbs
(1839–1903)

Contents

- Vectors, Types of Vectors.

INTRODUCTION

The word vector is derived from the Latin word "Veho" means "I carry". J. Willard Gobbs (1839–1903) was called the founder of Vector analysis. Physical quantities having only magnitude are called *Scalars*.

Example Distance, Speed, Density. Scalars are represented by numbers.
Physical quantities having both magnitude and direction are called *Vectors*.

Example: Velocity, Acceleration, Magnetic field intensity.

TYPES OF VECTORS

Null Vector or Zero Vector

A null vector is a vector with any direction and zero magnitude.

Equal Vector

Two vectors $\vec{a} = (x_1, y_1, z_1)$ and $\vec{b} = (x_2, y_2, z_2)$ are said to be equal if $x_1 = x_2, y_1 = y_2, z_1 = z_2$ and is written as $\vec{a} = \vec{b}$

Unit Vector

A vector whose magnitude is unity is called a unit vector and is generally denoted by $\hat{a}$. i.e., $\frac{\vec{a}}{|\vec{a}|} = \hat{a}$.

$$\therefore \hat{a} = \frac{\text{Vector } 'a'}{\text{Modulus of vector } \vec{a}}$$

Free Vector

A vector is said to be a free vector (or a sliding vector) if its magnitude and direction are fixed but position in space is not fixed.

Localised Vector

A vector whose position in space is fixed is called localised vector.

Co-initial Vector or Concurrent Vector

A vector having the same initial point are called co-initial vector or concurrent vector.

Collinear or Parallel Vectors

Vectors which are parallel to the same straight line are called collinear vectors.

Non-collinear Vectors

Vectors which are not parallel to the same line are called non-collinear vectors.

Coplanar and Non-coplanar Vectors

Three or more vectors are said to be coplanar when they are parallel to the same plane otherwise they are said to be non co-planar vectors.

Reciprocal Vector

A vector having the same direction as that of a given vector $\vec{a}$ but magnitude equal to the reciprocal of the magnitude of the given vector $\vec{a}$ is called the reciprocal of the vector $\vec{a}$ and is written as $\vec{a}^{-1}$

i.e., $\vec{a}^{-1} = \frac{1}{|\vec{a}|} = \frac{1}{a}$

Linear Combinations of Vectors

If $\vec{a}$ and $\vec{b}$ be any two vectors and l and m be any two scalars, then $l\vec{a} + m\vec{b}$ is called linear combination of $\vec{a}$ and $\vec{b}$.

Position Vector of a Point

If the vector $\overrightarrow{OA} = \vec{a}$ represents the position of a point A relative to a fixed point O, then $\overrightarrow{OA}$ is called the position vector of the point A with reference to the point O as origin $O \xrightarrow{\vec{a}} A$.

List of Formulae:

1. Unit vector $= \hat{a} = \dfrac{\vec{a}}{|\vec{a}|}$ where $\vec{a} = a_1\hat{i} + a_2\hat{j} + a_3\hat{k}$
2. Magnitude of $\vec{a} = |\vec{a}| = \sqrt{(a_1)^2 + (a_2)^2 + (a_3)^2}$
3. Inverse of $(a_1, a_2, a_3) = (-a_1, -a_2, -a_3)$
4. Scalar product or Dot product of $\vec{a}$ and $\vec{b}$

   $= (\vec{a}\cdot\vec{b})$

   $= (a_1\hat{i} + a_2\hat{j} + a_3\hat{k})\cdot(b_1\hat{i} + b_2\hat{j} + b_3\hat{k})$

   $= a_1b_1 + a_2b_2 + a_3b_3 \quad (\because \hat{i}\cdot\hat{i} = 1, \hat{j}\cdot\hat{j} = 1, \hat{k}\cdot\hat{k} = 1.)$
5. Vector product or Cross product of $\hat{a}$ and $\vec{b}$

   $= \hat{a}\times\vec{b}$

   $$= \begin{vmatrix} \hat{i} & \hat{j} & \hat{k} \\ a_1 & a_2 & a_3 \\ b_1 & b_2 & b_3 \end{vmatrix}$$

Worked Examples

Example 1 If $\vec{x} = \vec{a} + \vec{b}$, $\vec{y} = \vec{b} + \vec{c}$, $\vec{z} = \vec{a} - \vec{c}$ find

(a) $\vec{x} + 2\vec{y}$ and (b) $\vec{x} + \vec{y} - \vec{z}$.

Solution

(a) $\vec{x} + 2\vec{y} = (\vec{a} + \vec{b}) + 2(\vec{b} + \vec{c})$

$= \vec{a} + \vec{b} + 2\vec{b} + 2\vec{c}$

$\therefore\ \vec{x} + 2\vec{y} = a + 3\vec{b} + 2\vec{c}$

(b) $\vec{x} + \vec{y} - \vec{z} = (\vec{a} + \vec{b}) + (\vec{b} + \vec{c}) - (\vec{a} - \vec{c})$

$= \vec{a} + \vec{b} + \vec{b} + \vec{c} - \vec{a} + \vec{c}$

$\therefore\ \vec{x} + \vec{y} - \vec{z} = 2\vec{b} - 2\vec{c}$

$= 2(\vec{b} + \vec{c})$

Example 2 If $\vec{x} = 2\vec{a} + \vec{b}$, $\vec{y} = \vec{a} - 2\vec{b}$ and $\vec{z} = \vec{a} + \vec{b} + \vec{c}$ find the values of

(a) $2\vec{x} + \vec{y} - \vec{z}$ and (b) $\vec{x} + \vec{y} + 2\vec{z}$.

Solution:

(a) $2\vec{x} + \vec{y} + \vec{z} = 2(2\vec{a} + \vec{b}) + (\vec{a} - 2\vec{b}) - (\vec{a} + \vec{b} + \vec{c})$

$= 4\vec{a} + 2\vec{b} + \vec{a} - 2\vec{b} - \vec{a} - \vec{b} - \vec{c}$

$\therefore\ 2\vec{x} + \vec{y} - \vec{z} = 4\vec{a} - \vec{b} - \vec{c}$

(b) $\vec{a} + \vec{y} + 2\vec{z} = (2\vec{a} + \vec{b}) + (\vec{a} - 2\vec{b}) + 2(\vec{a} + \vec{b} + \vec{c})$

$= 2\vec{a} + \vec{b} + \vec{a} - 2\vec{b} + 2\vec{a} + 2\vec{b} + 2\vec{c}$

$\therefore\ \vec{x} + \vec{y} + 2\vec{z} = 5\vec{a} + \vec{b} + 2\vec{c}$

Example 3 If $\vec{a} = (2, 2, 1)$ $\vec{b} = (1, 4, 5)$ find

(a) $\vec{a} + \vec{b}$ (b) $2\vec{a} - 5\vec{b}$.

Solution:

(a) $\vec{a} + \vec{b} = (2, 2, 1) + (1, 4, 5)$

$= (2 + 1, 2 + 4, 1 + 5)$

$\therefore\ \vec{a} + \vec{b} = (3, 6, 6)$

(b) $2\vec{a} - 5\vec{b} = 2\,(2, 2, 1) + 5\,(1, 4, 51)$

$= (4, 4, 2) - (5, 20, 25)$

$= (4 - 5, 4 - 20, 2 - 25)$

$\therefore\ 2\vec{a} - 5\vec{b} = (-1, -16, -23)$

Example 4 If $\vec{a} = (2, 3, -1)$, $\vec{b} = (1, 0, 4)$ find

(a) $2\vec{a} + 3\vec{b}$ (b) $3\vec{a} - 2\vec{b}$.

Solution:

(a) $2\vec{a} + 3\vec{b} = 2\,(2, 3, -1) + 3\,(1, 0, 4)$

$= (4, 6, -2) + (3, 0, 12)$

$= (4 + 3, 6 + 0, -2 + 12)$

$\therefore\ 2\vec{a} + 3\vec{b} = (7, 6, 10)$

(b) $3\vec{a} - 2\vec{b} = 3\,(2, 3, -1) - 2\,(1, 0, 4)$

$= (6, 9, -3) - (2, 0, 8)$

$= (6 - 2, 9 - 0, -3 - 8)$

$\therefore\ 3\vec{a} - 2\vec{b} = (4, 9, -11)$

Example 5 If $\vec{a} = (2, 1, 2)$, $\vec{b} = (4, 5, 1)$ find

(a) $\vec{a} + \vec{b}$ (b) $\vec{a} - \vec{b}$.

Solution:

(a) $\vec{a} + \vec{b} = (2, 1, 2) + (4, 5, 1)$

$= (2 + 4, 1 + 5, 2 + 1)$

$\therefore\ \vec{a} + \vec{b} = (6, 6, 3)$

(b) $\vec{a} - \vec{b} = (2, 1, 2) - (4, 5, 1)$

$= (2 - 4, 1 - 5, 2 - 1)$

$\therefore\ \vec{a} - \vec{b} = (-2, -4, 1)$

Example 6 If $\vec{a} = (3, 2)$, $\vec{b} = (5, -1)$, $\vec{c} = (-4, 2)$ find

(a) $2\vec{a} + \vec{b} - 3\vec{c}$ (b) $3\vec{a} + 2\vec{b} - \vec{c}$.

Solution:

(a) $2\vec{a} + \vec{b} - 3\vec{c} = 2(3, 2) + (5 -1), -3(-4, 2)$

$= (6, 4) + (5, -1) - (-12, 6)$

$= (6 + 5 + 12, 4 - 1 - 6)$

$\therefore\ 2\vec{a} + \vec{b} - 3\vec{c} = (23, -3)$

(b) $3\vec{a} + 2\vec{b} - \vec{c} = 3(3, 2) + 2(5 -1), -(-4, 2)$

$= (9, 6) + (10, -2) - (-4, 2)$

$= (9 + 10 + 4, 6 - 2 - 2)$

$\therefore\ 3\vec{a} + 2\vec{b} - \vec{c} = (23, 2)$

Example 7 If $\vec{a} = (2, 1, 3)$, $\vec{b} = (2, 1, 4)$, $\vec{c} = (0, 1, 2)$ find

(a) $\vec{a} - \vec{b}$ (b) $4(\vec{c} - \vec{a})$.

Solution:

(a) $\vec{a} - \vec{b} = (2, 1, 3) - (2, 1\ 4)$

$= (2 - 2, 1 - 1, 3 - 4)$

$\therefore\ \vec{a} - \vec{b} = (0, 0, -1)$

(b) $4(\vec{c} - \vec{a}) = 4[(0, 1, 2) - (2, 1, 3)]$

$= 4[(0 - 2, 1 - 1, 2 - 3)]$

$= 4[-2, 0, -1]$

$\therefore\ 4(\vec{c} - \vec{a}) = (-8, 0, -4)$

Example 8 If $\vec{a} = (2, 4, -1)$, $\vec{b} = (3, 2, 0)$ find

(a) $2\vec{a} + 3\vec{b} - \vec{c}$ (b) $4\vec{a} - \vec{b} + 2\vec{c}$.

Solution:

(a) $2\vec{a} + 3\vec{b} - \vec{c} = 2(2, 4, -1) + 3(3, 2, 0) - \vec{c}$

$= (4, 8, -2) + (9, 6, 0) - \vec{c}$

$= (4 + 9, 8 + 6, -2 + 0) - \vec{c}$

$\therefore\ 2\vec{a} + 3\vec{b} - \vec{c} = (13, 14, -2) - \vec{c}$

(b) $4\vec{a} - \vec{b} + 2\vec{c} = 4\,(2, 4, -1) - (3, 2, 0) + 2\vec{c}$

$= (8, 16, -4) - (3, 2, 0) + 2\vec{c}$

$= (8 - 3, 16 - 2, -4 - 0) + 2\vec{c}$

$\therefore\ 4\vec{a} - \vec{b} + 2\vec{c} = (5, 14, -4) + 2\vec{c}$

Example 9 If $\vec{a} = (3, 2, -1)$, $\vec{b} = (2, 0, -3)$, $\vec{c} = (1, -2, 3)$ What is the value of $2\vec{a} + \vec{b} + 3\vec{c}$? And find its magnitude

or

what is the magnitude of the $2\vec{a} + \vec{b} + 3\vec{c}$.

Solution:

$2\vec{a} + \vec{b} + 3\vec{c} = 2\,(3, 2, -1) + (2, 0, -3) + 3\,(1, -2, 3)$

$= (6, 4 -2) + (2, 0 -3) + (3, -6, 9)$

$= (6 + 2 + 3, 4 + 0 - 6, -2 - 3 + 9)$

$\therefore\ 2\vec{a} + \vec{b} + 3\vec{c} = (11, -2, 4)$

Magnitude $= |2\vec{a} + \vec{b} + 3\vec{c}| = |(11, -2, 4)| = \sqrt{(11)^2 + (-2)^2 + (4)^2}$

$= \sqrt{121 + 4 + 16}$

$= \sqrt{141}$

Example 10 If $\vec{a} = (2, 3)$, $\vec{b} = (-1, 5)$ what are (a) $\vec{a} + \vec{b}$ and (b) $2\vec{a} + 3\vec{b}$.

Solution:

(a) $\vec{a} + \vec{b} = (2, 3) + (-1, 5)$

$= (2 - 1, 3 + 5)$

$\therefore\ \vec{a} + \vec{b} = (1, 8)$

(b) $2\vec{a} + 3\vec{b} = 2\,(2, 3) + 3\,(-1, 5)$

$= (4, 6) + (-3, 15)$

$= (4 - 3, 6 + 15)$

$\therefore\ 2\vec{a} + 3\vec{b} = (1, 21)$

Example 11 If $\vec{a} = (2, 3)$, $\vec{b} = (-2, 5)$ What are the $\vec{a} + \vec{b}$ and $2\vec{a} + 3\vec{b}$ find their inverses.

Solution

(a) $\vec{a} + \vec{b} = (2, 3) + (-2, 5)$

$= (2 - 2, 3 + 5)$

$\therefore\ \vec{a} + \vec{b} = (0, 8)$

(b) $2\vec{a} + 3\vec{b} = 2\,(2, 3) + 3\,(-2, 5)$

$= (4, 6) + (-6, 15)$

$= (4 - 6, 6 + 15)$

$\therefore\ 2\vec{a} + 3\vec{b} = (-2, 21)$

Inverse of $\vec{a} + \vec{b} = (0, -8)$

Inverse of $2\vec{a} + 3\vec{b} = (2, -21)$

Example 12 If $\vec{a} = (1, 2)$, $\vec{b} = (2, 3)$, $\vec{c} = (-2, 4)$ what is the magnitude of $2\vec{a} + 3\vec{b} - 4\vec{c}$.

Solution:

$$2\vec{a} + 3\vec{b} - 4\vec{c} = 2(1, 2) + 3(2, 3) - 4(-2, 4)$$
$$= (2, 4) + (6, 9) - (-8, 16)$$
$$= (2 + 6 + 8, 4 + 9 - 16)$$
$$= (16, -3)$$

Magnitude $= |2\vec{a} + 3\vec{b} - 4\vec{c}| = |(16, -3)|$

$$= \sqrt{(16)^2 + (-3)^2}$$
$$= \sqrt{256 + 9}$$
$$= \sqrt{265}$$

Example 13 If $\vec{a} = (1, 1, -1)$, $\vec{b} = (4, 1, 2)$ and $\vec{c} = (0, 1, -2)$ so that $(\vec{a} - \vec{b}) = 3(\vec{c} - \vec{a})$.

Solution

L.H.S $= (\vec{a} - \vec{b})$

$= (1, 1, -1) - (4, 1, 2)$

$= (1 - 4, 1 - 1, -1 - 2)$

L.H.S $= (-3, 0, -3)$

$= 3[-1, 0, -1]$

R.H.S $= 3(\vec{c} - \vec{a})$

$= 3[(0, 1, -2) - (1, 1, -1)]$

$= 3[0 - 1, 1 - 1, -2 - (-1)]$

$= 3[-1, 0, -2 + 1]$

$= 3[-1, 0, -1]$

R.H.S $= (-3, 0, -3)$

L.H.S = R.H.S

$\therefore\ (\vec{a} - \vec{b}) = 3(\vec{c} - \vec{a})$

Example 14 If $\vec{a} = (2, 3)$, $\vec{b} = (-1, 5)$ inverse of $3\vec{a} + 4\vec{b}$?

Solution:

$$3\vec{a} + 4\vec{b} = 3(2, 3) + 9(-1, 5)$$
$$= (6, 9) + (-4, 20)$$
$$= (6 + (-4), 9 + 20)$$

$3\vec{a} + 4\vec{b} = (2, 29)$

$\therefore$ Inverse of $3\vec{a} + 4\vec{b} = (-2, -29)$

Example 15 If $\vec{a} = (4, 6, -8)$, $\vec{b} = (-2, 6, -4)$ find the inverses of (a) $\vec{a} + \vec{b}$, and (b) $\vec{a} - \vec{b}$.

Solution:

(a) $\vec{a} + \vec{b} = (4, 6, -8) + (-2, 6, -4)$

$= [4 - 2, 6 + 6, -8 + (-4)]$

$\vec{a} + \vec{b} = (2, 12, -12)$

Inverse of $\vec{a} + \vec{b} = (-2, -12, 12)$

(b) $\vec{a} - \vec{b} = (4, 6, 8) - (-2, 6 -4)$

$= [4 - (-2), 6 - 6, -8 - (-4)]$

$\vec{a} - \vec{b} = (6, 0, -4)$

Inverse of $\vec{a} - \vec{b} = (-6, 0, 4)$

Example 16 Find the scalar product (dot product) of the following pairs of vectors.

(a) $\vec{a} = (2i + 3j - k)$ and $\vec{b} = (6i - j + 2k)$

Solution: Scalar product or Dot product = $\vec{a} \cdot \vec{b}$

$= (2i + 3j - k) \cdot (6i - j + 2k)$

$= (2)(6) + (3)(-1) + (-1)(+2)$

$= 12 - 3 - 2$

$= 7$

(b) $\vec{a} = (i - 3j - 7k)$ $\vec{b} = (-2i + j - 3k)$

Dot product $= \vec{a} \cdot \vec{b}$

Scalar product $= (i - 3j - 7k) \cdot (-2i + j - 3k)$

$= (1)(-2) + (-3)(1) + (-7)(-3)$

$= -2 - 3 + 21$

$= 16$

(c) $\vec{a} = (2i + j)$ $\vec{b} = (3j + 2k)$

Dot product $= \vec{a} \cdot \vec{b}$

$= (2i + j) . (3j + 2k)$

$= (2)(0) + (1)(3) + (0)(2)$

$= 0 + 3 + 0$

$= 3$

(d) $\vec{a} = (i + 3k)$, $\vec{b} = (4i - j)$

Dot product $= \vec{a} \cdot \vec{b}$

$= (i + 3k) . (4i - j)$

$= (1)(4) + (3)(0) + (0)(-1)$

$= 4 + 0 + 0$

$= 4$

(e) $\vec{a} = (2i + 4j - 6k)$, $\vec{b} = (i - j - 3k)$

Dot product $= (2i + 4j - 6k) . (i - j - 3k)$
$= (2)(1) + (4)(-1) + (-6)(-3)$
$= 2 + (-4) + 18$
$= 16$

Example 17 Find the vector product (cross product) of the following pairs of vectors.

(a) $\vec{a} = (3i - 2j + k)$ $\vec{b} = (2i + j - 4k)$

Solution:

$$\text{Vector product or cross product} = \vec{a} \times \vec{b} = \begin{vmatrix} + & - & + \\ i & j & k \\ 3 & -2 & 1 \\ 2 & 1 & -4 \end{vmatrix}$$

$= i\,[(-2)(-4) - (1)(1)] - j\,[(3)(-4) - (2)(1)] + k\,[(3)(1) - (2)(-2)]$
$= i\,(8 - 1) - j\,(-12 - 2) + k\,(3 + 4)$
$= 7i + 14j + 7k$

(b) $\vec{a} = (2i + 3j)$, $\vec{b} = (2j - k)$

Solution:

$$\text{Cross product} = \vec{a} \times \vec{b} = \begin{vmatrix} + & - & + \\ i & j & k \\ 2 & 3 & 0 \\ 0 & 2 & -1 \end{vmatrix}$$

$= i\,(-3 - 0) - j\,(-2) + k\,(4)$
$= -3i + 2j + 4k$

(c) $\vec{a} = (2i + j - 3k)$, $\vec{b} = (i - 2j + k)$

Solution:

$$\text{Cross product} = \vec{a} \times \vec{b} \begin{vmatrix} + & - & + \\ i & j & k \\ 2 & 1 & -3 \\ 1 & -2 & 1 \end{vmatrix}$$

$= i\,(6 - 1) - j\,(+2 - (-3)) + k\,(-4 - 1)$
$= -5i - 5j - 5k$

Example 18 If $\vec{a} = (2, -1)$, $\vec{b} = (3, 4)$, $\vec{c} = (-1, 0)$

find (a) $\vec{a} + \vec{b} + \vec{c}$, (b) $\vec{a} - \vec{b} + \vec{c}$, (c) $2\vec{a} + 3\vec{b} - \vec{c}$, (d) $3\vec{a} - 2\vec{b} + 4\vec{c}$.

Solution:

(a) $\vec{a} + \vec{b} + \vec{c} = (2, -1) + (3, 4) + (-1, 0)$

$= (2 + 3 - 1, -1 + 4)$

$\therefore \vec{a} + \vec{b} + \vec{c} = (4, 3)$

(b) $\vec{a} - \vec{b} + \vec{c} = (2, -1) - (3, 4) + (-1, 0)$

$= (2 - 3 - 1, -1 - 4)$

$\therefore \vec{a} - \vec{b} + \vec{c} = (-2, -5)$

(c) $2\vec{a} + 3\vec{b} - \vec{c} = 2\,(2, -1) + 3\,(3, 4) - (-1, 0)$

$= (4, -2) + (9, 12) - (-1, 0)$

$= (4 + 9 + 1, -2, + 12)$

$\therefore 2\vec{a} + 3\vec{b} - \vec{c} = (14, 10)$

(d) $3\vec{a} - 2\vec{b} + 4\vec{c} = 3\,(2, -1) - 2\,(3, 4) + 4\,(-1, 0)$

$= (6, -3) - (6, 8) + (-4, 0)$

$= (6 - 6 - 4, -3 - 8)$

$\therefore 3\vec{a} - 2\vec{b} + 4\vec{c} = (-4, -11)$

Example 19 If $\vec{a} = (1, 2)$, $\vec{b} = (2, 3)$, $\vec{c} = (-1, 4)$, $\vec{d} = (2, -4)$ find

(a) $\vec{a} + \vec{b} + \vec{c} + \vec{d}$, (b) $\vec{a} - \vec{b} + \vec{c} - \vec{d}$, (c) $2\vec{a} - \vec{b} + 2\vec{c} - \vec{d}$,

(d) $\vec{a} + 2\vec{b} - 3\vec{c} + 4\vec{d}$, (e) $3\vec{a} - 2\vec{b} + 3\vec{c} - 4\vec{d}$, (f) $\vec{a} - 3\vec{b} + 4\vec{c} - 5\vec{d}$,

(g) $3\vec{a} - \vec{b} + 2\vec{c} + 3\vec{d}$.

Solution:

(a) $\vec{a} + \vec{b} + \vec{c} + \vec{d} = (1, 2) + (2, 3) + (-1, 4) + (2, -4)$

$= (1 + 2 - 1 + 2, 2 + 3 + 4 - 4)$

$= (4, 5)$

(b) $\vec{a} - \vec{b} + \vec{c} - \vec{d} = (1, 2) - (2, 3) + (-1, 4) - (2, -4)$

$= (1 - 2 - 1 - 2, 2 - 3 + 4 + 4)$

$= (-4, 7)$

(c) $2\vec{a} - \vec{b} + 2\vec{c} - \vec{d} = 2\,(1, 2) - (2, 3) + 2\,(-1, 4) - (2, -4)$

$= (2, 4) - (2, 3) + (-2, 8) - (2, -4)$

$= (2 - 2 - 2 - 2, 4 - 3 + 8 + 4)$

$= (-4, 13)$

(d) $\vec{a} + 2\vec{b} - 3\vec{c} + 4\vec{d} = (1, 2) + 2\,(2, 3) - 3\,(-1, 4) + 4\,(2, -4)$

$= (1, 2) + (4, 6) - (-3, 12) + (8, -16)$

$= (1 + 4 + 3 + 8, 2 + 6 - 12 - 16)$

$= (16, -20)$

(e) $3\vec{a} - 2\vec{b} + 3\vec{c} - 4\vec{d} = 3\,(1, 2) - 2\,(2, 3) + 3\,(-1, 4) - 4\,(2, -4)$

$= (3, 6) - (4, 6) + (-3, 12) - (8, -16)$

$= (3 - 4 - 3 + 8, 6 - 6 + 12 + 16)$

$= (-12, 28)$

(f) $\vec{a} - 3\vec{b} + 4\vec{c} - 5\vec{d} = (1, 2) - 3\,(2, 3) + 4\,(-1, 4) - 5\,(2, -4)$
$= (1, 2) - (6, 9) + (-4, 16) - (10, -20)$
$= (1 - 6 - 4 - 10, 2 - 9 + 16 + 20)$
$= (-19, 29)$

(g) $3\vec{a} - \vec{b} + 2\vec{c} + 3\vec{d} = 3\,(1, 2) - (2, 3) + 2\,(-1, 4) + 3\,(2, -4)$
$= (3, 6) - (2, 3) + (-2, 8) + (6, -12)$
$= (3 - 2 - 2 + 6, 6 - 3 + 8 - 12)$
$= (5, -1)$

Example 20 If $\vec{a} = (1, 2, 3)$, $\vec{b} = (2, -3, 0)$, $\vec{c} = (1, 2, -1)$ find

(a) $\vec{a} + \vec{b} + \vec{c}$, (b) $2\vec{a} - 3\vec{b} + 2\vec{c}$, (c) $3\vec{a} - 2\vec{b} - \vec{c}$,

(d) $4\vec{a} + 2\vec{b} - 5\vec{c}$, (e) $\vec{a} + 3\vec{b} - 4\vec{c}$, (f) $2\vec{a} + 4\vec{b} + 2\vec{c}$.

Solution:

(a) $\vec{a} + \vec{b} + \vec{c} = (1, 2, 3) + (2, -3, 0) + 2\,(1, 2, -1)$
$= (1 + 2 + 1, 2 - 3 + 2, 3 - 1)$
$= (4, 1, 2)$

(b) $2\vec{a} - 3\vec{b} + 2\vec{c} = 2\,(1, 2, 3) - 3\,(2, -3, 0) + 2\,(1, 2, -1)$
$= (2, 4, 6) - (6, -9, 0) + (2, 4, -2)$
$= (2 - 6 + 2, 4 + 9 + 4, 6 - 2)$
$= (-2, 17, 4)$

(c) $3\vec{a} - 2\vec{b} - \vec{c} = 3\,(1, 2, 3) - 2\,(2, -3, 0) - (1, 2, -1)$
$= (3, 6, 9) - (4, -6, 0) - (1, 2, -1)$
$= (3 - 4 - 1, 6 + 6 - 2, 9 + 1)$
$= (-2, 10, 10)$

(d) $4\vec{a} + 2\vec{b} - 5\vec{c} = 4\,(1, 2, 3) + 2\,(2, -3, 0) - 5\,(1, 2, -1)$
$= (4, 8, 12) + (4, -6, 0) - (5, 10, -5)$
$= (4 + 4 - 5, 8 - 6 - 10, 12 + 5)$
$= (3, -8, 17)$

(e) $\vec{a} + 3\vec{b} - 4\vec{c} = (1, 2, 3) + 3\,(2, -3, 0) - 4\,(1, 2, -1)$
$= (1, 2, 3) + (6, -9, 0) - (4, 8, -4)$
$= (1 + 6 - 4, 2 - 9 - 8, 3 + 4)$
$= (3, -15, 7)$

(f) $2\vec{a} + 4\vec{b} + 2\vec{c} = 2\,(1, 2, 3) + 4\,(2, -3, 0) + 2\,(1, 2, -1)$
$= (2, 4, 6) + (8, -12, 0) + (2, 4, -2)$
$= (2 + 8 + 2, 4 - 12 + 4, 6 - 2)$
$= (12, -4, 4)$

Example 21 If $\vec{a} = (-1, 2, 0)$, $\vec{b} = (0, -2, 1)$, $\vec{c} = (1, 0, 3)$, $\vec{d} = (1, 2, 3)$ find

(a) $\vec{a} - \vec{b} + \vec{c} - \vec{d}$, (b) $2\vec{a} + 2\vec{b} - \vec{c} + 4\vec{d}$, (c) $2\vec{a} - 3\vec{b} + 4\vec{c} + 2\vec{d}$

(d) $3\vec{a} - \vec{b} + 3\vec{c} - \vec{d}$, (e) $-\vec{a} + 2\vec{b} + \vec{c} + 2\vec{d}$, (f) $-4\vec{a} + 3\vec{b} - \vec{c} - 2\vec{d}$

(g) $\vec{a} + \vec{b} + 2\vec{c} + 3\vec{d}$

Solution:

(a) $\vec{a} - \vec{b} + \vec{c} - \vec{d} = (-1, 2, 0) - (0, -2, 1) + (1, 0, 3) - (1, 2, 3)$
$= (-1 + 1 - 1, 2 + 2 - 2, 0 - 1 + 3 - 3)$
$= (-1, 2, -1)$

(b) $2\vec{a} + 2\vec{b} - \vec{c} + 4\vec{d} = 2(-1, 2, 0) + 2(0, -2, 1) - (1, 0, 3) + 4(1, 2, 3)$
$= (-2, 4, 0) + (0, -4, 2) - (1, 0, 3) + (4, 8, 12)$
$= (-2 - 1 + 4, 4 - 4 + 8, 0 + 2 - 3 + 12)$
$= (1, 8, 11)$

(c) $2\vec{a} - 3\vec{b} + 4\vec{c} + 2\vec{d} = 2(-1, 2, 0) - 3(0, -2, 1) + 4(1, 0, 3) + 2(1, 2, 3)$
$= (-2, 4, 0) - (0, -6, 3) + (4, 0, 12) + (2, 4, 6)$
$= (-2 + 4 + 2, 4 + 6 + 4, -3 + 12 + 6)$
$= (4, 14, 15)$

(d) $3\vec{a} - \vec{b} + 3\vec{c} - \vec{d} = 3(-1, 2, 0) - (0, -2, 1) + 3(1, 0, 3) - (1, 2, 3)$
$= (-3, 6, 0) - (0, -2, 1) + (3, 0, 9) - (1, 2, 3)$
$= (-3 + 3 - 1, 6 + 2 - 2, -1 + 9 - 3)$
$= (-1, 6, 5)$

(e) $-\vec{a} + 2\vec{b} + \vec{c} + 2\vec{d} = -(-1, 2, 0) + 2(0, -2, 1) + (1, 0, 3) + 2(1, 2, 3)$
$= (1, -2, 0) + (0, -4, 2) + (1, 0, 3) + (2, 4, 6)$
$= (1 + 0 + 1 + 2, -2 - 4 + 4, +2 + 3 + 6)$
$= (4, -2, 11)$

(f) $-4\vec{a} + 3\vec{b} - \vec{c} - 2\vec{d} = -4(-1, 2, 0) + 3(0, -2, 1) - (1, 0, 3) - 2(1, 2, 3)$
$= (4, -8, 0) + (0, -6, 3) - (1, 0, 3) - (2, 4, 6)$
$= (4 + 0 - 1 - 2, -8 - 6 - 4, 3 - 3 - 6)$
$= (1, -18, -6)$

(g) $\vec{a} + \vec{b} + 2\vec{c} + 3\vec{d} = (-1, 2, 0) + (0, -2, 1) + 2(1, 0, 3) + 3(1, 2, 3)$
$= (-1, 2, 0) + (0, -2, 1) + (2, 0, 6) + (3, 6, 9)$
$= (-1 + 2 + 3, + 2 - 2 + 6, 1 + 6 + 9)$
$= (4, 6, 16)$

Example 22 If $\vec{a} = (1, -1, 2)$, $\vec{b} = (0, 2, -3)$, $\vec{c} = (2, 0, -3)$, $\vec{d} = (-1, 2, 3)$ find

(a) $\vec{a} + \vec{b} - \vec{c} + \vec{d}$, (b) $3[2\vec{a} + \vec{b}] + 4[\vec{c} - 2\vec{d}]$, (c) $2[3\vec{a} + 3\vec{c}] - [4\vec{b} + 6\vec{d}]$,

(d) $[4\vec{a} + 2\vec{d}] + 3[\vec{b} - \vec{c}]$ and its magnitude,

(e) $2[\vec{d} - \vec{c}] + 3[\vec{b} - 2\vec{a}]$ and its magnitude.

Solution:

(a) $\vec{a} + \vec{b} - \vec{c} + \vec{d} = (1, -1, 2) + (0, 2, -3) - (2, 0, -3) + (-1, 2, 3)$
$= (1 - 2 - 1, -1 + 2 + 2, 2 - 3 + 3 + 3)$
$= (-2, 3, 5)$

(b) $3[2\vec{a} + \vec{b}] + 4[\vec{c} - 2\vec{d}] = 3\ [2\ (1, -1, 2) + (0, 2, -3)] + 4\ [(2, 0, -3) - 2\ (-1, 2, 3)]$
$= 3\ [(2, -2, 4) + (0, 2, -3)] + 4\ [(2, 0, -3) - (-2, 4, 6)]$
$= 3\ [2, 0, 1] + 4\ [4, -4, -9]$
$= (6, 0, 3) + (16, -16, -36)$
$= (6 + 16, 0 - 16, 3 - 36)$
$= (22, -16, -33)$

(c) $2[3\vec{a} + 3\vec{c}] - [4\vec{b} + 6\vec{d}] = 2\ [3\ (1, -1, 2) + 3\ (2, 0, -3)] - [4\ (0, 2, -3) + 6\ (-1, 2, 3)]$
$= 2\ [(3, -3, 6) + (6, 0, -9)] - [(0, 8, -12) + (-6, 12, 18)]$
$= 2\ [9, -3, -3] - [-6, 20, 6]$
$= (18, -6, -6) - (-6, 20, 6)$
$= (18 + 6, -6 - 20, -6 - 6)$
$= (24, -26, -12)$

(d) $[4\vec{a} + 2\vec{d}] + 3[\vec{b} - \vec{c}] = [4\ (1, -1, 2) + 2\ (-1, 2, 3)] + 3\ [(0, 2, -3) - (2, 0, -3)]$
$= [(4, -4, 8) + (-2, 4, 6)] + 3\ [(0, 2, -3) - (2, 0, -3)]$
$= [2, 0, 14] + 3\ [-2, 2, 0]$
$= (2, 0, 14) + (-6, 6, 0)$
$= (2 - 6, 0 + 6, 14 + 0)$
$= (-4, 6, 14)$

$\therefore$ Magnitudes $= \sqrt{(-4)^2 + (6)^2 + (14)^2}$

$= \sqrt{16 + 36 + 196}$

$= \sqrt{248}$

(e) $2\ [\vec{d} - \vec{c}] + 3[\vec{b} - 2\vec{a}] = 2\ [-3, 2, 6] + 3\ [-2, 4, -7]$
$= (-6, 4, 12) + (-6, 12, -21)$
$= (-12, 16, -9)$

$\therefore$ Magnitudes $= \sqrt{(-12)^2 + (16)^2 + (-9)^2}$

$= \sqrt{144 + 256 + 81}$

$= \sqrt{481}$

Example 23 Given the vectors $p = (-1, 4, 3)$, $q = (1, -2, 0)$, $r = (0, 4, -1)$ compute $(2p - q + r)$ and $\left(\frac{5}{2}p - \frac{3}{2}q + \frac{7}{2}r\right)$.

Solution:

(a) $2p - q + r = 2(-1, 4, 3) - (1, -2, 0) + (0, 4, -1)$
$= (-2, 8, 6) - (1, -2, 0) + (0, 4, -1)$
$= (-2 - 1 + 0, 8 + 2 + 4, 6 - 0 - 1)$

$\therefore\ 2p - q + r = (-3, 14, 5)$

(b) $\left(\frac{5}{2}p - \frac{3}{2}q + \frac{7}{2}r\right) = \frac{5}{2}(-1, 4, 3) - \frac{3}{2}(1, -2, 0) + \frac{7}{2}(0, 4, -1)$

$= \left(\frac{-5}{2}, 10, \frac{15}{2}\right) - \left(\frac{3}{2}, \frac{-6}{2}, 0\right) + \left(0, 14, \frac{-7}{2}\right)$

$= \left(\frac{-5}{2} - \frac{3}{2}, 10 + \frac{6}{2} + 14, \frac{15}{2} - \frac{7}{2}\right)$

$\therefore\ \left(\frac{5}{2}p - \frac{3}{2}q + \frac{7}{2}r\right) = (-4, 27, 4)$

Example 24 Given the vectors $u = (2, 3, 1)$, $v = (2, -1, 0)$ and $w = (0, -2, 2)$ compute $(u - v + 2w)$ and $\left(\frac{5}{2}u - \frac{7}{2}v + \frac{9}{2}w\right)$.

Solution:

(a) $u - v + 2w = (2, 3, 1) - (2, -1, 0) + 2(0, -2, 2)$
$= (2, 3, 1) - (2, -1, 0) + (0, -4, 4)$
$= (2 - 2, 3 - (1) - 4, 1 + 4)$

$\therefore\ u - v + 2w = (0, 0, 5)$

(b) $\frac{5}{2}u - \frac{7}{2}v + \frac{9}{2}w = \frac{5}{2}(2, 3, 1) - \frac{7}{2}(2, -1, 0) + \frac{9}{2}(0, -2, 2)$

$= \left(5, \frac{15}{2}, \frac{5}{2}\right) - \left(7, \frac{-7}{2}, 0\right)(0, -9, 9)$

$= \left(5 - 7, \frac{15}{2} + \frac{7}{2} - 9, \frac{5}{2} + 9\right)$

$\therefore\ \frac{5}{2}u - \frac{7}{2}v + \frac{9}{2}w = \left(-2, 2, \frac{23}{2}\right)$

Example 25 If $\vec{a} = (1, -1)$, $\vec{b} = (-2, 5)$ find,

(a) $\vec{a} + \vec{b}$; (b) $\vec{a} - \vec{b}$, (c) $2\vec{a} + 3\vec{b}$, (d) $3\vec{a} - 2\vec{b}$.

Solution:

(a) $\vec{a} + \vec{b} = (1, -1) + (-2, 5)$
$= (1 - 2, -1 + 5)$
$= (-1, 4)$

(b) $\vec{a} - \vec{b} = (1, -1) - (-2, 5)$
$= (1 + 2, -1 - 5)$

$= (3, -6)$

(c) $2\vec{a} + 3\vec{b} = 2(1, -1) + 3(-2, 5)$

$= (2, -2) + (-6, 15)$

$= (2 - 6, -2 + 15)$

$= (-4, 13)$.

(d) $3\vec{a} - 2\vec{b} = 3(1, -1) - 2(-2, 5)$

$= (3, -3) - (-4, 10)$

$= (3 + 4, -3 - 10)$

$= (7, -13)$

Problems on Co-planar Vectors

Example 26 Prove that the vectors $5\vec{a} + 6\vec{b} + 7\vec{c}$, $7\vec{a} - 8\vec{b} + 9\vec{c}$ and $3\vec{a} + 20\vec{b} + 5\vec{c}$ are co-planar vectors. Where $\vec{a}, \vec{b}, \vec{c}$ be any vectors.

Solution: Let $\vec{a}, \vec{b}$, and $\vec{c}$ be any vectors and where l and m are the unknown values.

$$5\vec{a} + 6\vec{b} + 7\vec{c} = l(7\vec{a} - 8\vec{b} + 9\vec{c}) + m(3\vec{a} + 20\vec{b} + 5\vec{c})$$

equating the coefficients of $\vec{a}, \vec{b}, \vec{c}$ me get

$$5 = 7l + 3m \quad \ldots (1)$$

$$6 = -8l + 20m \quad \ldots (2)$$

$$7 = 9l + 5m \quad \ldots (3)$$

Solving eqn. (1) and (2) we get

$$7l + 3m = 5 \times 8$$

$$-8l + 20m = 6 \times 7$$

$$\cancel{56l} + 24m = 40$$

$$\cancel{56l} + 140m = 42$$

$$164m = 82$$

$$\therefore\ m = \frac{1}{2}$$

Put $m = \frac{1}{2}$ in to (1) we get

$$7l = 5 - 3\left(\frac{1}{2}\right)$$

$$7l = \frac{10 - 3}{2}$$

$$7l = \frac{7}{2}$$

$$\therefore\ l = \frac{1}{2}$$

substituting the values $l = \frac{1}{2}$ and $m = \frac{1}{2}$ in eqn. (3) we get

$$7 = 9\left(\frac{1}{2}\right) + 5\left(\frac{1}{2}\right)$$

$$= \frac{9}{2} + \frac{5}{2}$$

$$7 = \frac{14}{2} = 7$$

$$\therefore\ 7 = 7$$

Therefore the given vectors are co-planar vectors.

Example 27 Show that the vectors $\vec{a} - 2\vec{b} + 3\vec{c}, \left(-2\vec{a} + 3\vec{b} - 4\vec{c}\right), \vec{a} - 3\vec{b} + 5\vec{c}$ are co-planar.

Solution: Let $\vec{a}, \vec{b},$ and $\vec{c}$ be any vectors and where *p* and *q* are unknown values.

$$\vec{a} - 2\vec{b} + 3\vec{c} = p\left(-2\vec{a} + 3\vec{b} - 4\vec{c}\right) + q\left(\vec{a} - 3\vec{b} + 5\vec{c}\right)$$

Equating the co-efficients of $\vec{a}, \vec{b},$ and $\vec{c}$ we get

$$1 = -2p + q \quad \ldots (1)$$

$$-2 = 3p - 3q \quad \ldots (2)$$

$$3 = -4p + 5q \quad \ldots (3)$$

Solving equations (1) and (2)

$$-2p + q = 1 \times 3$$

$$3p - 3q = -2 \quad \ldots (4)$$

$$-6p + 3q = 3 \quad \ldots (5)$$

(4) + (5)

$$-3p = 1$$

$$\therefore\ p = -\frac{1}{3}$$

P... $\therefore\ p = -\frac{1}{3}$ into (1) we get

$$1 + 2\left(-\frac{1}{3}\right) = q$$

$$\therefore\ q = \frac{1}{3}$$

substituting the values $p = -\frac{1}{3}$ and $q = \frac{1}{3}$ in equation (3) we get

$$3 = -4\left(-\frac{1}{3}\right) + 5\left(\frac{1}{3}\right)$$

$$3 = \frac{4}{3} + \frac{5}{3}$$

$$3 = \frac{9}{3}$$

$$3 = 3$$

Therefore the given vectors are co-planar.

Example 28 Show that the following vectors $4\vec{i} + 2\vec{j} + \hat{k}, (2\hat{i} - \hat{j} + 3\hat{k})$ and $(8\hat{i} + 7\hat{k})$ are co-planar.

Solution: Let $\hat{i}, \hat{j}, \hat{k}$ be any vectors and take l and m are unknown values.

$$4\vec{i} + 2\vec{j} + \hat{k} = l(2\hat{i} - \hat{j} + 3\hat{k}) + m(8\hat{i} + 7\hat{k})$$

equating the coefficients of i, j, k we get

$$4 = 2l + 8m \quad \dots (1)$$

$$2 = -l \quad \dots (2)$$

$$1 = 3l + 7m \quad \dots (3)$$

Solving (2) we get

$$-l = 2$$

$$\therefore l = -2$$

Put $l = -2$ in to (1) we get

$$4 - [2(-2)] = 8m$$

$$8 = 8m$$

$$\therefore m = 1$$

Substituting the values of $l = -2$, $m = 1$ in equation (3) we get

$$1 = 3(-2) + 7(1)$$

$$1 = -6 + 7$$

$$\therefore 1 = 1$$

Therefore the given vectors are coplanar.

Example 29 Show that the four vectors $4i + 5j + k, -(j + k), 5i + 9j + 4k, -i + k$ are coplanar.

Solution:

Let $\overrightarrow{OA} = 4i + 5j + k$

$$\overrightarrow{OB} = -(j + k) = -j - k$$

$$\overrightarrow{OC} = 5i + 9j + 4k$$

$$\overrightarrow{OD} = -i + k$$

$$\overrightarrow{AB} = \overrightarrow{OB} - \overrightarrow{OA}$$

$$= (-j - k) - (4i + 5j + k)$$

$$\therefore \overrightarrow{AB} = 4i - 6j - 2k$$

$$\vec{AC} = \vec{OC} - \vec{OA}$$

$$= (5i + 9j + 4k) - (4i + 5j + k)$$

$$\vec{AC} = i + 4j + 3k$$

$$\vec{AD} = \vec{OD} - \vec{OA}$$

$$= (-i + k) - (4i + 5j + k)$$

$$\therefore \vec{AD} = -5i - 5j$$

take l and m as the unknown values.

$$\vec{AB} = l\left(\vec{AC}\right) + m\left(\vec{AD}\right)$$

$$-4i - 6j - 2k = l\,(i + 4j + 3k) + m\,(-5i - 5j)$$

equating the coefficients of i, j, k values we get

$$l - 5m = -4 \quad \ldots (1)$$

$$4l - 5m = -6 \quad \ldots (2)$$

$$3l = -2 \quad \ldots (3)$$

from (3) we get

$$3l = -2$$

$$l = -\frac{2}{3}$$

Put this value in (2)

$$4\left(-\frac{2}{3}\right) - 5m = -6$$

$$-\frac{8}{3} + 6 = 5m$$

$$\frac{-8 + 18}{3} = 5m$$

$$\frac{10}{3} = 5m$$

$$m = \frac{2}{3}$$

Substituting l and m values in equation (1)

$$l - 5m = -4$$

$$-\frac{2}{3} - 5\left(\frac{2}{3}\right) = -4$$

$$-\frac{2}{3} - \frac{10}{3} = -4$$

$$-\frac{12}{3} = -4$$

$$-4 = -4$$

$\therefore$ The given values are coplanar

Exercises

I. Questions carrying THREE marks each (Short Answers):

1. If $\vec{x} = \vec{a} + \vec{b}$, $\vec{y} = \vec{b} + \vec{c}$ and $\vec{z} = \vec{a} - \vec{c}$ find (a) $\vec{x} + \vec{y} - \vec{z}$ and (b) $\vec{x} + 2\vec{y}$.

   [*Ans: (a)* $2\vec{b} + 2\vec{c}$, *(b)* $\vec{a} + 3\vec{b} + 2\vec{c}$]

2. If $\vec{x} = 2\vec{a} + \vec{b}$, $\vec{y} = \vec{a} - 2\vec{b}$, $\vec{z} = \vec{a} + \vec{b} + \vec{c}$ find the value of (a) $2\vec{x} + \vec{y} - \vec{z}$ and (b) $\vec{x} + \vec{y} - 2\vec{z}$.

   [*Ans: (a)* $4\vec{a} - \vec{b} - \vec{c}$, *(b)* $\vec{a} - 3\vec{b} - 2\vec{c}$]

3. If $\vec{a} = \vec{x} + \vec{y} - \vec{z}$, $\vec{b} = \vec{x} - \vec{y} + \vec{z}$ show that $\vec{x} = \dfrac{(\vec{a} + \vec{b})}{2}$.

4. If $\vec{a} = \hat{i} + \hat{j} + \hat{k}$, $\vec{b} = 2\hat{i} + 2\hat{j} - \hat{k}$ find (a) $\vec{a} + \vec{b}$ (b) $\vec{a} - \vec{b}$ (c) $\vec{a} + 3\vec{b}$.

   [*Ans: (a)* $3\hat{i} + 3\hat{j}$ *(b)* $-\hat{i} - \hat{j} + 2\hat{k}$ *(c)* $7\hat{i} + 7\hat{j} - 2\hat{k}$]

5. If $\vec{w} = (-4, 1, 3, 2)$, $\vec{x} = (2, 0, 0, 4)$, $\vec{y} = (4, 3, 2, 3)$ and $\vec{z} = (1, 1, 1, 1)$ then compute the following.

   (a) $\vec{w} + \vec{x} + \vec{y} - \vec{z}$, (b) $2\vec{w} - 5\vec{x} - \vec{z}$, (c) $2\vec{w} + \vec{x} - \vec{y} + 2\vec{z}$

   [Ans: (a) (1, 3, 4, 8) (b) (–19, 1, 5, –17) (c) (–8, 1, 6, 7)]

6. If $\vec{x} = (3, 1, -4)$, $\vec{y} = (2, 2, -3)$ and $\vec{z} = (0, -4, 1)$ then compute the vectors.

   (a) $2\vec{x} - 5\vec{y}$, (b) $2\vec{x} - 3\vec{y} - \vec{z}$, (c) $2\vec{x} - \vec{y}$, (d) $2\vec{x} + \vec{y} - \vec{z}$.

   [Ans: (a) (–4, –8, 7) (b) (0, 0, 0) (c) (4, 0, –5) (d) (8, 8, –12)]

7. If $\vec{x} = (1, -1, 0, 1)$, $\vec{y} = (-1, -1, -1, 2)$ and $\vec{z} = (2, 0, 1, -1)$ then compute the vectors

   (a) $2\vec{x} + 3\vec{y} + \vec{z}$, (b) $\vec{x} - 2\vec{y} + 3\vec{z}$.

   [Ans: (a) (1, –5, –2, 7) (b) (9, 1, 5, –6)]

8. Find $\vec{a}$ if $3\vec{a} + (-2, 1, 5) = 2\vec{a} - (1, 0, 1)$. ***[Ans: (a) (1, –1, –6)]***

9. Find the magnitude (or) modulus of the following vectors.

   (a) $2\hat{i} + 3\hat{j} + 6\hat{k}$, (b) $2\hat{i} - \hat{j} + 2\hat{k}$, (c) $\hat{i} + 2\hat{j} - 3\hat{k}$,

   (d) $5\hat{i} - 2\hat{j} - \hat{k}$, (e) $3\hat{i} - 4\hat{k}$.

   [*Ans: (a) 7, (b) 3, (c)* $\sqrt{14}$, *(d)* $\sqrt{30}$, *(e) 5*]

10. Find the scalar product (or) Dot product of the following pairs of vectors.

(a) $2\hat{i} + 3\hat{j} - \hat{k}$ and $6\hat{i} - \hat{j} + 2\hat{k}$

(b) $4\hat{i} + \hat{j} + 5\hat{k}$ and $\hat{i} - 3\hat{j} + 4\hat{k}$

(c) $\hat{i} - 3\hat{j} - 7\hat{k}$ and $-2\hat{i} + \hat{j} - 3\hat{k}$

(d) $2\hat{i} + \hat{j}$ and $3\hat{j} + 2\hat{k}$

(e) $\hat{i} + 3\hat{k}$ and $4\hat{i} - \hat{j}$

(f) $2\hat{i} - 4\hat{j} + \hat{k}$ and $2\hat{i} + 3\hat{j} - \hat{k}$

(g) (2, 1, 4) and (–1, 2, 1)

(h) $\hat{i} + 3\hat{j} - 2\hat{k}$ and $4\hat{i} - 2\hat{j} + 4\hat{k}$

(i) $\hat{i} + \hat{j} + \hat{k}$ and $\hat{i} - \hat{j} - \hat{k}$

(j) $3\hat{i} + \hat{j} + 2\hat{k}$ and $2\hat{i} - 2\hat{j} + 4\hat{k}$

[Ans: (a) 7, (b) 21, (c) 16, (d) 3, (e) 4, (f) –9, (g) 4, (h) –10, (i) –1, (j) 12]

11. Find the vector product (or) cross product of the following pairs of vectors. Also find their magnitude.

(a) $3i - 2j + k, 2i + j - 4k$, (b) $i - j + 2k, -3i + j - k$, (c) $2i + 3j - 4k, i + 2j + 2k$

(d) $3i + 2j + 5k, 4i + j + 3k$, (e) $2i + 3j, 2j - k$, (f) $2i - j + k, 3i + 4j - k$

[Ans: (a) $7i + 14j + 7k, 7\sqrt{6}$, (b) $-i - 5j - 2k, \sqrt{30}$, (c) $14i - 8j + k, \sqrt{261}$, (d) $i + 11j - 5k, \sqrt{147}$, (e) $-3i + 2j + 4k, \sqrt{29}$, (f) $-3i + 5j + 11k, \sqrt{155}$]

12. If $\vec{a} = (1, 2, 3)$, $\vec{b} = (1, 1, 2)$ find $\vec{a} \times \vec{b}$. ***[Ans: (1, 1, –1)]***

13. If $\vec{a} = 2i - j + 2k$, $\vec{b} = i - 2j + 7k$ find $\vec{a} \times \vec{b}$.

[Ans: $-3i - 12j - 3k$]

14. If $\vec{a} = 2i + 3j - 5k$ and $\vec{b} = -3i + j + 2k$ find the value of $\vec{a} \cdot \vec{b}$ and $\vec{a} \times \vec{b}$.

[Ans: $-13, 11i + 11j + 11k$]

II Questions carrying FIVE marks each (Essay Type Answers):

1. If $\vec{a} = (-1, 0)$, $\vec{b} = (2, -1)$, $\vec{c} = (2, 1)$ and $\vec{d} = (1, 2)$ find $\vec{a} + \vec{b}$, $\vec{c} - \vec{d}$, $3\vec{a}$, $4\vec{a} + 3\vec{d}$, $2\vec{c} - \vec{b}$, $2\vec{a} + \vec{b} - 3\vec{c}$ and $\frac{1}{2}(\vec{a} + \vec{b} + \vec{c} + \vec{d})$.

[Ans: (1, –1), (1, –1), (–3, 0), (–1, 6), (2, 3), (–6, –4), (2, 1)]

2. If $\vec{a} = (-1, 0, 2)$, $\vec{b} = (2, -1, 3)$ and $\vec{c} = (2, 3, -5)$ find $\vec{a} + \vec{b}$, $2\vec{a} - 3\vec{c}$, $2(\vec{a} + \vec{b} + \vec{c})$ and $2\vec{c} - 3\vec{b} + 4\vec{a}$.

[Ans: (1, –1, 5), (–8, –9, 19), (6, 4, 0), (–6, 9, –11)]

3. If $\vec{a} = 2i + 3j$ and $\vec{b} = 3i + 4j$ then find

   (a) The magnitude of $\vec{a}$.

   (b) $\vec{a} + \vec{b}$ and its magnitude.

   (c) $\vec{a} - \vec{b}$ and its unit vector.

[Ans: (a) $\sqrt{13}$, (b) $5i + 7j, \sqrt{74}$, (c) $-i - j, \frac{-(i+j)}{\sqrt{2}}$]

4. If $\vec{a} = 2i + 3j - k$ and $\vec{b} = 3i - j + 2k$ then find (a) $|2\vec{a} - 3\vec{b}|$ and (b) $|\vec{a} + \vec{b}|$.

[Ans: (a) $\sqrt{170}$, (b) $\sqrt{30}$]

5. Find $\vec{a}$ and $\vec{b}$ if $\vec{a} - \vec{b} = (3, 4, 6)$ and $\vec{a} + \vec{b} = (1, 2, 8)$.

[Ans: $\vec{a} = (2, 3, 7)$, $\vec{b} = (-1, -1, 1)$]

6. Find $\vec{a}$ and $\vec{b}$ if $\vec{a} + \vec{b} = (2, 3, 1)$ and $2\vec{a} - 3\vec{b} = (1, 2, 3)$.

[Ans: $\vec{a} = \left(\frac{7}{5}, \frac{11}{5}, \frac{6}{5}\right)$, $\vec{b} = \left(\frac{3}{5}, \frac{4}{5}, \frac{-1}{5}\right)$]

7. Solve for x and y:

   (a) $x(1, 2) + y(2, 1) = (8, 7)$

   (b) $x^2(x, 1) + 4(0, y) = (6, 10)$ **[Ans: (a) $x = 2, y = 3$, (b) $x = 3, y = 2$]**

8. If $\vec{a} = (-1, 0, 2)$, $\vec{b} = (2, 1, -1)$ and $\vec{c} = (1, 2, 2)$ then compute (a) $\vec{a} \times \vec{b}$ (b) $\vec{b} \times \vec{c}$ (c) $\vec{c} \times \vec{a}$.

[Ans: (a) $-2i + 3j - k$, (b) $4i - 5j + 3k$, (c) $4i - 4j + 2k$]

9. If $\vec{a} = 2i - 3j - k$ and $\vec{b} = i + 4j - 2k$ then find (a) $\vec{a} \times \vec{b}$ and (b) $\vec{b} \times \vec{a}$.

[Ans: (a) $10i + 3j + 11k$ (b) $-10i - 3j - 11k$]

10. If $\vec{a} = 2i - j + k$, $\vec{b} = i - j + k$ and $\vec{c} = i + 2j$ find $(2\vec{a} + 3\vec{c}) \cdot (\vec{a} - \vec{b})$.

[Ans: 7]

11. If $\vec{a} = (+1, 1, 1)$ and $\vec{b} = (-1, 1, -1)$ find $(\vec{a} + \vec{b}) \cdot (\vec{a} - \vec{b})$. **[Ans: 0]**

12. Find the unit vectors in the direction of the following vectors.

    (a) $2i + 2j + k$, (b) $i + j + k$, (c) $3i - 6j + 2k$, (d) $4i + j - k$, (e) $5i - j + 3k$.

[Ans: (a) $\frac{2i + 2j + k}{3}$, (b) $\frac{i + j + k}{\sqrt{3}}$, (c) $\frac{3i - 6j + 2k}{7}$, (d) $\frac{4i + j - k}{3\sqrt{2}}$, (e) $\frac{5i - j + 3k}{\sqrt{35}}$]

13. Find $\vec{a} \cdot (\vec{b} \times \vec{c})$ in the following:

(a) $\vec{a} = 2i + 3j - 4k, \vec{b} = i - j + 5k, \vec{c} = 2i + j + k$

(b) $\vec{a} = i + j - 3k, \vec{b} = i + 2j + 4k, \vec{c} = -2i - j + 3k$

(c) $\vec{a} = 4i + j - 2k, \vec{b} = 2i - j + k, \vec{c} = i - 3j + 2k$

(d) $\vec{a} = i + 2j, \vec{b} = j + 2k, \vec{c} = k + 2i$

[*Ans: (a) 3, (b) –10, (c) 11, (d) 9*]

14. Show that the vectors $2i - j + k$, $i - 3j - 5k$ and $3i - 4j - 4k$ are coplanar.

15. Show that the following points:

(a) With position vectors $-6\vec{a} + 3\vec{b} + 2\vec{c}$, $3\vec{a} - 2\vec{b} + 4\vec{c}$, $5\vec{a} + 7\vec{b} + 3\vec{c}$ and $-13\vec{a} + 17\vec{b} - \vec{c}$ are coplanar.

(b) With position vectors $i + j + k$, $2i + 3j + 4k$, $3i + j + 2k$ and $-i + j$ are coplanar.

16. Show that the points A, B, C, D are coplanar

(a) A = (4, 5, 1), B = (0, –1, –1), C = (3, 9, 4) D = (–4, 4, 4).

(b) A = (2, 3, –1), B = (1, –2, 3), C = (3, 4, –2), D = (1, –6, 6)

(c) $A = 2i + j - 5k$, $B = -3i - 2j - 4k$, $C = 5i + 3j - 5k$, $D = 6i + 7j + 5k$

17. Show that the following points:

(a) With positions vectors $2\vec{a} - \vec{b} + \vec{c}$, $\vec{a} + 2\vec{b} - \vec{c}$, $3\vec{a} + \vec{b} - \vec{c}$, $-\vec{a} - 2\vec{b} + 3\vec{c}$ are coplanar.

(b) With position vectors $3\vec{a} + 2\vec{b} - 5\vec{c}$, $\vec{a} + 4\vec{b} - 3\vec{c}$, $-3\vec{a} + 2\vec{b} + \vec{c}$, $3\vec{a} + 8\vec{b} + 5\vec{c}$ are not coplanar.

18. Show that the points with position vectors $6\vec{a} - 4\vec{b} + 4\vec{c}$, $-\vec{a} - 2\vec{b} - 3\vec{c}$, $\vec{a} + 2\vec{b} - 5\vec{c}$, $-4\vec{c}$ are coplanar.

3

Matrices

Arthur Cayley
(1821–1895)

J.J. Sylvester

Contents

- Matrices, Type of Matrices.

INTRODUCTION

Matrix theory was invented by British mathematicians Arthur Cayley and J.J. Sylvester in 1858.

The elements of the matrix are enclosed within the brackets of the type. [ ] or ( ) or ‖ ‖.

TYPES OF MATRICES

Row Matrix (or) Row Vector

A matrix having only one row is called a row matrix number of rows is denoted by 'm' rows. For example, $(5\ 3\ 2)$.

Column Matrix (or) Column Vector

A matrix having only one column is called a column matrix No. of columns is denoted by 'n' columns.

For example,

$$\begin{pmatrix} -3 \\ 2 \\ 0 \end{pmatrix}$$

Zero (or) Null Matrix

A matrix in which all the elements are zeroes is called a zero or null matrix and is denoted by '0'.

For example,

$$0 = \begin{pmatrix} 0 & 0 & 0 \\ 0 & 0 & 0 \end{pmatrix}$$

Square Matrix

A matrix in which the number of rows is equal to the number of columns is called a square matrix.

For example,

$$\begin{pmatrix} 2 & 7 \\ 5 & 3 \end{pmatrix}$$

Diagonal Elements

In a square matrix the diagonal from left to right is called the principle diagonal, the elements lying along the principle diagonal are called the diagonal elements.

For example,

$$A = \begin{pmatrix} a_{11} & a_{12} & a_{13} \\ a_{21} & a_{22} & a_{23} \\ a_{31} & a_{32} & a_{33} \end{pmatrix}$$

Diagonal Matrix

A square matrix in which all the non-diagonal elements are equal to zero is said to be a diagonal matrix.

For example,

$$\begin{pmatrix} a & 0 & 0 \\ 0 & b & 0 \\ 0 & 0 & c \end{pmatrix}$$

Scalar Matrix

A diagonal matrix in which all the diagonal elements are equal is called a scalar matrix.

For example,

$$\begin{pmatrix} a & 0 & 0 \\ 0 & a & 0 \\ 0 & 0 & a \end{pmatrix}$$

Unit Matrix (or) Identity Matrix

A scalar matrix in which each diagonal element is equal to 1 is called a unit matrix it is denoted by I.

For example,

$$I = \begin{pmatrix} 1 & 0 & 0 \\ 0 & 1 & 0 \\ 0 & 0 & 1 \end{pmatrix}_{3\times 3}, \quad I = \begin{pmatrix} 1 & 0 \\ 0 & 1 \end{pmatrix}_{2\times 2}$$

Transpose of a Matrix

Is denoted by A^T or A'.

For example,

$$A = \begin{pmatrix} 2 & 7 \\ 5 & 3 \end{pmatrix} \qquad A' = \begin{pmatrix} 2 & 5 \\ 7 & 3 \end{pmatrix}$$

Equal Matrices

Two matrices are said to be equal iff their orders are same and their corresponding elements are equal.

For example,

$$A = \begin{pmatrix} 2 & 3 \\ 5 & 7 \end{pmatrix} \text{ and } B = \begin{pmatrix} 2 & 3 \\ 5 & 7 \end{pmatrix}$$

$$\therefore\ A = B$$

Triangular Matrix

A square matrix (a_{ij}) whose elements $a_{ij} = 0$ whenever $i < j$ is called a lower triangular matrix (◺)

$|||_x$ a square matrix (a_{ij}) whose elements $a_{ij} = 0$ whenever $i > j$ is called an upper triangular matrix (◿).

For example,

$$\begin{pmatrix} 1 & 0 & 0 \\ 4 & 5 & 0 \\ 7 & 8 & 9 \end{pmatrix}, \begin{pmatrix} 1 & 0 \\ 2 & 0 \end{pmatrix} \text{ are Lower triangular matrices.}$$

$$\text{and } \begin{pmatrix} 1 & 2 & 3 \\ 0 & 4 & 5 \\ 0 & 0 & 6 \end{pmatrix}, \begin{pmatrix} 1 & 2 \\ 0 & 3 \end{pmatrix} \text{ are Upper triangular matrices.}$$

List of Formulae

1. Matrix multiplication is possible only when the number of columns of first matrix (prefactor) is equal to the number of rows of second matrix (postfactor).
2. Adjoint (or) Adjagate of a square matrix = adj A = [cofactors of A]′. Where, A is any given square matrix.
3. A square matrix A is said to be singular if $|A| = 0$
4. A square matrix A is said to be non-singular if $|A| \neq 0$.
5. Inverse of a square matrix $A = A' = \dfrac{I \cdot \text{adj } A}{|A|}$ where, I is identity square matrix.
6. In Cramer's rule, $x = \dfrac{\Delta_1}{\Delta}, y = \dfrac{\Delta_2}{\Delta}$ *and* $z = \dfrac{\Delta_3}{\Delta}$.

Worked Examples

Example 1. Evaluate $2A + 5B$ where $A = \begin{bmatrix} 1 & 5 \\ 2 & 6 \end{bmatrix}$ and $B = \begin{bmatrix} 0 & 1 \\ 5 & 0 \end{bmatrix}$

Solution:

$$2A + 5B = 2\begin{bmatrix} 1 & 5 \\ 2 & 6 \end{bmatrix} + 5\begin{bmatrix} 0 & 1 \\ 5 & 0 \end{bmatrix}$$

$$= \begin{bmatrix} 2 & 10 \\ 4 & 12 \end{bmatrix} + \begin{bmatrix} 0 & 5 \\ 25 & 0 \end{bmatrix}$$

$$= \begin{bmatrix} 2+0 & 10+5 \\ 4+25 & 12+0 \end{bmatrix}$$

$$\therefore\ 2A + 5B = \begin{bmatrix} 2 & 15 \\ 29 & 12 \end{bmatrix}$$

Example 2. If $A = \begin{bmatrix} 3 & 5 & 7 \\ 4 & 6 & 8 \end{bmatrix}$ and $B = \begin{bmatrix} 1 & 0 & 3 \\ 5 & 9 & 2 \end{bmatrix}$ then find $A + B$.

Solution:

$$A + B = \begin{bmatrix} 3 & 5 & 7 \\ 4 & 6 & 8 \end{bmatrix} + \begin{bmatrix} 1 & 0 & 3 \\ 5 & 9 & 2 \end{bmatrix}$$

$$= \begin{bmatrix} 4 & 5 & 10 \\ 9 & 15 & 10 \end{bmatrix}$$

Example 3. If $A = \begin{bmatrix} 9 & 7 \\ 8 & 2 \\ 3 & 5 \end{bmatrix}$ and $B = \begin{bmatrix} 2 & 4 \\ 0 & 7 \\ 6 & 1 \end{bmatrix}$ then find $A - B$.

Solution:

$$A - B = \begin{bmatrix} 9 & 7 \\ 8 & 2 \\ 3 & 5 \end{bmatrix} - \begin{bmatrix} 2 & 4 \\ 0 & 7 \\ 6 & 1 \end{bmatrix}$$

$$= \begin{bmatrix} 9-2 & 7-4 \\ 8-0 & 2-7 \\ 3-6 & 5-1 \end{bmatrix}$$

$$\therefore\ A - B = \begin{bmatrix} 7 & 3 \\ 8 & -5 \\ -3 & 4 \end{bmatrix}$$

Example 4. If $A = \begin{bmatrix} 0 & 2 & 3 \\ 2 & 1 & 4 \end{bmatrix}$ and $B = \begin{bmatrix} 7 & 6 & 0 \\ 0 & 4 & 1 \end{bmatrix}$ find $3B - 2A$.

Solution:

$$3B - 2A = 3\begin{bmatrix} 7 & 6 & 0 \\ 0 & 4 & 1 \end{bmatrix} - 2\begin{bmatrix} 0 & 2 & 3 \\ 2 & 1 & 4 \end{bmatrix}$$

$$= \begin{bmatrix} 21 & 18 & 0 \\ 0 & 12 & 3 \end{bmatrix} - \begin{bmatrix} 0 & 4 & 6 \\ 4 & 2 & 8 \end{bmatrix}$$

$$= \begin{bmatrix} 21-0 & 18-4 & 0-6 \\ 0-4 & 12-2 & 3-8 \end{bmatrix}$$

$$= \begin{bmatrix} 21 & 14 & -6 \\ -4 & 10 & -5 \end{bmatrix}$$

Example 5. If $A = \begin{bmatrix} 2 & -1 & 0 \\ 3 & 2 & 4 \\ -4 & 1 & 5 \end{bmatrix}$ find x, if $A + X$ equals unit matrix.

Solution:

$$A + X = I$$

$$X = I - A$$

$$X = \begin{bmatrix} 1 & 0 & 0 \\ 0 & 1 & 0 \\ 0 & 0 & 1 \end{bmatrix} - \begin{bmatrix} 2 & -1 & 0 \\ 3 & 2 & 4 \\ -4 & 1 & 5 \end{bmatrix}$$

$$= \begin{bmatrix} 1-2 & 0+1 & 0-0 \\ 0-3 & 1-2 & 0-4 \\ 0+4 & 0-1 & 1-5 \end{bmatrix}$$

$$\therefore X = \begin{bmatrix} -1 & 1 & 0 \\ -3 & -1 & -4 \\ 4 & -1 & -4 \end{bmatrix}$$

Example 6. Evaluate $4A - 5B - 6I$, where I is a unit matrix of the order 2×2 and $A = \begin{bmatrix} 3 & 4 \\ 1 & 7 \end{bmatrix}$ and $B = \begin{bmatrix} 3 & 4 \\ 5 & 6 \end{bmatrix}$

Solution:

$$4A - 5B - 6I$$

$$= 4\begin{bmatrix} 3 & 4 \\ 1 & 7 \end{bmatrix} - 5\begin{bmatrix} 3 & 4 \\ 5 & 6 \end{bmatrix} - 6\begin{bmatrix} 1 & 0 \\ 0 & 1 \end{bmatrix}$$

$$= \begin{bmatrix} 12 & 16 \\ 4 & 28 \end{bmatrix} - \begin{bmatrix} 15 & 20 \\ 25 & 30 \end{bmatrix} - \begin{bmatrix} 6 & 0 \\ 0 & 6 \end{bmatrix}$$

$$= \begin{bmatrix} 12-15-6 & 16-20-0 \\ 4-25-0 & 28-30-6 \end{bmatrix}$$

$$= \begin{bmatrix} -9 & -4 \\ -21 & -8 \end{bmatrix}$$

Example 7. If $A = \begin{bmatrix} 3 & -2 & 1 \\ 4 & -3 & -4 \\ -3 & 2 & -1 \end{bmatrix}$ and $A + X$ equals a unit matrix find X.

Solution:

$$A + X = I$$

$$X = I - A$$

$$X = \begin{bmatrix} 1 & 0 & 0 \\ 0 & 1 & 0 \\ 0 & 0 & 1 \end{bmatrix} - \begin{bmatrix} 3 & -2 & 1 \\ 4 & -3 & -4 \\ -3 & 2 & -1 \end{bmatrix}$$

$$X = \begin{bmatrix} 1-3 & 0+2 & 0-1 \\ 0-4 & 1+3 & 0+4 \\ 0+3 & 0-2 & 1+1 \end{bmatrix}$$

$$\therefore\ X = \begin{bmatrix} -2 & 2 & -1 \\ -4 & 4 & 4 \\ 3 & -2 & 2 \end{bmatrix}$$

Example 8. If $A = \begin{bmatrix} 1 & 3 & -2 & 5 \\ 3 & 1 & 2 & 6 \end{bmatrix}$ and $B = \begin{bmatrix} 4 & 7 & 1 & 3 \\ 2 & 5 & 2 & -3 \end{bmatrix}$ find $3A + 4B$ and $4A - 3B$.

Solution:

$$3A + 4B = 3\begin{bmatrix} 1 & 3 & -2 & 5 \\ 3 & 1 & 2 & 6 \end{bmatrix} + 4\begin{bmatrix} 4 & 7 & 1 & 3 \\ 2 & 5 & 2 & -3 \end{bmatrix}$$

$$= \begin{bmatrix} 3 & 9 & -6 & 15 \\ 9 & 3 & 6 & 18 \end{bmatrix} + \begin{bmatrix} 16 & 28 & 4 & 12 \\ 8 & 20 & 8 & -12 \end{bmatrix}$$

$$= \begin{bmatrix} 3+16 & 9+28 & -6+4 & 15+12 \\ 9+8 & 3+20 & 6+8 & 18-12 \end{bmatrix}$$

$$\therefore\ 3A + 4B = \begin{bmatrix} 19 & 37 & -2 & 27 \\ 17 & 23 & 14 & 6 \end{bmatrix}$$

$$4A - 3B = 4\begin{bmatrix} 1 & 3 & -2 & 5 \\ 3 & 1 & 2 & 6 \end{bmatrix} - 3\begin{bmatrix} 4 & 7 & 1 & 3 \\ 2 & 5 & 2 & -3 \end{bmatrix}$$

$$= \begin{bmatrix} 4 & 12 & -8 & 20 \\ 12 & 4 & 8 & 24 \end{bmatrix} - \begin{bmatrix} 12 & 21 & 3 & 9 \\ 6 & 15 & 6 & -9 \end{bmatrix}$$

$$= \begin{bmatrix} 4-12 & 12-21 & -8-3 & 20-9 \\ 12-6 & 4-15 & 8-6 & 24+9 \end{bmatrix}$$

$$\therefore\ 4A - 3B = \begin{bmatrix} -8 & -9 & -11 & 11 \\ 6 & -11 & 2 & 33 \end{bmatrix}$$

Example 9. If $A = \begin{bmatrix} 2 & 1 & 3 \\ 3 & 4 & 5 \end{bmatrix}$ and $B = \begin{bmatrix} 4 & 3 \\ 0 & 6 \\ 5 & 7 \end{bmatrix}$ find $A + B'$ and $A' + B$. Verify that $(A + B')'$ $= (A' + B)$.

Solution:

$$A + B' = \begin{bmatrix} 2 & 1 & 3 \\ 3 & 4 & 5 \end{bmatrix} + \begin{bmatrix} 4 & 3 \\ 0 & 6 \\ 5 & 7 \end{bmatrix}'$$

$$= \begin{bmatrix} 2 & 1 & 3 \\ 3 & 4 & 5 \end{bmatrix} + \begin{bmatrix} 4 & 0 & 5 \\ 3 & 6 & 7 \end{bmatrix}$$

$$\therefore\ A + B' = \begin{bmatrix} 6 & 1 & 8 \\ 6 & 10 & 12 \end{bmatrix}$$

$$A' + B = \begin{bmatrix} 2 & 1 & 3 \\ 3 & 4 & 5 \end{bmatrix}' + \begin{bmatrix} 4 & 3 \\ 0 & 6 \\ 5 & 7 \end{bmatrix}$$

$$A' + B = \begin{bmatrix} 2 & 3 \\ 1 & 4 \\ 3 & 5 \end{bmatrix} + \begin{bmatrix} 4 & 3 \\ 0 & 6 \\ 5 & 7 \end{bmatrix}$$

$$\therefore\ A' + B = \begin{bmatrix} 6 & 6 \\ 1 & 10 \\ 8 & 12 \end{bmatrix}$$

Take L.H.S = (A + B′)′

$$= \begin{bmatrix} 6 & 1 & 8 \\ 6 & 10 & 12 \end{bmatrix}'$$

$$= \begin{bmatrix} 6 & 6 \\ 1 & 10 \\ 8 & 12 \end{bmatrix}$$

= (A′ + B) = R.H.S.

Example 10. If $A = \begin{bmatrix} 5 & 3 \\ 4 & 6 \end{bmatrix}$ and $B = \begin{bmatrix} 6 & 8 \\ 9 & 1 \end{bmatrix}$ find $2A + 3B$.

Solution:

$$2A + 3B = 2\begin{bmatrix} 5 & 3 \\ 4 & 6 \end{bmatrix} + 3\begin{bmatrix} 6 & 8 \\ 9 & 1 \end{bmatrix}$$

$$= \begin{bmatrix} 10 & 6 \\ 8 & 12 \end{bmatrix} + \begin{bmatrix} 18 & 24 \\ 27 & 3 \end{bmatrix}$$

$$\therefore\ 2A + 3B = \begin{bmatrix} 28 & 30 \\ 35 & 15 \end{bmatrix}$$

Example 11. If $A = \begin{bmatrix} 1 & 0 & 4 \\ 2 & 4 & 3 \end{bmatrix}$ and $B = \begin{bmatrix} 4 & 0 & 1 \\ 3 & 4 & 2 \end{bmatrix}$ find matrix Y such that $A + 2Y = 4B$.

Solution:

$$A + 2Y = 4B$$

$$2Y = 4B - A$$

$$\therefore\ Y = \frac{1}{2}[4B - A]$$

$$Y = \frac{1}{2} = \left\{4\begin{bmatrix} 4 & 0 & 1 \\ 3 & 4 & 2 \end{bmatrix} - \begin{bmatrix} 1 & 0 & 4 \\ 2 & 4 & 3 \end{bmatrix}\right\}$$

$$= \frac{1}{2}\left\{\begin{bmatrix} 16 & 0 & 4 \\ 12 & 16 & 8 \end{bmatrix} - \begin{bmatrix} 1 & 0 & 4 \\ 2 & 4 & 3 \end{bmatrix}\right\}$$

$$\therefore\ Y = \frac{1}{2}\begin{bmatrix} 15 & 0 & 0 \\ 10 & 12 & 5 \end{bmatrix}$$

Example 12. If $A = \begin{bmatrix} 5 & 6 & 7 \\ 8 & 9 & 0 \\ 1 & 2 & 3 \end{bmatrix}$, $B = \begin{bmatrix} 3 & 2 & 1 \\ 9 & 0 & 8 \\ 7 & 6 & 5 \end{bmatrix}$ find (a) $3A + 5B$ (b) $5A + 3B$.

Solution:

$$\text{(a)}\ 3A + 5B = 3\begin{bmatrix} 5 & 6 & 7 \\ 8 & 9 & 0 \\ 1 & 2 & 3 \end{bmatrix} + 5\begin{bmatrix} 3 & 2 & 1 \\ 9 & 0 & 8 \\ 7 & 6 & 5 \end{bmatrix}$$

$$= \begin{bmatrix} 15 & 18 & 21 \\ 24 & 27 & 0 \\ 3 & 6 & 9 \end{bmatrix} + \begin{bmatrix} 15 & 10 & 5 \\ 45 & 0 & 40 \\ 35 & 30 & 25 \end{bmatrix}$$

$$\therefore\ 3A + 5B = \begin{bmatrix} 30 & 28 & 26 \\ 69 & 27 & 40 \\ 38 & 36 & 34 \end{bmatrix}$$

(b) $5A + 3B = 5\begin{bmatrix} 5 & 6 & 7 \\ 8 & 9 & 0 \\ 1 & 2 & 3 \end{bmatrix} + 3\begin{bmatrix} 3 & 2 & 1 \\ 9 & 0 & 8 \\ 7 & 6 & 5 \end{bmatrix}$

$$= \begin{bmatrix} 25 & 30 & 35 \\ 40 & 45 & 0 \\ 5 & 10 & 15 \end{bmatrix} + \begin{bmatrix} 9 & 6 & 3 \\ 27 & 0 & 24 \\ 21 & 18 & 15 \end{bmatrix}$$

$$\therefore 5A + 3B = \begin{bmatrix} 34 & 36 & 38 \\ 67 & 45 & 24 \\ 26 & 28 & 30 \end{bmatrix}$$

Example 13. If $A = \begin{bmatrix} 3 & 2 \\ 1 & 5 \end{bmatrix}$ and $B = \begin{bmatrix} 6 & 7 \\ 8 & 9 \end{bmatrix}$ find $5A + 3B$.

Solution:

$$5A + 3B = 5\begin{bmatrix} 3 & 2 \\ 1 & 5 \end{bmatrix} + 3\begin{bmatrix} 6 & 7 \\ 8 & 9 \end{bmatrix}$$

$$= \begin{bmatrix} 15 & 10 \\ 5 & 25 \end{bmatrix} + \begin{bmatrix} 18 & 21 \\ 24 & 27 \end{bmatrix}$$

$$\therefore 5A + 3B = \begin{bmatrix} 33 & 31 \\ 29 & 52 \end{bmatrix}$$

Example 14. If $A = \begin{bmatrix} 8 & 7 \\ 6 & 5 \end{bmatrix}$ and $B = \begin{bmatrix} 5 & 6 \\ 7 & 8 \end{bmatrix}$ find matrix 'X' such that $3A + 2B + X = 0$.

Solution:

$$3A + 2B + X = 0$$

$$X = -3A - 2B$$

$$= -3\begin{bmatrix} 8 & 7 \\ 6 & 5 \end{bmatrix} - 2\begin{bmatrix} 5 & 6 \\ 7 & 8 \end{bmatrix}$$

$$\therefore X = \begin{bmatrix} -24 & -21 \\ -18 & -15 \end{bmatrix} + \begin{bmatrix} -10 & -12 \\ -14 & -16 \end{bmatrix}$$

$$\therefore X = \begin{bmatrix} -34 & -33 \\ -32 & -31 \end{bmatrix}$$

Example 15. If $A = \begin{bmatrix} 2 & 6 \\ 4 & 5 \end{bmatrix}$ and $B = \begin{bmatrix} 4 & 3 \\ 2 & 4 \end{bmatrix}$ find matrix X such that $2A + B + 4X = 0$

Solution:

$$2A + B + 4X = 0$$

$$4X = -2A - B$$

$$X = \frac{1}{4}[-2A - B]$$

$$X = \frac{1}{4}\left[-2\begin{pmatrix} 2 & 6 \\ 4 & 5 \end{pmatrix} - \begin{pmatrix} 4 & 3 \\ 2 & 4 \end{pmatrix}\right]$$

$$= \frac{1}{4}\left[\begin{pmatrix} -4 & -12 \\ -8 & -10 \end{pmatrix} + \begin{pmatrix} -4 & -3 \\ -2 & -4 \end{pmatrix}\right]$$

$$\therefore\ X = \frac{1}{4}\begin{bmatrix} -8 & -15 \\ -10 & -14 \end{bmatrix}$$

Example 16. If $A = \begin{bmatrix} -3 & -2 & -1 \\ 0 & -4 & 2 \\ -1 & -2 & -3 \end{bmatrix}$ and $B = \begin{bmatrix} 0 & -1 & -2 \\ -3 & -4 & -5 \\ -1 & 0 & -2 \end{bmatrix}$ show that $(A + B)' = A' + B'$.

Solution:

$$A + B = \begin{bmatrix} -3 & -2 & -1 \\ 0 & -4 & 2 \\ -1 & -2 & -3 \end{bmatrix} + \begin{bmatrix} 0 & -1 & -2 \\ -3 & -4 & -5 \\ -1 & 0 & -2 \end{bmatrix}$$

$$A + B = \begin{bmatrix} -3 & -3 & -3 \\ -3 & -8 & -3 \\ -2 & -2 & -5 \end{bmatrix}$$

$$\therefore\ (A + B)' = \begin{bmatrix} -3 & -3 & -3 \\ -3 & -8 & -3 \\ -2 & -2 & -5 \end{bmatrix}'$$

$$\therefore\ (A + B)' = \begin{bmatrix} -3 & -3 & -2 \\ -3 & -8 & -2 \\ -3 & -3 & -5 \end{bmatrix} \quad \text{... (1)}$$

$$A' + B' = \begin{bmatrix} -3 & -2 & -1 \\ 0 & -4 & 2 \\ -1 & -2 & -3 \end{bmatrix}' + \begin{bmatrix} 0 & -1 & -2 \\ -3 & -4 & -5 \\ -1 & 0 & -2 \end{bmatrix}'$$

$$= \begin{bmatrix} -3 & 0 & -1 \\ -2 & -4 & -2 \\ -1 & 2 & -3 \end{bmatrix} + \begin{bmatrix} 0 & -3 & -1 \\ -1 & -4 & 0 \\ -2 & -5 & -2 \end{bmatrix}$$

$$\therefore A' + B' = \begin{bmatrix} -3 & -3 & -2 \\ -3 & -8 & -2 \\ -3 & -3 & -5 \end{bmatrix} \qquad \text{... (2)}$$

from (1) and (2) we get

$$\therefore (A + B)' = A' + B'$$

Example 17. If $A = \begin{bmatrix} 2 & 4 \\ 5 & 6 \end{bmatrix}$ and $B = \begin{bmatrix} 4 & 2 \\ 2 & 4 \end{bmatrix}$ find matrix such that $2A + B + 4X = 0$

Solution:

$$2A + B + 4X = 0$$

$$4X = -2A - B$$

$$X = \frac{1}{4}[-2A - B]$$

$$X = \frac{1}{4}\left[-2\begin{pmatrix} 2 & 4 \\ 5 & 6 \end{pmatrix} - \begin{pmatrix} 4 & 2 \\ 2 & 4 \end{pmatrix}\right]$$

$$= \frac{1}{4}\left[\begin{pmatrix} -4 & -8 \\ -10 & -12 \end{pmatrix} + \begin{pmatrix} -4 & -2 \\ -2 & -4 \end{pmatrix}\right]$$

$$= \frac{1}{4}\begin{bmatrix} -8 & -10 \\ -12 & -16 \end{bmatrix}$$

$$\therefore X = \begin{bmatrix} -2 & -5/2 \\ -3 & -4 \end{bmatrix}$$

Example 18. Solve for A and B, $2A + 3B = \begin{bmatrix} 1 & 2 \\ 2 & 0 \end{bmatrix}$ and $A + 2B = \begin{bmatrix} 2 & 1 \\ 1 & -1 \end{bmatrix}$.

Solution: Solving

$$2A + 3B = \begin{bmatrix} 1 & 2 \\ 2 & 0 \end{bmatrix}$$

$$A + 2B = \begin{bmatrix} 2 & 1 \\ 1 & -1 \end{bmatrix} \times -2$$

$$\cancel{2A} + 3B = \begin{bmatrix} 1 & 2 \\ 2 & 0 \end{bmatrix}$$

$$\cancel{-2A} - 4B = \begin{bmatrix} -4 & -2 \\ -2 & +2 \end{bmatrix}$$

$$-B = \begin{bmatrix} 1 & 2 \\ 2 & 0 \end{bmatrix} + \begin{bmatrix} -4 & -2 \\ -2 & 2 \end{bmatrix}$$

$$--B = \begin{bmatrix} -3 & 0 \\ 0 & 2 \end{bmatrix}$$

$$\therefore\ B = \begin{bmatrix} 3 & 0 \\ 0 & -2 \end{bmatrix}$$

$$A + 2B = \begin{bmatrix} 2 & 1 \\ 1 & -1 \end{bmatrix}$$

$$A = \begin{bmatrix} 2 & 1 \\ 1 & -1 \end{bmatrix} - 2B$$

$$= \begin{bmatrix} 2 & 1 \\ 1 & -1 \end{bmatrix} - 2\begin{bmatrix} 3 & 0 \\ 0 & -2 \end{bmatrix}$$

$$= \begin{bmatrix} 2 & 1 \\ 1 & -1 \end{bmatrix} + \begin{bmatrix} -6 & 0 \\ 0 & +4 \end{bmatrix}$$

$$\therefore\ A = \begin{bmatrix} -4 & 1 \\ 1 & 3 \end{bmatrix}$$

Example 19. If $A = \begin{bmatrix} 1 & 2 & 0 & 4 \\ 2 & 4 & -1 & 3 \end{bmatrix}$ and $B = \begin{bmatrix} 2 & 1 & 0 & 3 \\ 1 & -1 & 2 & 3 \end{bmatrix}$ find matrix 'X' such that $A - X = 3B$.

Solution:

$$A - X = 3B$$

$$X = \mathrm{A} - 3B$$

$$= \begin{bmatrix} 1 & 2 & 0 & 4 \\ 2 & 4 & -1 & 3 \end{bmatrix} - 3\begin{bmatrix} 2 & 1 & 0 & 3 \\ 1 & -1 & 2 & 3 \end{bmatrix}$$

$$= \begin{bmatrix} 1-6 & 2-3 & 0-0 & 4-9 \\ 2-3 & 4+3 & -1-6 & 3-9 \end{bmatrix}$$

$$= \begin{bmatrix} -5 & -1 & 0 & -5 \\ -1 & 7 & -7 & -6 \end{bmatrix}$$

Example 20. If $A = \begin{bmatrix} 1 & 3 \\ 4 & 5 \end{bmatrix}$ and $B = \begin{bmatrix} 4 & 5 \\ 7 & 6 \end{bmatrix}$ find $3A - 2B + I$ where I is a unit matrix.

Solution:

$$3A - 2B + I = 3\begin{bmatrix} 1 & 3 \\ 4 & 5 \end{bmatrix} - 2\begin{bmatrix} 4 & 5 \\ 7 & 6 \end{bmatrix} + \begin{bmatrix} 1 & 0 \\ 0 & 1 \end{bmatrix}$$

$$= \begin{bmatrix} 3 & 9 \\ 12 & 15 \end{bmatrix} - \begin{bmatrix} 8 & 10 \\ 14 & 12 \end{bmatrix} + \begin{bmatrix} 1 & 0 \\ 0 & 1 \end{bmatrix}$$

$$= \begin{bmatrix} 3 - 8 + 1 & 9 - 10 + 0 \\ 12 - 14 + 0 & 15 - 12 + 1 \end{bmatrix}$$

$$= \begin{bmatrix} -4 & -1 \\ -2 & 4 \end{bmatrix}$$

Example 21. If $A = \begin{bmatrix} 0 & 2 & 3 \\ 2 & 1 & 9 \end{bmatrix}$ and $B = \begin{bmatrix} 7 & 6 & 3 \\ 1 & 4 & 5 \end{bmatrix}$ find the value of $2A - 9B$.

Solution:

$$2A - 9B = 2\begin{bmatrix} 0 & 2 & 3 \\ 2 & 1 & 9 \end{bmatrix} - 9\begin{bmatrix} 7 & 6 & 3 \\ 1 & 4 & 5 \end{bmatrix}$$

$$= \begin{bmatrix} 0 & 4 & 6 \\ 4 & 2 & 18 \end{bmatrix} - \begin{bmatrix} 63 & 54 & 27 \\ 9 & 36 & 45 \end{bmatrix}$$

$$= \begin{bmatrix} -63 & -50 & -21 \\ -5 & -34 & -27 \end{bmatrix}$$

Example 22. Find matrix C if $A = \begin{bmatrix} 3 & 7 \\ 2 & 5 \end{bmatrix}$, $B = \begin{bmatrix} -3 & 2 \\ 4 & -1 \end{bmatrix}$ and $5C + 2B = A$.

Solution:

$$5C = A - 2B$$

$$C = \frac{1}{5}[A - 2B]$$

$$= \frac{1}{5}\left[\begin{pmatrix} 3 & 7 \\ 2 & 5 \end{pmatrix} - 2\begin{pmatrix} -3 & 2 \\ 4 & -1 \end{pmatrix}\right]$$

$$= \frac{1}{5}\left[\begin{pmatrix} 3 & 7 \\ 2 & 5 \end{pmatrix} + \begin{pmatrix} 6 & -4 \\ -8 & 2 \end{pmatrix}\right]$$

$$\therefore C = \frac{1}{5}\begin{bmatrix} 9 & 3 \\ -6 & 7 \end{bmatrix}$$

Example 23. If $3A + B = \begin{bmatrix} 3 & 2 \\ 8 & 3 \\ 10 & 4 \end{bmatrix}$ and $A - B = \begin{bmatrix} 1 & -2 \\ 0 & -3 \\ 2 & -4 \end{bmatrix}$ find A and B.

Solution:

$$3A + \cancel{B} = \begin{bmatrix} 3 & 2 \\ 8 & 3 \\ 10 & 4 \end{bmatrix}$$

$$A - \cancel{B} = \begin{bmatrix} 1 & -2 \\ 0 & -3 \\ 2 & -4 \end{bmatrix}$$

$$4A = \begin{bmatrix} 4 & 0 \\ 8 & 0 \\ 12 & 0 \end{bmatrix}$$

$$\therefore \; A = \begin{bmatrix} 1 & 0 \\ 2 & 0 \\ 3 & 0 \end{bmatrix}$$

$$B = A - \begin{bmatrix} 1 & -2 \\ 0 & -3 \\ 2 & -4 \end{bmatrix}$$

$$= \begin{bmatrix} 1 & 0 \\ 2 & 0 \\ 3 & 0 \end{bmatrix} - \begin{bmatrix} 1 & -2 \\ 0 & -3 \\ 2 & -4 \end{bmatrix}$$

$$\therefore \; B = \begin{bmatrix} 0 & 2 \\ 2 & +3 \\ 1 & 4 \end{bmatrix}$$

Example 24. If $A = \begin{bmatrix} 4 & 3 \\ 1 & 2 \end{bmatrix}$ and $B = \begin{bmatrix} -4 \\ 3 \end{bmatrix}$ find AB.

Solution:

$$AB = \overset{\rightarrow}{\begin{pmatrix} 4 & 3 \\ 1 & 2 \end{pmatrix}} \begin{bmatrix} -4 \\ 3 \end{bmatrix} \downarrow$$

$$= \begin{bmatrix} 4 \times -4 + 3 \times 3 \\ 1 \times -4 + 2 \times 3 \end{bmatrix}$$

$$= \begin{bmatrix} -16 + 9 \\ -4 + 6 \end{bmatrix}$$

$$\therefore \; AB = \begin{bmatrix} -7 \\ 2 \end{bmatrix}$$

Example 25. If $A = \begin{pmatrix} 1 \\ 3 \end{pmatrix}$, $B = (1\ 6\ 7)$ find AB.

Solution:

$$AB = \begin{bmatrix} 1 \\ 3 \end{bmatrix} [1 \quad 6 \quad 7]$$

$$\therefore \; AB = \begin{bmatrix} 1 & 6 & 7 \\ 3 & 18 & 21 \end{bmatrix}$$

Example 26. If $U = (1\ 5\ 6)$ and $V = \begin{bmatrix} 3 \\ 4 \\ 8 \end{bmatrix}$ find UV.

Solution:

$$UV = [1 \quad 5 \quad 6] \begin{bmatrix} 3 \\ 4 \\ 8 \end{bmatrix}$$

$$= (3 + 20 + 48)$$

$$\therefore \; UV = (71)$$

Example 27. If $A = \begin{bmatrix} 1 & -1 \\ -2 & 3 \end{bmatrix}$ and $B = \begin{bmatrix} 4 & 1 \\ 3 & -2 \end{bmatrix}$ verify that $(AB)^T = B^T A^T$, where A^T is the transpose of A.

Solution:

$$AB = \begin{bmatrix} 1 & -1 \\ -2 & 3 \end{bmatrix} \begin{bmatrix} 4 & 1 \\ 3 & -2 \end{bmatrix}$$

$$AB = \begin{bmatrix} 4 - 3 & 1 + 2 \\ -8 + 9 & -2 - 6 \end{bmatrix}$$

$$AB = \begin{bmatrix} +1 & 3 \\ 1 & -8 \end{bmatrix}$$

$$\therefore (AB)^T = \begin{bmatrix} 1 & 3 \\ 1 & -8 \end{bmatrix}^T = \begin{bmatrix} 1 & 1 \\ 3 & -8 \end{bmatrix}$$

$$B^T = \begin{bmatrix} 4 & 1 \\ 3 & -2 \end{bmatrix}^T = \begin{bmatrix} 4 & 3 \\ 1 & -2 \end{bmatrix}$$

$$A^T = \begin{bmatrix} 1 & -1 \\ -2 & 3 \end{bmatrix}^T = \begin{bmatrix} 1 & -2 \\ -1 & 3 \end{bmatrix}$$

$$\therefore B^T A^T = \begin{bmatrix} 4 & 3 \\ 1 & -2 \end{bmatrix}\begin{bmatrix} 1 & -2 \\ -1 & 3 \end{bmatrix}$$

$$= \begin{bmatrix} 4-3 & -8+9 \\ 1+2 & -2-6 \end{bmatrix}$$

$$\therefore B^T A^T = \begin{bmatrix} 1 & 1 \\ 3 & -8 \end{bmatrix}$$

$$\therefore (AB)^T = B^T A^T$$

Example 28. If $A = \begin{bmatrix} 1 & 3 & 4 \\ 5 & -2 & 1 \\ 0 & 1 & 3 \end{bmatrix}$ and $B = \begin{bmatrix} 5 & -1 & 0 \\ 0 & 1 & 2 \\ 3 & -2 & -1 \end{bmatrix}$ find AB and BA.

Solution:

$$AB = \begin{bmatrix} 1 & 3 & 4 \\ 5 & -2 & 1 \\ 0 & 1 & 3 \end{bmatrix}\begin{bmatrix} 5 & -1 & 0 \\ 0 & 1 & 2 \\ 3 & -2 & -1 \end{bmatrix}$$

$$= \begin{bmatrix} 5+0+12 & -1+3-8 & 0+6-4 \\ 25+0+3 & -5-2-2 & 0-4-1 \\ 0+0+9 & 0+1-6 & 0+2-3 \end{bmatrix}$$

$$\therefore AB = \begin{bmatrix} 17 & -6 & 2 \\ 28 & -9 & -5 \\ 9 & -5 & -1 \end{bmatrix}$$

$$BA = \begin{bmatrix} 5 & -1 & 0 \\ 0 & 1 & 2 \\ 3 & -2 & -1 \end{bmatrix}\begin{bmatrix} 1 & 3 & 4 \\ 5 & -2 & 1 \\ 0 & 1 & 3 \end{bmatrix}$$

$$= \begin{bmatrix} 5-5+0 & 15+2+0 & 20-1+0 \\ 0+5+0 & 0-2+2 & 0+1+6 \\ 3-10+0 & 9+4-1 & 12-2-3 \end{bmatrix}$$

$$\therefore\ BA = \begin{bmatrix} 0 & 17 & 19 \\ 5 & 0 & 7 \\ -7 & 12 & 7 \end{bmatrix}$$

Example 29. If $X = [5\ 4\ 3]$ and $Y = \begin{bmatrix} 4 \\ 6 \\ 3 \end{bmatrix}$ find XY and YX.

Solution:

$$XY = [5\quad 4\quad 3]\begin{bmatrix} 4 \\ 6 \\ 3 \end{bmatrix}$$

$$\therefore\ XY = [20 + 24 + 9] = [53]$$

$$YX = \begin{bmatrix} 4 \\ 6 \\ 3 \end{bmatrix}[5\quad 4\quad 3]$$

$$\therefore\ YX = \begin{bmatrix} 20 & 16 & 12 \\ 30 & 24 & 18 \\ 15 & 12 & 9 \end{bmatrix}$$

Example 30. If $A = \begin{bmatrix} 2 & -3 \\ -4 & 7 \end{bmatrix}$ and $B = \begin{bmatrix} 8 & -2 \\ 3 & -1 \end{bmatrix}$ verify that $(AB)' = B'A'$.

Solution:

$$AB = \begin{bmatrix} 2 & -3 \\ -4 & 7 \end{bmatrix}\begin{bmatrix} 8 & -2 \\ 3 & -1 \end{bmatrix}$$

$$= \begin{bmatrix} 16 - 9 & -4 + 3 \\ -32 + 21 & 8 - 7 \end{bmatrix}$$

$$AB = \begin{bmatrix} 7 & -1 \\ -11 & 1 \end{bmatrix}$$

$$\therefore\ (AB)' = \begin{bmatrix} 7 & -1 \\ -11 & 1 \end{bmatrix}' = \begin{bmatrix} 7 & -11 \\ -1 & 1 \end{bmatrix}$$

$$B'A' = \begin{bmatrix} 8 & -2 \\ 3 & -1 \end{bmatrix}'\begin{bmatrix} 2 & -3 \\ -4 & 7 \end{bmatrix}'$$

$$= \begin{bmatrix} 8 & 3 \\ -2 & -1 \end{bmatrix}\begin{bmatrix} 2 & -4 \\ -3 & 7 \end{bmatrix}$$

$$= \begin{bmatrix} 16-9 & -32+21 \\ -4+3 & 8-7 \end{bmatrix}$$

$$B'A' = \begin{bmatrix} 7 & -11 \\ -1 & 1 \end{bmatrix}$$

$$\therefore (AB)' = B'A'$$

Example 31. If $P\,[1\ 2]$, $Q = \begin{bmatrix} 2 & 3 \\ 4 & 5 \end{bmatrix}$, $R = \begin{bmatrix} 3 \\ 5 \end{bmatrix}$ form as many products as you can.

Solution:

$$PQ = [1\quad 2]\begin{bmatrix} 2 & 3 \\ 4 & 5 \end{bmatrix} = [2+8\quad 3+10]$$

$$\therefore PQ = [10\quad 13]$$

$$PR = [1\quad 2]\begin{bmatrix} 3 \\ 5 \end{bmatrix}$$

$$= [3+10]$$

$$\therefore PR = [13]$$

$$QR = \begin{bmatrix} 2 & 3 \\ 4 & 5 \end{bmatrix}\begin{bmatrix} 3 \\ 5 \end{bmatrix}$$

$$= \begin{bmatrix} 6+15 \\ 12+25 \end{bmatrix}$$

$$\therefore QR = \begin{bmatrix} 21 \\ 37 \end{bmatrix}$$

$$\therefore QP = \begin{bmatrix} 2 & 3 \\ 4 & 5 \end{bmatrix}[1\quad 2]$$

Here multiplication is not possible. ($\because$ No. of columns of $Q \neq$ No. of rows of P)

$$RP = \begin{bmatrix} 3 \\ 5 \end{bmatrix}[1\quad 2]$$

$$\therefore RP = \begin{bmatrix} 3 & 6 \\ 5 & 10 \end{bmatrix}$$

$$RQ = \begin{bmatrix} 3 \\ 5 \end{bmatrix}\begin{bmatrix} 2 & 3 \\ 4 & 5 \end{bmatrix}$$

Here multiplication is not possible. ($\because$ No. of column of $R \neq$ No. of rows of Q)

Example 32: Given $A = \begin{bmatrix} 2 & 1 & 5 \\ 4 & 1 & 2 \\ 2 & 2 & 3 \end{bmatrix}$, $B = \begin{bmatrix} 4 & 7 & --2 \\ 2 & 3 & -4 \\ 1 & 1 & 3 \end{bmatrix}$ verify whether $(AB)' = B'A'$.

Solution:

$$AB = \begin{bmatrix} 2 & 1 & 5 \\ 4 & 1 & 2 \\ 2 & 2 & 3 \end{bmatrix}\begin{bmatrix} 4 & 7 & -2 \\ 2 & 3 & -4 \\ 1 & 1 & 3 \end{bmatrix}$$

$$= \begin{bmatrix} 8+2+5 & 14+3+5 & -4-4+15 \\ 16+2+2 & 28+3+2 & -8-4+6 \\ 8+4+3 & 14+6+3 & -4-8+9 \end{bmatrix}$$

$$\therefore\ AB = \begin{bmatrix} 15 & 22 & 7 \\ 20 & 33 & -6 \\ 15 & 23 & -3 \end{bmatrix}$$

$$(AB)' = \begin{bmatrix} 15 & 22 & +7 \\ 20 & 33 & -6 \\ 15 & 23 & -3 \end{bmatrix}'$$

$$\therefore\ (AB)' = \begin{bmatrix} 15 & 20 & 15 \\ 22 & 33 & 23 \\ 7 & -6 & -3 \end{bmatrix}'$$

$$B'A' = \begin{bmatrix} 4 & 7 & -2 \\ 2 & 3 & -4 \\ 1 & 1 & 3 \end{bmatrix}'\begin{bmatrix} 2 & 1 & 5 \\ 4 & 1 & 2 \\ 2 & 2 & 3 \end{bmatrix}'$$

$$= \begin{bmatrix} 4 & 2 & 1 \\ 7 & 3 & 1 \\ -2 & -4 & 3 \end{bmatrix}\begin{bmatrix} 2 & 4 & 2 \\ 1 & 1 & 2 \\ 5 & 2 & 3 \end{bmatrix}$$

$$= \begin{bmatrix} 8+2+5 & 16+2+2 & 8+4+3 \\ 14+3+5 & 28+3+2 & 14+6+3 \\ -4-4+15 & -8-4+6 & -4-8+9 \end{bmatrix}$$

$$\therefore\ B'A' = \begin{bmatrix} 15 & 20 & 15 \\ 22 & 33 & 23 \\ 7 & -6 & -3 \end{bmatrix}$$

$$\therefore\ (AB)' = B'A'$$

Example 33. If $A = \begin{bmatrix} 1 & 1 & -1 \\ 2 & -3 & 4 \\ 3 & -2 & 3 \end{bmatrix}$, $B = \begin{bmatrix} -1 & -2 & -1 \\ 6 & 12 & 6 \\ 5 & 10 & 5 \end{bmatrix}$ and $C = \begin{bmatrix} -1 & -1 & 1 \\ 2 & 2 & -2 \\ -3 & -3 & 3 \end{bmatrix}$ show that AB is a null matrix but AC is not a null matrix.

Solution:

$$AB = \begin{bmatrix} 1 & 1 & -1 \\ 2 & -3 & 4 \\ 3 & -2 & 3 \end{bmatrix} \begin{bmatrix} -1 & -2 & -1 \\ 6 & 12 & 6 \\ 5 & 10 & 5 \end{bmatrix}$$

$$= \begin{bmatrix} -1+6-5 & -2+12-10 & -1+6-5 \\ -2-18+20 & -4-36+40 & -2-18+20 \\ -3-12+15 & -6-24+30 & -3-12-15 \end{bmatrix}$$

$$\therefore\ AB = \begin{bmatrix} 0 & 0 & 0 \\ 0 & 0 & 0 \\ 0 & 0 & 0 \end{bmatrix} = \text{A null matrix}$$

$$AC = \begin{bmatrix} 1 & 1 & -1 \\ 2 & -3 & 4 \\ 3 & -2 & 3 \end{bmatrix} \begin{bmatrix} -1 & -1 & 1 \\ 2 & 2 & -2 \\ -3 & -3 & 3 \end{bmatrix}$$

$$= \begin{bmatrix} -1+2+3 & -1+2+3 & 1-2-3 \\ -2-6-12 & -2-6-12 & 2+6+12 \\ -3-4-9 & -3-4-9 & 3+4+9 \end{bmatrix}$$

$$AC = \begin{bmatrix} 4 & 4 & -4 \\ -20 & -20 & 20 \\ -16 & -16 & 16 \end{bmatrix}$$

$\therefore\ AC =$ not a null matrix.

Example 34. If $A = \begin{bmatrix} 2 & 3 & 4 \\ 1 & 2 & 3 \\ 1 & 1 & 2 \end{bmatrix}$ and $B = \begin{bmatrix} 1 & 3 & 0 \\ 1 & 2 & 1 \\ 0 & 0 & 2 \end{bmatrix}$ find AB and BA.

Solution:

$$AB = \begin{bmatrix} 2 & 3 & 4 \\ 1 & 2 & 3 \\ 1 & 1 & 2 \end{bmatrix} \begin{bmatrix} 1 & 3 & 0 \\ 1 & 2 & 1 \\ 0 & 0 & 2 \end{bmatrix}$$

$$= \begin{bmatrix} 2+3+0 & 6+6+0 & 0+3+8 \\ 1+2+0 & 3+4+0 & 0+2+6 \\ 1+1+0 & 3+2+0 & 0+1+4 \end{bmatrix}$$

$$\therefore\ AB = \begin{bmatrix} 5 & 12 & 11 \\ 3 & 7 & 8 \\ 2 & 5 & 5 \end{bmatrix}$$

$$BA = \begin{bmatrix} 1 & 3 & 0 \\ 1 & 2 & 1 \\ 0 & 0 & 2 \end{bmatrix} \begin{bmatrix} 2 & 3 & 4 \\ 1 & 2 & 3 \\ 1 & 1 & 2 \end{bmatrix}$$

$$= \begin{bmatrix} 2+3+0 & 3+6+0 & 4+9+0 \\ 2+2+1 & 3+4+1 & 4+6+2 \\ 0+0+2 & 0+0+2 & 0+0+4 \end{bmatrix}$$

$$\therefore\ BA = \begin{bmatrix} 5 & 9 & 13 \\ 5 & 8 & 12 \\ 2 & 2 & 4 \end{bmatrix}$$

Example 35. If $A = \begin{bmatrix} 1 & 2 & 3 \\ 2 & 4 & 6 \\ 1 & 2 & 3 \end{bmatrix}$ verify $(5A)' = 5A'$.

Solution:

$$5A = 5\begin{bmatrix} 1 & 2 & 3 \\ 2 & 4 & 6 \\ 1 & 2 & 3 \end{bmatrix} = \begin{bmatrix} 5 & 10 & 15 \\ 10 & 20 & 30 \\ 5 & 10 & 15 \end{bmatrix}$$

$$(5A)' = \begin{bmatrix} 5 & 10 & 15 \\ 10 & 20 & 30 \\ 5 & 10 & 15 \end{bmatrix}' = \begin{bmatrix} 5 & 10 & 5 \\ 10 & 20 & 10 \\ 15 & 30 & 15 \end{bmatrix}$$

$$\therefore 5A' = 5\begin{bmatrix}1 & 2 & 3\\2 & 4 & 6\\1 & 2 & 3\end{bmatrix}' = 5\begin{bmatrix}1 & 2 & 1\\2 & 4 & 2\\3 & 6 & 3\end{bmatrix} = \begin{bmatrix}5 & 10 & 5\\10 & 20 & 10\\15 & 30 & 15\end{bmatrix}$$

$$\therefore (5A)' = 5A'$$

Example 36. If $A = \begin{bmatrix}1 & 2 & 3\\4 & 5 & 6\\7 & 8 & 9\end{bmatrix}$ and $B = \begin{bmatrix}9 & 8 & 7\\6 & 5 & 4\\3 & 2 & 1\end{bmatrix}$ find (a) AB (b) BA.

Solution:

$$\text{(a) } AB = \begin{bmatrix}1 & 2 & 3\\4 & 5 & 6\\7 & 8 & 9\end{bmatrix}\begin{bmatrix}9 & 8 & 7\\6 & 5 & 4\\3 & 2 & 1\end{bmatrix}$$

$$= \begin{bmatrix}9+12+9 & 8+10+6 & 7+8+3\\36+30+18 & 32+25+12 & 28+20+6\\63+48+27 & 56+40+18 & 49+32+9\end{bmatrix}$$

$$\therefore AB = \begin{bmatrix}30 & 24 & 18\\84 & 69 & 54\\138 & 114 & 90\end{bmatrix}$$

$$\text{(b) } BA = \begin{bmatrix}9 & 8 & 7\\6 & 5 & 4\\3 & 2 & 1\end{bmatrix}\begin{bmatrix}1 & 2 & 3\\4 & 5 & 6\\7 & 8 & 9\end{bmatrix}$$

$$= \begin{bmatrix}9+32+49 & 18+40+56 & 27+48+63\\6+20+28 & 12+25+32 & 18+30+36\\3+8+7 & 6+10+8 & 9+12+9\end{bmatrix}$$

$$\therefore BA = \begin{bmatrix}90 & 114 & 138\\54 & 69 & 84\\18 & 24 & 30\end{bmatrix}$$

Example 37. If $A = \begin{bmatrix}1 & 2 & 3\\3 & -2 & 1\\4 & 2 & 1\end{bmatrix}$ find $A^2 - 23A - 40I$.

Solution:

$$A^2 = A \cdot A = \begin{bmatrix} 1 & 2 & 3 \\ 3 & -2 & 1 \\ 4 & 2 & 1 \end{bmatrix} \begin{bmatrix} 1 & 2 & 3 \\ 3 & -2 & 1 \\ 4 & 2 & 1 \end{bmatrix}$$

$$= \begin{bmatrix} 1+6+12 & 2-4+6 & 3+2+3 \\ 3-6+4 & 6+4+2 & 9-2+1 \\ 4+6+4 & 8-4+2 & 12+2+1 \end{bmatrix}$$

$$A^2 = \begin{bmatrix} 19 & 4 & 8 \\ 1 & 12 & 8 \\ 14 & 6 & 15 \end{bmatrix}$$

$$\therefore A^2 - 23A - 40I$$

$$= \begin{bmatrix} 19 & 4 & 8 \\ 1 & 12 & 8 \\ 14 & 6 & 15 \end{bmatrix} - 23 \begin{bmatrix} 1 & 2 & 3 \\ 3 & -2 & 1 \\ 4 & 2 & 1 \end{bmatrix} - 40 \begin{bmatrix} 1 & 0 & 0 \\ 0 & 1 & 0 \\ 0 & 0 & 1 \end{bmatrix}$$

$$= \begin{bmatrix} 19 & 4 & 8 \\ 1 & 12 & 8 \\ 14 & 6 & 15 \end{bmatrix} - \begin{bmatrix} 23 & 46 & 69 \\ 69 & -46 & 23 \\ 92 & 46 & 23 \end{bmatrix} - \begin{bmatrix} 40 & 0 & 0 \\ 0 & 40 & 0 \\ 0 & 0 & 40 \end{bmatrix}$$

$$= \begin{bmatrix} 19-23-40 & 4-46-0 & 8-69-0 \\ 1-69-0 & 12+46-40 & 8-23-0 \\ 14-92-0 & 6-46-0 & 15-23-40 \end{bmatrix}$$

$$\therefore A^2 - 23A - 40I = \begin{bmatrix} -44 & -42 & -61 \\ -68 & 18 & -15 \\ -78 & -40 & -48 \end{bmatrix}$$

Example 38. If $A = [1 \quad 0 \quad 9]$, $B = \begin{bmatrix} 3 \\ 4 \\ 5 \end{bmatrix}$ find AB and BA.

Solution:

$$AB = [1 \quad 0 \quad 9] \begin{bmatrix} 3 \\ 4 \\ 5 \end{bmatrix}$$

$$\therefore AB = [3 + 0 + 45] = [48]$$

Example 39. If $A = \begin{bmatrix} 1 & 2 \\ 4 & 1 \end{bmatrix}$ and $B = \begin{bmatrix} 5 & 2 \\ -1 & 3 \end{bmatrix}$ find $A^2 - B^2$.

Solution:

$$A + B = \begin{bmatrix} 1 & 2 \\ 4 & 1 \end{bmatrix} + \begin{bmatrix} 5 & 2 \\ -1 & 3 \end{bmatrix}$$

$$\therefore\ A + B = \begin{bmatrix} 6 & 4 \\ 3 & 4 \end{bmatrix}$$

$$A - B = \begin{bmatrix} 1 & 2 \\ 4 & 1 \end{bmatrix} - \begin{bmatrix} 5 & 2 \\ -1 & 3 \end{bmatrix}$$

$$\therefore\ A - B = \begin{bmatrix} -4 & 0 \\ 5 & -2 \end{bmatrix}$$

$$\therefore\ A^2 - B^2 = (A + B)(A - B)$$

$$= \begin{bmatrix} 6 & 4 \\ 3 & 4 \end{bmatrix} \begin{bmatrix} -4 & 0 \\ 5 & -2 \end{bmatrix}$$

$$= \begin{bmatrix} -24 + 20 & 0 - 8 \\ -12 + 20 & 0 - 8 \end{bmatrix}$$

$$\therefore\ A^2 - B^2 = \begin{bmatrix} -4 & -8 \\ +8 & -8 \end{bmatrix}$$

Example 40. If $A = \begin{bmatrix} -1 & -2 \\ -3 & -4 \end{bmatrix}$ and $B = \begin{bmatrix} 3 & 4 \\ 5 & 2 \end{bmatrix}$ S.T $AB \neq BA$.

Solution:

$$AB = \begin{bmatrix} -1 & -2 \\ -3 & -4 \end{bmatrix} \begin{bmatrix} 3 & 4 \\ 5 & 2 \end{bmatrix}$$

$$\therefore\ AB = \begin{bmatrix} -3 - 10 & -4 - 4 \\ -9 - 20 & -12 - 8 \end{bmatrix} = \begin{bmatrix} -13 & -8 \\ -29 & -20 \end{bmatrix}$$

$$BA = \begin{bmatrix} 3 & 4 \\ 5 & 2 \end{bmatrix} \begin{bmatrix} -1 & -2 \\ -3 & -4 \end{bmatrix}$$

$$= \begin{bmatrix} -3 - 12 & -6 - 16 \\ -5 - 6 & -10 - 8 \end{bmatrix}$$

$$\therefore\ BA = \begin{bmatrix} -15 & -22 \\ -11 & -18 \end{bmatrix}$$

$$\therefore\ AB \neq BA$$

Example 41. If $A = \begin{bmatrix} 4 & -2 \\ 3 & -1 \end{bmatrix}$ and $B = \begin{bmatrix} 2 & 4 \\ 3 & 6 \end{bmatrix}$ find $AB - B$.

Solution:

$$AB = \begin{bmatrix} 4 & -2 \\ 3 & -1 \end{bmatrix}\begin{bmatrix} 2 & 4 \\ 3 & 6 \end{bmatrix}$$

$$= \begin{bmatrix} 8-6 & 16-12 \\ 6-3 & 12-6 \end{bmatrix}$$

$$AB = \begin{bmatrix} 2 & 4 \\ 3 & 6 \end{bmatrix}$$

$$\therefore\ AB - B = \begin{bmatrix} 2 & 4 \\ 3 & 6 \end{bmatrix} - \begin{bmatrix} 2 & 4 \\ 3 & 6 \end{bmatrix}$$

$$\therefore\ AB - B = \begin{bmatrix} 0 & 0 \\ 0 & 0 \end{bmatrix}$$

Example 42. Solve for X : $\begin{bmatrix} x^2 & 1 \\ 2 & 3 \end{bmatrix} + \begin{bmatrix} 2x & 3 \\ 1 & 4 \end{bmatrix} = \begin{bmatrix} 3 & 4 \\ 3 & 7 \end{bmatrix}$.

Solution:

$$\begin{bmatrix} x^2 + 2x & 4 \\ 3 & 7 \end{bmatrix} = \begin{bmatrix} 3 & 4 \\ 3 & 7 \end{bmatrix}$$

$$\Rightarrow x^2 + 2x = 3$$

$$\Rightarrow x^2 + 2x - 3 = 0$$

$$\Rightarrow x^2 + 3x - x - 3 = 0$$

-3 ↙ ↘ $+3$ -1

$$\Rightarrow x(x+3) - (x+3) = 0$$

$$\Rightarrow (x+3)(x-1) = 0$$

$$\therefore\ x = -3 \text{ or } x = 1$$

Example 43. Solve for x and y, $x\begin{pmatrix} 2 \\ 1 \end{pmatrix} + y\begin{pmatrix} 3 \\ 5 \end{pmatrix} + \begin{pmatrix} 4 \\ 6 \end{pmatrix} = \begin{pmatrix} 12 \\ 17 \end{pmatrix}$.

Solution:

$$\begin{pmatrix} 2x \\ x \end{pmatrix} + \begin{pmatrix} 3y \\ 5y \end{pmatrix} + \begin{pmatrix} 4 \\ 6 \end{pmatrix} = \begin{pmatrix} 12 \\ 17 \end{pmatrix}$$

$$\begin{pmatrix} 2x + 3y + 4 \\ x + 5y + 6 \end{pmatrix} = \begin{pmatrix} 12 \\ 17 \end{pmatrix}$$

$\Rightarrow 2x + 3y + 4 = 12,$ $\qquad x + 5y + 6 = 17$

$\Rightarrow 2x + 3y = 8,$ $\qquad x + 5y = 11$

$2x + 3y = 8$

$x + 5y = 11 \times 2$ $\qquad x = 11 - 5y$

$\cancel{2x} + 3y = 8$ $\qquad = 11 - 10$

$\cancel{2x} + 10y = 22$ $\qquad \therefore x = 1$

$(-)\ (-)\ (-)$

$-7y = -14$ $\qquad \therefore x = 1$ and $y = 2$

$\therefore y = 2$

Example 44. Find x and y if $(x \quad y)\begin{pmatrix} 2 & 3 \\ 4 & -1 \end{pmatrix} = (2 \quad -4)$.

Solution:

$(2x + 4y \quad 3x - y) = (2 \quad -4)$

$\Rightarrow 2x + 4y = 2,$ $\qquad 3x - y = -4$

$\Rightarrow 2x + 4y = 2$

$3x - y = -4 \times 4$

$2x + \cancel{4y} = 2$

$12x - \cancel{4y} = -16$ $\qquad 2x + 4y = 2$

$14x = -14$ $\qquad 4y = 2 - 2x$

$\therefore x = -1$ $\qquad 4y = 2 - 2(-1)$

$4y = 2 + 2$

$\therefore y = \frac{4}{4} = 1$

$\therefore x = -1$ and $y = 1$

Example 45. If $\begin{bmatrix} 1 & 0 & 0 \\ 0 & y & 0 \\ 0 & 0 & 1 \end{bmatrix}\begin{bmatrix} x \\ 1 \\ z \end{bmatrix} = \begin{bmatrix} 1 \\ 0 \\ 1 \end{bmatrix}$ find x, y and z.

Solution:

$$\begin{bmatrix} x+0+0 \\ 0+y+0 \\ 0+0+z \end{bmatrix} = \begin{bmatrix} 1 \\ 0 \\ 1 \end{bmatrix}$$

$$\begin{bmatrix} x \\ y \\ z \end{bmatrix} = \begin{bmatrix} 1 \\ 0 \\ 1 \end{bmatrix}$$

$\therefore\ x = 1,\ y = 0$ and $z = 1$

Example 46. If $A = \begin{bmatrix} -2 & 1 \\ 1 & -2 \end{bmatrix}$, $B = \begin{bmatrix} 4 \\ 5 \end{bmatrix}$ and $AX = B$, find X.

Solution:

$AX = B$

$$\begin{bmatrix} -2. & 1 \\ 1 & -2 \end{bmatrix} \begin{bmatrix} x \\ y \end{bmatrix} = \begin{bmatrix} 4 \\ 5 \end{bmatrix}$$

$-2x + y = 4$

$x - 2y = 5 \times 2$

$-2x + y = 4$ (with $-2x$ struck out)

$2x - 4y = 10$ (with $2x$ struck out)

$-3y = 14$

$\therefore\ y = \dfrac{-14}{3}$

$x = 5 + 2y$

$= 5 + 2\left(\dfrac{-14}{3}\right)$

$= 5 - \dfrac{28}{3}$

$x = \dfrac{15 - 28}{3}$

$\therefore\ x = \dfrac{-13}{3}$

$\therefore\ x = \dfrac{-13}{3}$ and $y = \dfrac{-14}{3}$

i.e., $X = \begin{bmatrix} x \\ y \end{bmatrix} = \begin{bmatrix} -13/3 \\ -14/3 \end{bmatrix}$

Example 47. Find the adjoint of the matrix.

(a) $A = \begin{bmatrix} 2 & 3 \\ 4 & 8 \end{bmatrix}$ $\qquad \begin{bmatrix} + & - \\ - & + \end{bmatrix}$

Solution:

$$\therefore\ \text{adj } A = \begin{bmatrix} 8 & -4 \\ -3 & 2 \end{bmatrix}' = \begin{bmatrix} 8 & -3 \\ -4 & 2 \end{bmatrix}$$

(b) $A = \begin{bmatrix} 3 & -2 \\ 2 & 3 \end{bmatrix}$

Solution:

$$\therefore \text{ adj } A = \begin{bmatrix} 3 & -2 \\ -(-2) & 3 \end{bmatrix}' = \begin{bmatrix} 3 & 2 \\ -2 & 3 \end{bmatrix}$$

(c) $A = \begin{bmatrix} 2 & 3 \\ 2 & 5 \end{bmatrix}$

Solution:

$$\therefore \text{ adj } A = \begin{bmatrix} 5 & -2 \\ -3 & 2 \end{bmatrix}' = \begin{bmatrix} 5 & -3 \\ -2 & 2 \end{bmatrix}$$

Example 48. Find the adjoint of the matrix $\begin{bmatrix} 0 & 1 & 3 \\ 1 & 0 & 4 \\ 1 & 2 & -1 \end{bmatrix}$

Solution: Cofactors of A:-

For 0, = + [0 – 8] = –8
For 1, = – [–1 – 4] = +5
For 3, = + [2 – 0] = 2
For 1, = – [–1 – 6] = 7
For 0, = + [0 – 3] = –3
For 4, = – [0 – 1] = 1
For 1, = + [4 – 0] = 4
For 2, = – [0 – 3] = 3
For –1, = + [0 – 1] = –1

$$\therefore \text{ adj } A = [\text{Cofactors of } A]'$$

$$= \begin{bmatrix} -8 & +5 & 2 \\ 7 & -3 & 1 \\ 4 & 3 & -1 \end{bmatrix}'$$

$$\therefore \text{ adj } A = \begin{bmatrix} -8 & 7 & 4 \\ +5 & -3 & 3 \\ 2 & 1 & -1 \end{bmatrix}$$

Determinants

Example 49. Find $\begin{vmatrix} 1 & 3 \\ 5 & -8 \end{vmatrix}$.

Solution:

$$= (1)(-8) - (3)(5)$$
$$= -8 - 15$$
$$= -23$$

Example 50. Find $\begin{vmatrix} 2 & 4 \\ 3 & 6 \end{vmatrix}$.

Solution:

$$= (2)(6) - (3)(4)$$
$$= 12 - 12$$
$$= 0$$

Example 51. Find $\begin{vmatrix} \overset{+}{2} & \overset{-}{3} & \overset{+}{4} \\ -4 & x & -8 \\ 5 & 6 & 7 \end{vmatrix} = 0$

Solution:

$$\Rightarrow 2\begin{vmatrix} x & -8 \\ 6 & 7 \end{vmatrix} - 3\begin{vmatrix} -4 & -8 \\ 5 & 7 \end{vmatrix} + 4\begin{vmatrix} -4 & x \\ 5 & 6 \end{vmatrix} = 0$$

$$2\,[(x)(7) - (6)(-8)] - 3\,[(-4)(7) - (5)(-8)] + 4\,[(-4)(6) - (5)(x)] = 0$$
$$2\,[7x + 48] - 3\,[-28 + 40] + 4\,[-24 - 5x] = 0$$
$$14x + 96 + 84 - 120 - 96 - 20x = 0$$
$$-6x + 180 - 216 = 0$$
$$-6x - 36 = 0$$
$$6x = -36$$
$$\therefore\ x = \frac{-36}{6} = -6$$

Example 52. Solve $|A| = \begin{vmatrix} 2 & x & 4 \\ 3 & 2 & 1 \\ 1 & 2 & 3 \end{vmatrix} = -3$.

Solution:

$$|A| = \begin{vmatrix} 2 & x & 4 \\ 3 & 2 & 1 \\ 1 & 2 & 3 \end{vmatrix} = -3$$

$$2\begin{vmatrix} 2 & 1 \\ 2 & 3 \end{vmatrix} - x\begin{vmatrix} 3 & 1 \\ 1 & 3 \end{vmatrix} + 4\begin{vmatrix} 3 & 2 \\ 1 & 2 \end{vmatrix} = -3$$

2 [6 – 2] – x [9 – 1] + 4 [6 – 2] = –3

2 [4] –x [8] + 4 [4] = –3

8 – 8x + 16 = –3

24 – 8x = –3

24 = –3 + 8x

8x = 24 + 3

8x = 27

$$\therefore\ x = \frac{27}{8}$$

Example 53. Evaluate $\begin{vmatrix} 3 & 8 & 6 \\ 9 & 8 & 2 \\ 6 & 12 & 2 \end{vmatrix}$.

Solution:

$$= 3\begin{vmatrix} 8 & 2 \\ 12 & 2 \end{vmatrix} - 8\begin{vmatrix} 9 & 2 \\ 6 & 2 \end{vmatrix} + 6\begin{vmatrix} 9 & 8 \\ 6 & 12 \end{vmatrix}$$

= 3 [16 – 24] – 8 [18 – 12] + 6 [108 – 48]

= 3 [–8] –8 [6] + 6 [60]

= –24 – 48 + 360

= –72 + 360

= 288

or

$$= (3)(4)(2)\begin{vmatrix} 1 & 2 & 3 \\ 3 & 2 & 1 \\ 2 & 3 & 1 \end{vmatrix}$$

$$= 24\begin{vmatrix} 1 & 2 & 3 \\ 3 & 2 & 1 \\ 2 & 3 & 1 \end{vmatrix}$$

= 24 {1 [2 – 3] – 2 [3 – 2] + 3 [9 – 4]}

= 24 {1 [–1] – 2 [1] + 3 [5]}

= 24 {–1 – 2 + 15}

= 24 {12}

= 288

Example 54. Find the value $\begin{vmatrix} 1 & 2 & 3 \\ 4 & 5 & 6 \\ 7 & 8 & 9 \end{vmatrix}$.

Solution:

$$= 1\begin{vmatrix}5 & 6\\8 & 9\end{vmatrix} - 2\begin{vmatrix}4 & 6\\7 & 9\end{vmatrix} + 3\begin{vmatrix}4 & 5\\7 & 8\end{vmatrix}$$

$= 1\,[45 - 48] - 2\,[36 - 42] + 3\,[32 - 35]$

$= 1\,[-3] - 2\,[-6] + 3\,[-3]$

$= -3 + 12 - 9$

$= -12 + 12$

$= 0$

Example 55. Find the value of $\begin{vmatrix}1 & 2 & 3\\4 & 5 & 6\\7 & 8 & 9\end{vmatrix}$.

Solution: Refer problem No. (54).

Example 56. Evaluate $\begin{vmatrix}2 & 0 & -1\\0 & 1 & 5\\1 & 3 & -4\end{vmatrix}$.

Solution:

$$= 2\begin{vmatrix}1 & 5\\3 & -4\end{vmatrix} - 0\begin{vmatrix}0 & 5\\1 & -4\end{vmatrix} - 1\begin{vmatrix}0 & 1\\1 & 3\end{vmatrix}$$

$= 2\,[-4 - 15] - 0 - 1\,[0 - 1]$

$= 2\,[-19] + 1$

$= -38 + 1$

$= -37$

Example 57. Find the value $|X| = \begin{vmatrix}5 & 1 & 2\\3 & 2 & 2\\6 & 5 & 3\end{vmatrix}$

Solution:

$$|X| = \begin{vmatrix}5 & 1 & 2\\3 & 2 & 2\\6 & 5 & 3\end{vmatrix}$$

$= 5\,[6 - 10] - 1\,[9 - 12] + 2\,[15 - 12]$

$= 5\,[-4] - 1\,[-3] + 2\,[3]$

$= -20 + 3 + 6$

$= -20 + 9$

$= -11$

Example 58. Show that $\begin{vmatrix} 3 & 4 & 7 \\ 2 & 1 & 3 \\ -5 & -1 & 2 \end{vmatrix} = -40$.

Solution:

$$\begin{aligned} \text{L.H.S} &= 3\,[2+3] - 4\,[4+15] + 7\,[-2+5] \\ &= 3\,[5] - 4\,[19] + 7\,[3] \\ &= 15 - 76 + 21 \\ &= 36 - 76 \\ &= -40 = \text{R.H.S} \end{aligned}$$

Example 59. Find the value of $|A| = \begin{vmatrix} 5 & -1 & 2 \\ -2 & -3 & -2 \\ 3 & 2 & 1 \end{vmatrix}$.

Solution:

$$|A| = \begin{vmatrix} 5 & -1 & 2 \\ -2 & -3 & -2 \\ 3 & 2 & 1 \end{vmatrix}$$

$$\begin{aligned} |A| &= 5\,[-3+4] - (-1)\,[-2+6] + 2\,[-4+9] \\ &= 5\,[1] + 1\,[4] + 2\,[5] \\ &= 5 + 4 + 10 \end{aligned}$$

$$\therefore\ |A| = 19$$

Example 60. Evaluate $\begin{vmatrix} 7 & 6 & 1 \\ 5 & 3 & 8 \\ 3 & 2 & 4 \end{vmatrix}$.

Solution:

$$\begin{aligned} &= 7\,[12-16] - 6\,[20-24] + 1\,[10-9] \\ &= 7\,[-4] - 6\,[-4] + 1\,[1] \\ &= -28 + 24 + 1 \\ &= -28 + 25 \\ &= -3 \end{aligned}$$

Example 61. Find the adjoint of $A = \begin{bmatrix} -5 & 7 \\ -2 & 3 \end{bmatrix}$ and show that A (adj. A) = $|A|\ I$.

Solution:

$$A = \begin{bmatrix} -5 & 7 \\ -2 & 3 \end{bmatrix}$$

$$|A| = \begin{bmatrix} -5 & 7 \\ -2 & 3 \end{bmatrix}$$

$$|A| = (-5)(3) - (-2)(7)$$
$$= -15 + 14$$
$$\therefore |A| = -1$$

$$\text{adj } A = \begin{bmatrix} 3 & -(-2) \\ -(7) & -5 \end{bmatrix}'$$

$$= \begin{bmatrix} 3 & +2 \\ -7 & -5 \end{bmatrix}'$$

$$\therefore \text{adj } A = \begin{bmatrix} 3 & -7 \\ 2 & -5 \end{bmatrix}$$

$\therefore$ L.H.S $= A$ (adj A)

$$= \begin{bmatrix} -5 & 7 \\ -2 & 3 \end{bmatrix} \begin{bmatrix} 3 & -7 \\ 2 & -5 \end{bmatrix}$$

$$= \begin{bmatrix} -15 + 14 & +35 + (-35) \\ -6 + 6 & 14 - 5 \end{bmatrix}$$

$$= \begin{bmatrix} -1 & 0 \\ 0 & -1 \end{bmatrix}$$

R.H.S $= |A|\, I$

$$= -1 \begin{bmatrix} 1 & 0 \\ 0 & 1 \end{bmatrix} = \begin{bmatrix} -1 & 0 \\ 0 & -1 \end{bmatrix}$$

$\therefore$ L.H.S = R.H.S

Example 62. Evaluate $\begin{vmatrix} 7 & 6 & 1 \\ 5 & 3 & 8 \\ 3 & 2 & 4 \end{vmatrix}$

Solution:

$$= 7\,[12 - 16] - 6\,[20 - 24] + 1\,[10 - 9]$$
$$= 7\,[-4] - 6\,[-4] + 1\,[1]$$
$$= -28 + 24 + 1$$
$$= -28 + 25$$
$$= -3$$

Example 63. Find 'y' if $\begin{vmatrix} -3 & -6 & 1 \\ 5 & y & -2 \\ 2 & -3 & 5 \end{vmatrix} = -7$.

Solution:

$$-3\,[5y - 6] - (-6)\,[25 + 4] + 1\,[-15 - 2y] = -7$$

$$-3\,[5y-6]+6\,[29]+[-15-2y]=-7$$
$$-15y+18+174-15-2y=-7$$
$$-17y+3+174=-7$$
$$-17y=-7-3-174$$
$$-17y=-184$$

$$y=\frac{-184}{-17}$$

$$\therefore\ y=\frac{184}{17}$$

Example 64. Find the value of $\begin{vmatrix}1 & 1 & 1\\ 1 & (1+a) & 1\\ 1 & 1 & (1+b)\end{vmatrix}$.

Solution:

$$=1[(1+a)(1+b)-1]-1[(1+b)-1]+1[1-(1+a)]$$
$$=1[1+a+b+ab-1]-[1+b-1]+[1-1-a]$$
$$=[a+b+ab]-b-a$$
$$=[a+b+ab-b-a]$$
$$=ab$$

Inverse of a Matrix

Example 65. Find the inverse of $\begin{vmatrix}2 & 3\\ 2 & 5\end{vmatrix}$.

Solution:

Let $A=\begin{vmatrix}2 & 3\\ 2 & 5\end{vmatrix}$.

$|A|=10-6=4\neq 0\ \therefore\ A^{-1}$ exists.

$$\text{adj } A=\begin{bmatrix}5 & -2\\ -3 & 2\end{bmatrix}'=\begin{bmatrix}5 & -3\\ -2 & 2\end{bmatrix}$$

$$\therefore\ A^{-1}=\frac{\text{adj}}{|A|}=\begin{bmatrix}5 & -3\\ -2 & 2\end{bmatrix}$$

$$=\frac{1}{4}\begin{bmatrix}5 & -3\\ -2 & 2\end{bmatrix}$$

Example 66. If $A = \begin{vmatrix} 1 & 3 & 4 \\ 1 & 4 & 5 \\ 2 & 5 & 7 \end{vmatrix}$ show A^{-1} does not exist.

Solution:

$$A = \begin{vmatrix} 1 & 3 & 4 \\ 1 & 4 & 5 \\ 2 & 5 & 7 \end{vmatrix}$$

$$|A| = 1\begin{vmatrix} 4 & 5 \\ 5 & 7 \end{vmatrix} - 3\begin{vmatrix} 1 & 5 \\ 2 & 7 \end{vmatrix} + 4\begin{vmatrix} 1 & 4 \\ 2 & 5 \end{vmatrix}$$

$$= 1[28-25] - 3\,[7-10] + 4\,[5-8]$$

$$= 1\,[3] - 3\,[-3] + 4\,[-3]$$

$$= 3 + 9 - 12$$

$$= 12 - 12$$

$$\therefore\ |A| = 0$$

$\therefore$ A^{-1} does not exist.

Example 67. If $A = \begin{vmatrix} 2 & 3 \\ 3 & 5 \end{vmatrix}$ verify that $A^2 - 7A + I = 0$ and hence obtain A^{-1}.

Solution:

$$A = \begin{vmatrix} 2 & 3 \\ 3 & 5 \end{vmatrix}$$

$$A^2 = A \cdot A = \begin{bmatrix} 2 & 3 \\ 3 & 5 \end{bmatrix}\begin{bmatrix} 2 & 3 \\ 3 & 5 \end{bmatrix}$$

$$= \begin{bmatrix} 4+9 & 6+15 \\ 6+15 & 9+25 \end{bmatrix}$$

$$\therefore\ A^2 = \begin{bmatrix} 13 & 21 \\ 21 & 34 \end{bmatrix}$$

$$\text{L.H.S} = A^2 - 7A + I = \begin{bmatrix} 13 & 21 \\ 21 & 34 \end{bmatrix} - 7\begin{bmatrix} 2 & 3 \\ 3 & 5 \end{bmatrix} + \begin{bmatrix} 1 & 0 \\ 0 & 1 \end{bmatrix}$$

$$= \begin{bmatrix} 13 & 21 \\ 21 & 34 \end{bmatrix} - \begin{bmatrix} 14 & 21 \\ 21 & 35 \end{bmatrix} + \begin{bmatrix} 1 & 0 \\ 0 & 1 \end{bmatrix}$$

$$= \begin{bmatrix} 13-14+1 & 21-21+0 \\ 21-21+0 & 34-35+1 \end{bmatrix}$$

$$= \begin{bmatrix} 0 & 0 \\ 0 & 0 \end{bmatrix}$$

$= 0 =$ R.H.S

Now take $A^2 - 7A + I = 0$

Multiply this by A^{-1} we get

$$A^2 \cdot A^{-1} - 7A \cdot A^{-1} + I \cdot A^{-1} = 0$$

$$A^1 - 7A^0 + A^{-1} = 0$$

$$A - 7I + A^{-1} = 0$$

$$A^{-1} = 7I - A \qquad (\because I = 1)$$

$$= 7\begin{bmatrix} 1 & 0 \\ 0 & 1 \end{bmatrix} - \begin{bmatrix} 2 & 3 \\ 3 & 5 \end{bmatrix}$$

$$= \begin{bmatrix} 7 & 0 \\ 0 & 7 \end{bmatrix} - \begin{bmatrix} 2 & 3 \\ 3 & 5 \end{bmatrix}$$

$$= \begin{bmatrix} 7-2 & -3 \\ 0-3 & 7-5 \end{bmatrix}$$

$$\therefore A^{-1} = \begin{bmatrix} 5 & -3 \\ -3 & 2 \end{bmatrix}$$

Example 68. Find the inverse of $A = \begin{bmatrix} 1 & 1 & 2 \\ 2 & -1 & 1 \\ 2 & 2 & 1 \end{bmatrix}$

Solution:

$$A = \begin{bmatrix} 1 & 1 & 2 \\ 2 & -1 & 1 \\ 2 & 2 & 1 \end{bmatrix}$$

$|A| = 1\,[-1-2] - 1\,[2-2] + 2\,[4+2]$
$= -3 - 0 + 2\,[6]$
$= -3 + 12$

$\therefore |A| = 9 \neq 0$

$\therefore$ A^{-1} exists.

$$\text{adj } A = \begin{bmatrix} +(-3) & -(0) & (6) \\ -(-3) & (-3) & -(0) \\ (3) & -(-3) & (-3) \end{bmatrix}$$

$$= \begin{bmatrix} -3 & 0 & 6 \\ 3 & -3 & 0 \\ 3 & 3 & -3 \end{bmatrix}'$$

$$\text{adj } A = \begin{bmatrix} -3 & 3 & 3 \\ 0 & -3 & 3 \\ 6 & 0 & -3 \end{bmatrix}$$

$$\therefore A^{-1} = \frac{1}{|A|} \text{adj } A$$

$$= \frac{1}{9} \begin{bmatrix} -3 & 3 & 3 \\ 0 & -3 & 3 \\ 6 & 0 & -3 \end{bmatrix}$$

$$\therefore A^{-1} = \frac{1}{3} \begin{bmatrix} -1 & 1 & 1 \\ 0 & -1 & 1 \\ 2 & 0 & -1 \end{bmatrix}$$

Example 69. If $A = \begin{bmatrix} 2 & -1 \\ -1 & 2 \end{bmatrix}$ show that $A^2 - 4A + 3I = 0$ and hence find A^{-1}.

Solution:

$$A^2 = A \cdot A = \begin{bmatrix} 2 & -1 \\ -1 & 2 \end{bmatrix} \begin{bmatrix} 2 & -1 \\ -1 & 2 \end{bmatrix}$$

$$= \begin{bmatrix} 4+1 & -2-2 \\ -2-2 & +1+4 \end{bmatrix}$$

$$\therefore A^2 = \begin{bmatrix} 5 & -4 \\ -4 & 5 \end{bmatrix}$$

$$\therefore \text{L.H.S} = A^2 - 4A + 3I$$

$$= \begin{bmatrix} 5 & -4 \\ -4 & 5 \end{bmatrix} - 4 \begin{bmatrix} 2 & -1 \\ -1 & 2 \end{bmatrix} + 3 \begin{bmatrix} 1 & 0 \\ 0 & 1 \end{bmatrix}$$

$$= \begin{bmatrix} 5 & -4 \\ -4 & 5 \end{bmatrix} - \begin{bmatrix} 8 & -4 \\ -4 & 8 \end{bmatrix} + \begin{bmatrix} 3 & 0 \\ 0 & 3 \end{bmatrix}$$

$$= \begin{bmatrix} 5-8+3 & -4+4+0 \\ -4+4+0 & 5-8+3 \end{bmatrix}$$

$$= \begin{bmatrix} 0 & 0 \\ 0 & 0 \end{bmatrix} = 0 = \text{R.H.S.}$$

Now take $A^2 - 4A + 3I = 0$

Multiplying this by A^{-1}, we get

$$A^2 \cdot A^{-1} - 4A \cdot A^{-1} + 3I \cdot A^{-1} = 0$$

$$A - 4I + 3A^{-1} = 0 \qquad [\because I \cdot A^{-1} = A^{-1}, A \cdot A^{-1} = I]$$

$$3A^{-1} = 4I - A$$

$$A^{-1} = \frac{1}{3}[4I - A]$$

$$= \frac{1}{3}\left\{\begin{bmatrix} 4 & 0 \\ 0 & 4 \end{bmatrix} - \begin{bmatrix} 2 & -1 \\ -1 & 2 \end{bmatrix}\right\}$$

$$= \frac{1}{3}\begin{bmatrix} 4-2 & 0+1 \\ 0+1 & 4-2 \end{bmatrix}$$

$$\therefore A^{-1} = \frac{1}{3}\begin{bmatrix} 2 & 1 \\ 1 & 2 \end{bmatrix}$$

Example 70. Show that the matrix $P = \begin{bmatrix} 2 & 3 \\ 3 & 5 \end{bmatrix}$ satisfies the relation $P^2 + I = 7P$ where I is the unit matrix of the order 2 and hence find the inverse of P.

Solution:

$$P = \begin{bmatrix} 2 & 3 \\ 3 & 5 \end{bmatrix}$$

$$P^2 = P \cdot P = \begin{bmatrix} 2 & 3 \\ 3 & 5 \end{bmatrix}\begin{bmatrix} 2 & 3 \\ 3 & 5 \end{bmatrix}$$

$$= \begin{bmatrix} 4+9 & 6+15 \\ 6+15 & 9+25 \end{bmatrix}$$

$$\therefore P^2 = \begin{bmatrix} 13 & 21 \\ 21 & 34 \end{bmatrix}$$

So L.H.S $= P^2 + I$

$$= \begin{bmatrix} 13 & 21 \\ 21 & 34 \end{bmatrix} + \begin{bmatrix} 1 & 0 \\ 0 & 1 \end{bmatrix}$$

$$= \begin{bmatrix} 14 & 21 \\ 21 & 35 \end{bmatrix}$$

$$\text{R.H.S} = 7P = 7\begin{bmatrix} 2 & 3 \\ 3 & 5 \end{bmatrix} = \begin{bmatrix} 14 & 21 \\ 21 & 35 \end{bmatrix}$$

$\therefore$ L.H.S = R.H.S.

$\therefore$ Now take $P^2 + I = 7P$

$P^2 + I - 7P = 0$

Multiplying P^{-1}

$$P^2 \cdot P^{-1} - 7P \cdot P^{-1} + I \cdot P^{-1} = 0$$

$$P - 7I + P^{-1} = 0$$

$$\therefore P^{-1} = 7I - P$$

$$= 7\begin{bmatrix} 1 & 0 \\ 0 & 1 \end{bmatrix} - \begin{bmatrix} 2 & 3 \\ 3 & 5 \end{bmatrix}$$

$$= \begin{bmatrix} 7 & 0 \\ 0 & 7 \end{bmatrix} - \begin{bmatrix} 2 & 3 \\ 3 & 5 \end{bmatrix}$$

Example 71. Find the inverse of $A = \begin{bmatrix} 1 & 3 & 2 \\ 0 & 2 & 1 \\ 0 & 5 & 3 \end{bmatrix}$

Solution:

$$\text{Let } A = \begin{bmatrix} 1 & 3 & 2 \\ 0 & 2 & 1 \\ 0 & 5 & 3 \end{bmatrix}$$

$$|A| = 1[6-5] - 3[0-0] + 2[0-0]$$
$$= 1 - 0 + 0$$

$\therefore |A| = 1 \neq 0 \qquad \therefore A^{-1}$ exists.

$$\text{adj } A = \begin{bmatrix} (1) & -(0) & (0) \\ -(-1) & (3) & -(5) \\ (-1) & -(1) & +(2) \end{bmatrix}' = \begin{bmatrix} 1 & 0 & 0 \\ 1 & 3 & -5 \\ -1 & -1 & 2 \end{bmatrix}'$$

$$\text{adj } A = \begin{bmatrix} 1 & +1 & -1 \\ 0 & 3 & -1 \\ 0 & -5 & 2 \end{bmatrix}$$

$$\therefore A^{-1} = \frac{\text{adj } A}{|A|}$$

$$\therefore A^{-1} = (1)\begin{bmatrix} 1 & 1 & -1 \\ 0 & 3 & -1 \\ 0 & -5 & 2 \end{bmatrix}$$

Example 72. If $C = \begin{bmatrix} 1 & -3 \\ -6 & 20 \end{bmatrix}$ find C' and C^{-1}.

Solution:

$$\text{Take } C = \begin{bmatrix} 1 & -3 \\ -6 & 20 \end{bmatrix}$$

$$C' = \begin{bmatrix} 1 & -3 \\ -6 & 20 \end{bmatrix}'$$

$$\therefore C' = \begin{bmatrix} 1 & -6 \\ -3 & 20 \end{bmatrix}$$

$$|C| = 20 - 18 = 2 \neq 0$$

$$\therefore C^{-1} \text{ exists}$$

$$\text{adj } C = \begin{bmatrix} 20 & -(-6) \\ -(-3) & 1 \end{bmatrix}'$$

$$\therefore \text{adj } C = \begin{bmatrix} 20 & +6 \\ 3 & 1 \end{bmatrix}' = \begin{bmatrix} 20 & 3 \\ 6 & 1 \end{bmatrix}$$

$$\therefore C^{-1} = \frac{1}{|C|} \text{adj } C$$

$$\therefore C^{-1} = \frac{1}{2}\begin{bmatrix} 20 & 3 \\ 6 & 1 \end{bmatrix}$$

Example 73. Solve the equations using the matrix method:

$3x - y + 2z = 13$

$2x + y - z = 3$

$x + 3y - 5z = -8$

Solution:

$$A = \begin{bmatrix} 3 & -1 & 2 \\ 2 & 1 & -1 \\ 1 & 3 & -5 \end{bmatrix}, X = \begin{bmatrix} x \\ y \\ z \end{bmatrix}, B = \begin{bmatrix} 13 \\ 3 \\ -8 \end{bmatrix}$$

$$|A| = \begin{vmatrix} 3 & -1 & 2 \\ 2 & 1 & -1 \\ 1 & 3 & -5 \end{vmatrix}$$

$$= 3[-5+3] + 1[-10+1] + 2[6-1]$$
$$= 3[-2] + 1[-9] + 2[5]$$
$$= -6 + [-9] + 10$$
$$= -15 + 10$$
$$|A| = -5 \neq 0$$

$\therefore A^{-1}$ exists.

Co-factors are:

$$A_{11} = \begin{vmatrix} 1 & -1 \\ 3 & -5 \end{vmatrix} = -5 + 3 = -2$$

$$A_{12} = -\begin{vmatrix} 2 & -1 \\ 1 & -5 \end{vmatrix} = -(-10+1) = -(-9) = 9$$

$$A_{13} = \begin{vmatrix} 2 & 1 \\ 1 & 3 \end{vmatrix} = 6 - 1 = 5$$

$$A_{21} = -\begin{vmatrix} -1 & 2 \\ 3 & -5 \end{vmatrix} = -(5-6) = -(-1) = 1$$

$$A_{22} = \begin{vmatrix} 3 & 2 \\ 1 & -5 \end{vmatrix} = -15 - 2 = -17$$

$$A_{23} = -\begin{vmatrix} 3 & -1 \\ 1 & 3 \end{vmatrix} = -(9+1) = -10$$

$$A_{31} = \begin{vmatrix} -1 & 2 \\ 1 & -1 \end{vmatrix} = 1 - 2 = -1$$

$$A_{32} = -\begin{vmatrix} 3 & 2 \\ 2 & -1 \end{vmatrix} = -(-3-4) = 7$$

$$A_{33} = \begin{vmatrix} 3 & -1 \\ 2 & 1 \end{vmatrix} = 3 + 2 = 5$$

$\therefore$ adj A = [cofactors of A]′

$$= \begin{bmatrix} -2 & 9 & 5 \\ 1 & -17 & -10 \\ -1 & 7 & 5 \end{bmatrix}'$$

$$\text{adj } A = \begin{bmatrix} -2 & 1 & -1 \\ 9 & -17 & 7 \\ 5 & -10 & 5 \end{bmatrix}$$

$$\therefore \; A^{-1} = \frac{1}{|A|} \text{adj } A$$

$$A^{-1} = \frac{-1}{5} \begin{bmatrix} -2 & 1 & -1 \\ 9 & -17 & 7 \\ 5 & -10 & 5 \end{bmatrix}$$

$AX = B$

$\therefore \; X = A^{-1} \cdot B$

$$\begin{bmatrix} x \\ y \\ z \end{bmatrix} = \frac{-1}{5} \begin{bmatrix} -2 & 1 & -1 \\ 9 & -17 & 7 \\ 5 & -10 & 5 \end{bmatrix} \begin{bmatrix} 13 \\ 3 \\ -8 \end{bmatrix}$$

$$= \frac{-1}{5} \begin{bmatrix} -26 \; +3 \; +8 \\ 117 \; -51 \; -36 \\ 65 \; -30 \; -40 \end{bmatrix}$$

$$= \frac{-1}{5} \begin{bmatrix} -15 \\ 10 \\ -5 \end{bmatrix}$$

$$= \frac{1}{5} \begin{bmatrix} 15 \\ -10 \\ 5 \end{bmatrix}$$

$$\therefore \; \begin{bmatrix} x \\ y \\ z \end{bmatrix} = \begin{bmatrix} 15/5 \\ -10/5 \\ 5/5 \end{bmatrix} = \begin{bmatrix} 3 \\ -2 \\ 1 \end{bmatrix}$$

$\therefore \; x = 3, \; y = -2$ and $z = 1$

Verification:

Put $x = 3, y = -2$ and $z = 1$

In $3x - y + 2z = 13$.

We get L.H.S $= 3(3) - (-2) + 2\,(1) = 9 + 2 + 2$

$= 9 + 4$

$= 13 =$ R.H.S

Cramer's Rule:

To apply the rule, remember the following points:

1. First, write the given equations in order so that the constant terms are all on right side.
2. Take Δ as the determinant formed by the co-efficients of x, y, z.
3. In Δ, replace the first column by constant term of the equations call this Δ_1.
4. In Δ, replace the second column by constant terms and call it Δ_2.
5. In Δ, replace the third column by constant terms and call it Δ_3.
6. The Solution is $x = \frac{\Delta_1}{\Delta}, y = \frac{\Delta_2}{\Delta}, z = \frac{\Delta_3}{\Delta} (\Delta \neq 0)$.

In case $\Delta = 0$ there is no unique Solution.

Example 74. Solve the following using Cramer's rule.

(a) $2x + 3y = 7$, $3x + 5y = 9$

Solution:

$$\Delta = \begin{vmatrix} 2 & 3 \\ 3 & 5 \end{vmatrix} = 10 - 9 = 1$$

$$\Delta_1 = \begin{vmatrix} 7 & 3 \\ 9 & 5 \end{vmatrix} = 35 - 27 = 8$$

$$\Delta_2 = \begin{vmatrix} 2 & 7 \\ 3 & 9 \end{vmatrix} = 18 - 21 = -3$$

$$x = \frac{\Delta_1}{\Delta} = \frac{8}{1} = 8$$

$$y = \frac{\Delta_2}{\Delta} = \frac{-3}{1} = -3$$

$\therefore$ $x = 8$ and $y = -3$

(b) $10x + 5y = 125$

$9x + 12y = 150$

Solution:

$$\Delta = \begin{vmatrix} 10 & 5 \\ 9 & 12 \end{vmatrix} = 120 - 45 = 75$$

$$\Delta_1 = \begin{vmatrix} 125 & 5 \\ 150 & 12 \end{vmatrix} = 1500 - 750 = 750$$

$$\Delta_2 = \begin{vmatrix} 10 & 125 \\ 9 & 150 \end{vmatrix} = 1500 - 1125 = 375$$

$$x = \frac{\Delta_1}{\Delta} = \frac{750}{75} = 10$$

$$y = \frac{\Delta_2}{\Delta} = \frac{375}{75} = 5$$

$\therefore\ x = 10$ and $y = 5$

(c) $2x + y - z = 3, x + y + z = 1, x - 2y - 3z = 4$

Solution:

$$\Delta = \begin{vmatrix} 2 & 1 & -1 \\ 1 & 1 & 1 \\ 1 & -2 & -3 \end{vmatrix}$$

$$= 2[-3 - (-2)] - 1\,[-3 - 1] - 1\,[-2 - 1]$$
$$= -2 + 4 + 3$$
$$= 5$$

$$\Delta_1 = \begin{vmatrix} 3 & 1 & -1 \\ 1 & 1 & 1 \\ 4 & -2 & -3 \end{vmatrix}$$

$$= 3[-3 + 2] - 1\,[-3 - 4] - 1\,[-2 - 4]$$
$$= -3 + 7 + 6$$
$$= 10$$

$$\Delta_2 = \begin{vmatrix} 2 & 3 & -1 \\ 1 & 1 & 1 \\ 1 & 4 & -3 \end{vmatrix}$$

$$= 2[-3 - 4] - 3\,[-3 - 1] - 1\,[4 - 1]$$
$$= -14 + 12 - 3$$
$$= -5$$

$$\Delta_3 = \begin{vmatrix} 2 & 1 & 3 \\ 1 & 1 & 1 \\ 1 & -2 & 4 \end{vmatrix}$$

$$= 2[4 + 2] - 1\,[4 - 1] + 3\,[-2 - 1]$$
$$= 12 + 3 - 9$$
$$= 0$$

$$x = \frac{\Delta_1}{\Delta} = \frac{10}{5} = 2 \qquad z = \frac{\Delta_3}{\Delta} = \frac{0}{5} = 0$$

$$y = \frac{\Delta_2}{\Delta} = \frac{-5}{5} = -1 \qquad \therefore\ x = 2,\ y = -1 \text{ and } z = 0$$

(d) $4x + y + 2z = 7$, $7x - y + z = 7$, $3x + 4y + z = 8$

Solution:

$$\Delta = \begin{vmatrix} 4 & 1 & 2 \\ 7 & -1 & 1 \\ 3 & 4 & 1 \end{vmatrix}$$

$$= 4[-1-4] - 1\,[7-3] + 2\,[28+3]$$

$$= -20 - 4 + 62$$

$$= 38$$

$$\Delta_1 = \begin{vmatrix} 7 & 1 & 2 \\ 7 & -1 & 1 \\ 8 & 4 & 1 \end{vmatrix}$$

$$= 7[-1-4] - 1\,[7-8] + 2\,[28+8]$$

$$= -35 + 1 + 72$$

$$= 38$$

$$\Delta_2 = \begin{vmatrix} 4 & 7 & 2 \\ 7 & 7 & 1 \\ 3 & 8 & 1 \end{vmatrix}$$

$$= 4[7-8] - 1\,[7-3] + 2\,[56-21]$$

$$= -4 - 28 + 70$$

$$= 38$$

$$\Delta_3 = \begin{vmatrix} 4 & 1 & 7 \\ 7 & -1 & 7 \\ 3 & 4 & 8 \end{vmatrix}$$

$$= 4[-8-28] - 1\,[56-21] + 7\,[28+3]$$

$$= -144 - 35 + 217$$

$$= 38$$

$$x = \frac{\Delta_1}{\Delta} = \frac{38}{38} = 1 \qquad y = \frac{\Delta_2}{\Delta} = \frac{38}{38} = 1$$

$$z = \frac{\Delta_3}{\Delta} = \frac{38}{38} = 1 \qquad \therefore\ x = 1,\ y = 1 \text{ and } z = 1$$

Example 75. A commodity was produced by using 3 units of labour and 2 units of capital the total cost comes to Rs 62. If the commodity had been produced by using 4 units of labour and one unit of capital, the cost would have come to Rs 56. What is the cost per unit of labour and capital? Use matrix method.

Solution:

Let 'x' be cost/unit of labour.

Let 'y' be cost/unit of capital.

$3x + 2y = 62$
$4x + 1y = 56$

$$A = \begin{bmatrix} 3 & 2 \\ 4 & 1 \end{bmatrix}, X = \begin{bmatrix} x \\ y \end{bmatrix}, B = \begin{bmatrix} 62 \\ 56 \end{bmatrix}$$

$$|A| = \begin{vmatrix} 3 & 2 \\ 4 & 1 \end{vmatrix} = 3(1) - 4(2)$$

$$= 3 - 8 = -5 \neq 0$$

$\therefore$ A^{-1} exists

adj A = [Cofactors of A]′

$$= \begin{bmatrix} 1 & -4 \\ -2 & 3 \end{bmatrix}'$$

$$\therefore \text{adj } A = \begin{bmatrix} 1 & -2 \\ -4 & 3 \end{bmatrix}$$

$$A^{-1} = \frac{\text{adj } A}{|A|} = \frac{1}{-5}\begin{bmatrix} 1 & -2 \\ -4 & 3 \end{bmatrix} = \frac{1}{5}\begin{bmatrix} -1 & +2 \\ +4 & -3 \end{bmatrix}$$

$AX = B.$

$X = A^{-1} . B.$

$$\begin{bmatrix} x \\ y \end{bmatrix} = \frac{1}{5}\begin{bmatrix} -4 & 2 \\ 4 & -3 \end{bmatrix}\begin{bmatrix} 62 \\ 56 \end{bmatrix}$$

$$= \frac{1}{5}\begin{bmatrix} -62 + 112 \\ 248 - 168 \end{bmatrix}$$

$$\begin{bmatrix} x \\ y \end{bmatrix} = \frac{1}{5}\begin{bmatrix} 50 \\ 80 \end{bmatrix} = \begin{bmatrix} 10 \\ 16 \end{bmatrix}$$

$\therefore$ Cost/unit of labour = x = 10.

and cost/unit of capital = y =16.

Example 76. The price of 4 Accountancy books, 2 Commerce books, 3 Economics book is Rs 134. the cost of one Accountancy book, 3 Commerce books and 2 Economics book is Rs 81. The cost of 2 Accountancy books, one Commerce book and 5 Economics book is Rs 130. Find the rate per book of each using matrix method.

Solution

Let 'x' be the cost of Accountancy books,

Let 'y' be the cost of Commerce books,

and let 'z' be the cost of Economics books.

$\therefore$ $4x + 2y + 3z = 134$

$1x + 3y + 2z = 81$

$2x + 1y + 5z = 130$

$$A = \begin{bmatrix} 4 & 2 & 3 \\ 1 & 3 & 2 \\ 2 & 1 & 5 \end{bmatrix}, X = \begin{bmatrix} x \\ y \\ z \end{bmatrix}, B = \begin{bmatrix} 134 \\ 81 \\ 130 \end{bmatrix}$$

$$|A| = \begin{vmatrix} 4 & 2 & 3 \\ 1 & 3 & 2 \\ 2 & 1 & 5 \end{vmatrix}$$

$= 4\,[15 - 2] - 2\,[5 - 4] + 3\,[1 - 6]$

$= 4\,[13] - 2\,[1] + 3\,[-5]$

$\therefore |A| = 52 - 2 - 15 = 52 - 17 = 35 \neq 0$

$\therefore A^{-1}$ exists

adj A = [Cofactors of A]′

$$= \begin{bmatrix} (13) & -(1) & (-5) \\ -(7) & (14) & -(0) \\ (-5) & -(5) & (10) \end{bmatrix}'$$

$$= \begin{bmatrix} 13 & -1 & -5 \\ -7 & 14 & 0 \\ -5 & -5 & 10 \end{bmatrix}'$$

$$\text{adj } A = \begin{bmatrix} 13 & -7 & -5 \\ -1 & 14 & -5 \\ -5 & 0 & 10 \end{bmatrix}$$

$$A^{-1} = \frac{\text{adj } A}{|A|} = \frac{1}{35}\begin{bmatrix} 13 & -7 & -5 \\ -1 & 14 & -5 \\ -5 & 0 & 10 \end{bmatrix}$$

$AX = B$

$X = A^{-1}.\,B$

$$\begin{bmatrix} x \\ y \\ z \end{bmatrix} = \frac{1}{35}\begin{bmatrix} 13 & -7 & -5 \\ -1 & 14 & -5 \\ -5 & 0 & 10 \end{bmatrix}\begin{bmatrix} 134 \\ 81 \\ 130 \end{bmatrix}$$

$$= \frac{1}{35}\begin{bmatrix} 1742 - 567 - 650 \\ -134 + 1134 - 650 \\ -670 + 0 + 1300 \end{bmatrix}$$

$$\begin{bmatrix} x \\ y \\ z \end{bmatrix} = \frac{1}{35}\begin{bmatrix} 525 \\ 350 \\ 630 \end{bmatrix} = \begin{bmatrix} 15 \\ 10 \\ 18 \end{bmatrix}$$

$\therefore$ The cost of Accountancy books $= x = 15$ Rs

The cost of Commerce books $= y = 10$ Rs

And the cost of Economics books $= z = 18$ Rs

Exercises

I. Questions carrying THREE marks each (Short Answers):

1. If $A = \begin{bmatrix} 3 & 4 & 1 \\ 1 & 2 & 3 \end{bmatrix}$ and $B = \begin{bmatrix} 2 & 3 & 1 \\ 3 & 4 & 2 \end{bmatrix}$ find $(A + B)$ and $(A - B)'$

   [*Ans:* $A + B = \begin{bmatrix} 5 & 7 & 2 \\ 6 & 4 & 5 \end{bmatrix}, A - B = \begin{bmatrix} 1 & 1 & 0 \\ -2 & -2 & 1 \end{bmatrix}$]

2. If $A = \begin{bmatrix} 4 & 3 & 1 \\ 5 & 6 & 2 \end{bmatrix}, B = \begin{bmatrix} 1 & 2 & 3 \\ 3 & 2 & 1 \end{bmatrix}$ find $2A + 3B$.

   [*Ans:* $2A + 3B = \begin{bmatrix} 11 & 12 & 11 \\ 19 & 18 & 7 \end{bmatrix}$]

3. If $A = \begin{bmatrix} 2 & 1 & 3 \\ 4 & -1 & 2 \end{bmatrix}$ and $B = \begin{bmatrix} -3 & 1 \\ 2 & 5 \\ 0 & 2 \end{bmatrix}$ then find the value of AB.

   [*Ans:* $AB = \begin{bmatrix} -4 & 13 \\ -14 & 3 \end{bmatrix}$]

4. If $A = \begin{bmatrix} 2 & 3 & -1 \\ 2 & 0 & 2 \end{bmatrix}$ and $B = \begin{bmatrix} 1 & -2 \\ -1 & 2 \\ 2 & -4 \end{bmatrix}$ find AB.

   [*Ans:* $AB = \begin{bmatrix} -3 & 6 \\ 6 & -12 \end{bmatrix}$]

5. If $A = \begin{bmatrix} 1 \\ 2 \\ 3 \end{bmatrix}$ and $B = \begin{bmatrix} 1 & -1 & 1 \end{bmatrix}$ find AB.

   [*Ans:* $AB = \begin{bmatrix} 1 & -1 & 1 \\ 2 & -2 & 2 \\ 3 & -3 & 3 \end{bmatrix}$]

6. If $A = \begin{bmatrix} 1 & 4 \\ 2 & -1 \end{bmatrix}$ and $B = \begin{bmatrix} -3 & 0 \\ 5 & 7 \end{bmatrix}$ find matrix 'y' such that $2A + y = 3B$.

[**Ans:** $Y = \begin{bmatrix} -11 & -8 \\ 11 & 23 \end{bmatrix}$]

7. If $A = \begin{bmatrix} 9 & 1 \\ 4 & 3 \end{bmatrix}$ and $B = \begin{bmatrix} 1 & 5 \\ 7 & 12 \end{bmatrix}$ find matrix 'x' so that $3A + 5B + 2X = 0$.

[**Ans:** $X = \begin{bmatrix} -16 & -14 \\ -47/2 & -69/2 \end{bmatrix}$]

8. Find x and y if $(x \quad 7) + (5 \quad y) = (15 \quad 11)$. [**Ans:** $x = 10, y = 4$]

9. Solve for x if $\begin{bmatrix} x^2 & 4 \\ 6 & 2 \end{bmatrix} + \begin{bmatrix} -2x & 4 \\ 3 & 3 \end{bmatrix} = \begin{bmatrix} -1 & 8 \\ 9 & 5 \end{bmatrix}$. [**Ans:** $x = 1$]

10. Given that $\begin{bmatrix} x & a \\ b & y \end{bmatrix}\begin{bmatrix} 1 & 2 \\ 2 & 4 \end{bmatrix} = \begin{bmatrix} 1 & 2 \\ 2 & 4 \end{bmatrix}$ determine x, y in terms of a, b.

[**Ans:** $x = 1 - 2a,\ y = \frac{1}{2}(2 - b)$]

11. If matrix $\begin{bmatrix} 4 \\ 1 \\ 3 \end{bmatrix}\begin{bmatrix} x & y & z \end{bmatrix} = \begin{bmatrix} -4 & 8 & 4 \\ -1 & 2 & 1 \\ -3 & 6 & 3 \end{bmatrix}$ find the values of x, y and z.

[**Ans:** $x = -1, y = 2, z = 1$]

12. If $\begin{bmatrix} 2 & -1 \\ 2 & 0 \end{bmatrix} + 2A = \begin{bmatrix} -4 & 5 \\ 6 & 8 \end{bmatrix}$ find A^2. [**Ans:** $A^2 = \begin{bmatrix} 15 & 3 \\ 2 & 22 \end{bmatrix}$]

13. If $A = \begin{bmatrix} 4 & 2 \\ 0 & 6 \end{bmatrix}$ and $B = \begin{bmatrix} 2 & 4 \\ 4 & 6 \end{bmatrix}$ verify that $(A + B)' = A' + B'$.

14. Express the following in matrix notation.

$5x_1 + 3x_2 = 35$
$4x_1 - 5x_2 = 55$

[**Ans:** $\begin{bmatrix} 5 & 3 \\ 4 & -5 \end{bmatrix}\begin{bmatrix} x_1 \\ x_2 \end{bmatrix} = \begin{bmatrix} 35 \\ 5 \end{bmatrix}$]

15. Express the linear equation $4x_1 + 5x_2 - 3x_2$ as the inner product of two vectors.

[**Ans:** $\begin{bmatrix} 4 & 5 & -3 \end{bmatrix}\begin{bmatrix} x_1 \\ x_2 \\ x_3 \end{bmatrix}$]

16. Express the following equations as matrix equations:

$y + z = 3$

$x + 3y + 2z = 4$

$2x + 3z = 6$

$$\left[Ans: \begin{bmatrix} x & y & z \end{bmatrix} \begin{bmatrix} 0 & 1 & 2 \\ 1 & 3 & 0 \\ 1 & 2 & 3 \end{bmatrix} = \begin{bmatrix} 3 \\ 4 \\ 6 \end{bmatrix}\right]$$

17. Solve for 'y' using determinants:

$3x - y + z = 19$

$x + y + z = 21$

$2x - 2y + 3z = 22$

$$\left[Ans: y = \frac{\Delta y}{\Delta} = 7\right]$$

II. Questions carrying FIVE marks each (Essay Type Answers):

1. If $A = \begin{bmatrix} 1 & 2 & 3 \\ 5 & 0 & 2 \\ 1 & -1 & 1 \end{bmatrix}$, $A = \begin{bmatrix} 3 & -1 & 2 \\ 4 & 2 & 5 \\ 2 & 0 & 3 \end{bmatrix}$ and $C = \begin{bmatrix} 4 & 1 & 2 \\ 0 & 3 & 2 \\ 1 & -2 & 3 \end{bmatrix}$

   then verify $A + (B - C) = (A + B) - C$.

2. If $A = \begin{bmatrix} 1 & -2 & 3 \\ -4 & 2 & 5 \end{bmatrix}$ and $B = \begin{bmatrix} 2 & 3 \\ 4 & 5 \\ 2 & 1 \end{bmatrix}$ show that $AB \neq BA$.

3. Verify $(AB)' = B'A'$ for the matrices $A = \begin{bmatrix} 3 & -4 \\ 1 & 1 \\ 2 & 0 \end{bmatrix}$ and $B = \begin{bmatrix} 2 & 1 & 2 \\ 1 & 3 & 4 \end{bmatrix}$.

4. Find the product AB and BA of two matrices $A = \begin{bmatrix} 2 & 3 & 1 \\ 0 & 2 & -1 \end{bmatrix}$, $B = \begin{bmatrix} 3 & -4 \\ 1 & 4 \\ -2 & 2 \end{bmatrix}$ and

   prove that $AB \neq BA$.

5. If $A + B = \begin{bmatrix} 3 & 5 \\ 7 & 9 \end{bmatrix}$ and $A - B = \begin{bmatrix} -1 & -1 \\ -1 & -1 \end{bmatrix}$ find A and B.

$$\left[Ans: A = \begin{bmatrix} 1 & 2 \\ 3 & 4 \end{bmatrix}, B = \begin{bmatrix} 2 & 3 \\ 4 & 5 \end{bmatrix}\right]$$

6. Find the adjoint of

(a) $\begin{bmatrix} 1 & -2 & 3 \\ 0 & 2 & -1 \\ -4 & 5 & 2 \end{bmatrix}$ (b) $\begin{bmatrix} 1 & 2 & 3 \\ 2 & 3 & 2 \\ 3 & 3 & 4 \end{bmatrix}$ (c) $\begin{bmatrix} 2 & 0 & 0 \\ 0 & 2 & 0 \\ 0 & 2 & 2 \end{bmatrix}$

[*Ans: (a)* $\begin{bmatrix} 9 & 19 & -4 \\ 4 & 14 & 1 \\ 8 & 3 & 2 \end{bmatrix}$, *(b)* $\begin{bmatrix} 6 & 1 & -5 \\ -2 & -5 & 4 \\ -3 & 3 & -1 \end{bmatrix}$, *(c)* $\begin{bmatrix} 4 & 0 & 0 \\ 0 & 4 & 0 \\ 0 & -4 & 4 \end{bmatrix}$]

7. Find the adjoint and inverse of the matrix $A = \begin{bmatrix} 1 & 0 & -1 \\ 3 & 4 & 5 \\ 0 & -6 & -7 \end{bmatrix}$.

[*Ans: adj* = $\begin{bmatrix} 2 & 6 & 5 \\ 21 & -7 & -8 \\ -18 & 6 & 4 \end{bmatrix}$, *Inverse* = $\frac{1}{20}\begin{bmatrix} 2 & 6 & 5 \\ 21 & -7 & -8 \\ -18 & 6 & 4 \end{bmatrix}$]

8. If $A = \begin{bmatrix} 1 & 1 & -1 \\ 1 & 2 & 1 \\ -1 & 1 & 3 \end{bmatrix}$, find the inverse of A. [*Ans:* $\begin{bmatrix} -5/2 & 2 & -3/2 \\ 2 & -1 & 1 \\ -3/2 & 1 & -1/2 \end{bmatrix}$]

9. Find the inverse of:

(a) $\begin{bmatrix} 1 & 2 \\ 3 & -1 \end{bmatrix}$ (b) $\begin{bmatrix} 1 & -3 & 2 \\ 0 & 2 & 8 \\ -1 & 5 & 7 \end{bmatrix}$ (c) $\begin{bmatrix} 1 & 2 & 3 \\ 2 & 4 & 5 \\ 3 & 5 & 6 \end{bmatrix}$

[*Ans: (a)* $\frac{-1}{7}\begin{bmatrix} -1 & -2 \\ -3 & 1 \end{bmatrix}$, *(b)* $\begin{bmatrix} -13 & -4 & 1 \\ 3/2 & 9/2 & -1 \\ -14 & -4 & 1 \end{bmatrix}$, *(c)* $\begin{bmatrix} 1 & -3 & 2 \\ -3 & 3 & -1 \\ 2 & -1 & 0 \end{bmatrix}$]

10. If $A = \begin{bmatrix} 1 & 2 & 2 \\ 2 & 1 & 2 \\ 2 & 2 & 1 \end{bmatrix}$ show that $A^2 - 4A - 5I_3 = 0$

11. If $A = \begin{bmatrix} 2 & 3 \\ 3 & 5 \end{bmatrix}$, then verify that $A^2 = 7A - I$ where $I = \begin{bmatrix} 1 & 0 \\ 0 & 1 \end{bmatrix}$ and hence cbtain A^{-1} from the relationship.

[*Ans:* $A^{-1} = \begin{bmatrix} 5 & -3 \\ -3 & 2 \end{bmatrix}$]

12. Show that $A = \begin{bmatrix} 1 & 2 & 2 \\ 2 & 1 & 2 \\ 2 & 2 & 1 \end{bmatrix}$ satisfies the equations $A^2 - 4A - 5I = 0$ where I is the identity matrix and 0 denotes the zero matrix, hence find the inverse of A.

[*Ans:* $A^{-1} = \frac{1}{5}\begin{bmatrix} -3 & 2 & 2 \\ 2 & -3 & 2 \\ 2 & 2 & -3 \end{bmatrix}$]

13. Solve using matrix method:
    (a) $4x + 3y = 6$
    $5x - 2y = 19$
    (b) $x + y - z = -2$
    $3x + 2y + 3z = 13$
    $2x + 7y + 4z = 31$
    (c) $x - 3y + 2z = -2$
    $y + 4z = -2$
    $-x + 5y + 7z = 0$

[*Ans: (a)* $x = 3, y = -2$, *(b)* $x = \frac{-19}{18}, y = \frac{8}{3}, z = \frac{65}{18}$ *(c)* $x = -36, y = -10, z = 2$]

14. Solve the following equations by using Cramer's Rule:
    (a) $5x - y = 9$
    $3x + y = 7$
    (b) $4x + y = 24$
    $3x - 5y = 8$
    (c) $3x + 2y + z = 3$
    $2x - y + z = 5$
    $5x - y - 3z = -2$
    (d) $2x + 3y = 1$
    $5x + 6y = 4$
    (e) $4x + y + 2z = 7$
    $7x - y + z = 7$
    $3x + 4y + z = 8$
    (f) $x + y + z = 6$
    $2x + 3y - z = 5$
    $6x - 2y - 3z = -7$

[*Ans: (a)* $x = 2, y = 1$ *(b)* $x = \frac{128}{23}, y = \frac{40}{23}$ *(c)* $x = 1, y = -1, z = 2$ *(d)* $x = 2, y = -1$ *(e)* $x = 1, y = 1, z = 1$ *(f)* $x = 1, y = 2, z = 3$]

4

Differentiation

Gottfried Wilhelm Leibnitz
(1646–1716)

Sir Isaac Newton
(1642–1727)

Joseph Louis Lagrange
(1736–1813)

Contents

- Types of Derivatives, Functions, Limits, Max ima and Minima.

INTRODUCTION

The fundamental limiting processes of calculus are differentiation and integration. Differential calculus is concerned with the notion of the derivative. The symbol $\frac{dy}{dx}$ as a notation for the derivative was first used by the German mathematician Leibnitz (co-inventor of calculus). Newton used the symbol Y. The use of the symbol f' for the derivative of 'f' was first introduced by the French mathematician Lagrange in the eighteenth century.

Note:

1. $\frac{dy}{dx}$ is called the differential coefficient of 'y' $w.r.t$ 'x' or the derived function of 'y' $w.r.t$ 'x'.

2. $\frac{dy}{dx}$ may be indicated by any one of the symbols $\frac{d}{dx}(y)$, $\frac{d}{dx}[f(x)]$, Dy, y_1, y', y, $f'(x)$.

3. $\frac{d}{dx}$ is called the differential operator meaning "the derivative of ... *w.r.t* '*x*' and it is denoted by the symbol "*D*".

DERIVATIVES

1. Derivative of x^n

Derivative of x^n *w.r.t* '*x*' where *n* is any rational number.

i.e., $\frac{d}{dx}\left(x^n\right) = nx^{n-1}$ where *n* is any rational number.

Examples

(a) $\frac{d}{dx}(x) = 1 \cdot x^{1-1} = x^0 = 1$

(b) $\frac{d}{dx}\left(x^2\right) = 2 \cdot x^{2-1} = 2x$

(c) $\frac{d}{dx}\left(x^3\right) = 3 \cdot x^{3-1} = 3x^2$

(d) $\frac{d}{dx}\left(\sqrt{x}\right) = \frac{d}{dx}\left(x^{1/2}\right) = \frac{1}{2}x^{1/2-1} = \frac{1}{2x^{1/2}} = \frac{1}{2\sqrt{x}}$

(e) $\frac{d}{dx}\left(\frac{1}{x}\right) = \frac{d}{dx}\left(x^{-1}\right) = -1 \cdot x^{-1-1} = -x^{-2} = \frac{-1}{x^2}$

(f) $\frac{d}{dx}\left(\frac{1}{\sqrt[3]{x^2}}\right) = \frac{d}{dx}\left(\frac{1}{x^{2/3}}\right)$

$$\frac{d}{dx}\left(x^{-2/3}\right) = \frac{-2}{3} \cdot x^{-2/3-1}$$

$$= \frac{-2}{3} \cdot x^{-5/3}$$

$$= \frac{-2}{3x^{5/3}}$$

$$= \frac{-2}{3\sqrt[3]{x^5}}$$

2. Derivative of a Constant

Derivative of a constant. i.e., $\frac{d}{dx}$ (constant) = 0 (zero)

Examples.

(a) $\frac{d}{dx}(2) = 0$

(b) $\frac{d}{dx}\left(\frac{-5}{2}\right) = 0$

(c) $\frac{d}{dx}(ab) = 0$ If a is a constant.

3. Derivative of a Product of a Constant and a Function

Derivative of a product of a constant and a function. i.e., $\frac{d}{dx}$ (constant × function) = constant × $\frac{d}{dx}$ (function)

Examples

(a) $\frac{d}{dx}\left(4x^3\right) = 4\frac{d}{dx}\left(x^3\right)$

$= 4 \cdot 3x^{3-1} = 4 \cdot 3x^2$

$= 12x^2$

(b) $\frac{d}{dx}\left(ax^3\right) = a \cdot \frac{d}{dx}\left(x^3\right)$

$= a \cdot 3x^{3-1} = 3ax^2$

(c) $\frac{d}{dx}(5 \log x) = 5\frac{d}{dx}(\log x)$

$= 5\frac{1}{x} = \frac{5}{x}$

(d) $\frac{d}{dx}\left(\frac{-3}{4x^3}\right) = \frac{-3}{4}\frac{d}{dx}\left(\frac{1}{x^3}\right)$

$= \frac{-3}{4}\frac{d}{dx}\left(x^{-3}\right)$

$$= \frac{-3}{4}(-3)x^{-3-1}$$

$$= \frac{+9}{4}x^{-4}$$

$$= \frac{9}{4x^4}$$

(e) $\frac{d}{dx}\left(\frac{2}{\sqrt{x}}\right) = \left(2 \cdot x^{-1/2}\right)$

$$= 2 \cdot \left(\frac{-1}{2}\right)x^{-1/2-1} = -x^{-3/2}$$

$$= \frac{-1}{x^{3/2}}$$

(f) $\frac{d}{dx}\left(\frac{a}{x^n}\right) = a \cdot \frac{d}{dx}\left(x^{-n}\right)$

$$= a(-n)x^{-n-1}$$

$$= -an\,x^{-(n+1)}$$

$$= \frac{-an}{x^{n+1}}$$

4. Derivative of a Sum

Derivative of a sum i.e., $\frac{d}{dx}(u + v) = \frac{du}{dx} + \frac{dv}{dx}$ where u and v are functions of 'x'. In general if u, v, w, ... are functions of 'x' then

$$\frac{d}{dx}(u + v + w + ...) = \frac{du}{dx} + \frac{dv}{dx} + \frac{dw}{dx} + \dots .$$

Examples.

(a) $\frac{d}{dx}\left(8x^4 + 7x^3 + 6x^2 + 4x + 7\right)$

$$= \frac{d}{dx}\left(8x^4\right) + \frac{d}{dx}\left(7x^3\right) + \frac{d}{dx}\left(6x^2\right) + \frac{d}{dx}(4x) + \frac{d}{dx}(4)$$

$$= 32x^3 + 21x^2 + 12x + 4 + 0$$

$$= 32x^3 + 21x^2 + 12x + 4$$

(b) $\frac{d}{dx}\left(ax^3 + 3bx^2 + 3cx + d\right)$

$$= \frac{d}{dx}\left(ax^3\right) + \frac{d}{dx}\left(3bx^2\right) + \frac{d}{dx}(3cx) + \frac{d}{dx}(d)$$

$$= 3ax^2 + 6bx + 3c + 0$$

$$= 3ax^2 + 6bx + 3c$$

(c) $\frac{d}{dx}\left(x^n + \frac{1}{x^n}\right)$

$$= \frac{d}{dx}\left(x^n\right) + \frac{d}{dx}\left(\frac{1}{x^n}\right)$$

$$= n \cdot x^{n-1} + \frac{d}{dx}\left(x^{-n}\right)$$

$$= n \cdot x^{n-1} + (-n)x^{-n-1}$$

$$= n \cdot x^{n-1} + (-n)x^{-(n+1)}$$

$$= nx^{n-1} - \frac{n}{x^{n+1}}$$

(d) $\frac{d}{dx}\left(\frac{a}{x} + \frac{b}{x^2} + \frac{c}{x^3} + \frac{d}{x^4}\right)$

$$= \frac{d}{dx}\left(\frac{a}{x}\right) + \frac{d}{dx}\left(\frac{b}{x^2}\right) + \frac{d}{dx}\left(\frac{c}{x^3}\right) + \frac{d}{dx}\left(\frac{d}{x^4}\right)$$

$$= a\frac{d}{dx}\left(\frac{1}{x}\right) + b\frac{d}{dx}\left(\frac{1}{x^2}\right) + c\frac{d}{dx}\left(\frac{1}{x^3}\right) + d\frac{d}{dx}\left(\frac{1}{x^4}\right)$$

$$= \frac{ad}{dx}\left(x^{-1}\right) + \frac{bd}{dx}\left(x^{-2}\right) + \frac{cd}{dx}\left(x^{-3}\right) + \frac{dd}{dx}\left(x^{-4}\right)$$

$$= -ax^{-2} + b(-2)x^{-3} + (-3c)x^{-4} + d(-4)x^{-5}$$

$$= \frac{-a}{x^2} - \frac{2b}{x^3} - \frac{3c}{x^4} - \frac{4d}{x^5}$$

(e) $\frac{d}{dx}\left(\sqrt{x} + e^x + \log x\right)$

$$= \frac{d}{dx}\left(\sqrt{x}\right) + \frac{d}{dx}\left(e^x\right) + \frac{d}{dx}(\log x)$$

$$= \frac{1}{2\sqrt{x}} + e^x + \frac{1}{x}$$

5. Derivative of a Difference

Derivative of a difference. i.e., $\frac{d}{dx}(u - v) = \frac{du}{dx} - \frac{dv}{dx}$ where u and v are functions of 'x'. In general if $u, v, w, \ldots$ are functions of 'x', then

$$\frac{d}{dx}(u - v - w \ldots) = \frac{du}{dx} - \frac{dv}{dx} - \frac{dw}{dx} \ldots .$$

Examples

(a) $\frac{d}{dx}\left(9x^5 - 4x^3 - 12\right)$

$$= \frac{d}{dx}\left(9x^5\right) - \frac{d}{dx}\left(4x^3\right) - \frac{d}{dx}(12)$$

$$= 9\frac{d}{dx}\left(x^5\right) - 4\frac{d}{dx}\left(x^3\right) - \frac{d}{dx}(12)$$

$$= 9 \cdot 5x^4 - 4 \cdot 3x^2 - 0$$

$$= 45x^4 - 12x^2$$

(b) $\frac{d}{dx}\left(5x^4 + 3x^3 - 7x^2 + 9x + 8\right)$

$$= \frac{d}{dx}\left(5x^4\right) + \frac{d}{dx}\left(3x^3\right) + \frac{d}{dx}\left(-7x^2\right) + \frac{d}{dx}(9x) + \frac{d}{dx}(8)$$

$$= 5\frac{d}{dx}\left(x^4\right) + 3\frac{d}{dx}\left(x^3\right) - 7\frac{d}{dx}\left(x^2\right) + 9\frac{d}{dx}(x) + \frac{d}{dx}(8)$$

$$= 5 \cdot 4x^3 + 3 \cdot 3x^2 - 7 \cdot 2x + 9 + 0$$

$$= 20x^3 + 9x^2 - 14x + 9$$

6. Derivative of a Product

Derivative of a product i.e., $\frac{d}{dx}(uv) = u\frac{dv}{dx} + v\frac{du}{dx}$ where u and v are functions of 'x'.

The formula can be remembered as follows:

$$\frac{d}{dx}(\text{First function} \cdot \text{Second function})$$

$$= (\text{First function}) \cdot \frac{d}{dx}(\text{Second function}) +$$

$$(\text{Second function}) \cdot \frac{d}{dx}(\text{First function})$$

Examples

(a) $\frac{d}{dx}\left(xe^x\right) = x\frac{d}{dx}\left(e^x\right) + e^x\frac{d}{dx}(x)$

$= xe^x + e^x \cdot 1$

$= xe^x + e^x$

$= (1 + x)\,e^x$

(b) $\frac{d}{dx}\left(\sqrt{x}\log x\right) = \sqrt{x}\frac{d}{dx}(\log x) + \log x\frac{d}{dx}\left(\sqrt{x}\right)$

$= \sqrt{x}\left(\frac{1}{x}\right) + \log x\left(\frac{1}{2\sqrt{x}}\right)$

$= \frac{\sqrt{x}}{\sqrt{x}\cdot\sqrt{x}} + \frac{\log x}{2\sqrt{x}} = \frac{1}{\sqrt{x}} + \frac{\log x}{2\sqrt{x}}$

$= \frac{2 + \log x}{2\sqrt{x}}$

(c) $\frac{d}{dx}\left[\left(x^2 + 1\right)\left(1 - 2x^3\right)\right]$

$= \left(x^2 + 1\right)\frac{d}{dx}\left(1 - 2x^3\right) + \left(1 - 2x^3\right)\frac{d}{dx}\left(x^2 + 1\right)$

$= \left(x^2 + 1\right)\left(-6x^2\right) + \left(1 - 2x^3\right)(2x)$

(d) $\frac{d}{dx}\left[\left(2x^3 - 1\right)\left(x^4 + 7x^2 - 1\right)\right]$

$= \left(2x^3 - 1\right)\frac{d}{dx}\left(x^4 + 7x^2 - 1\right) + \left(x^4 + 7x^2 + 1\right)\frac{d}{dx}\left(2x^3 - 1\right)$

$= \left(2x^3 - 1\right)\left(4x^3 + 14x\right) + \left(x^4 + 7x^2 - 1\right)\left(6x^2\right)$

If $y = u\,v\,w$ where u, v, w are functions of x then $\frac{dy}{dx} = \frac{d}{dx}(u\,v\,w)$.

$$= uv\frac{dw}{dx} + uw\frac{dv}{dx} + vw\frac{du}{dx}$$

Example. $\frac{d}{dx}\left[\left(x^2+1\right)\left(1-x^3\right)\left(2-3x^4\right)\right]$

$$= \left(x^2+1\right)\left(1-x^3\right)\frac{d}{dx}\left(2-3x^4\right) + \left(x^2+1\right)\left(2-3x^4\right)\frac{d}{dx}\left(1-x^3\right)$$

$$+ \left(1-x^3\right)\left(2-3x^4\right)\frac{d}{dx}\left(x^2+1\right)$$

$$= \left(x^2+1\right)\left(1-x^3\right)\left(-12x^3\right) + \left(x^2+1\right)\left(2-3x^4\right)\left(-3x^2\right)$$

$$+ \left(1-x^3\right)\left(2-3x^4\right)(2x)$$

7. Derivative of a Quotient

Derivative of a Quotient i.e., $\frac{d}{dx}\left(\frac{u}{v}\right) = \frac{v \cdot \frac{du}{dx} - u \cdot \frac{dv}{dx}}{v^2}$ where 'u' and 'v' are functions of 'x' the above formula is remembered as:

$$\frac{d}{dx}\left(\frac{u}{v}\right) = \frac{(\text{Denominator})\,\frac{d}{dx}(\text{Numerator}) - (\text{Numerator})\,\frac{d}{dx}(\text{Denominator})}{(\text{Denominator})^2}$$

$$= \frac{(\text{Dr})\frac{d}{dx}(\text{Nr}) - (\text{Nr})\frac{d}{dx}(\text{Dr})}{(\text{Dr})^2}$$

Examples.

(a) $\frac{d}{dx}\left(\frac{x}{1+x^2}\right)$

$$= \frac{\left(1+x^2\right)\frac{d}{dx}(x) - x \cdot \frac{d}{dx}\left(1+x^2\right)}{\left(1+x^2\right)^2}$$

$$= \frac{(1+x^2)\cdot 1 - x(2x)}{(1+x^2)^2}$$

$$= \frac{1+x^2-2x^2}{(1+x^2)^2} = \frac{1-x^2}{(1+x^2)^2}$$

(b) $$\frac{d}{dx}\left(\frac{\log x}{x^2}\right) = \frac{(x^2)\frac{d}{dx}(\log x) - (\log x)\frac{d}{dx}(x^2)}{(x^2)^2}$$

$$= \frac{x^2\left(\frac{1}{x}\right) - (\log x)(2x)}{x^4}$$

$$= \frac{x - 2x \log x}{x^4}$$

$$= \frac{1 - 2\log x}{x^3}$$

(c) $$\frac{d}{dx}\left(\frac{x^2 - x + 1}{x^2 + x + 1}\right)$$

$$= \frac{(x^2+x+1)\frac{d}{dx}(x^2-x+1) - (x^2-x+1)\frac{d}{dx}(x^2+x+1)}{(x^2+x+1)^2}$$

$$= \frac{(x^2+x+1)(2x-1) - (x^2-x+1)(2x+1)}{(x^2+x+1)^2}$$

8. Differentiation of Parametric Functions

Differentiation of parametric functions. i.e., $\dfrac{dy}{dx} = \dfrac{\left(\dfrac{dy}{dt}\right)}{\left(\dfrac{dx}{dt}\right)}$

Example. Find $\frac{dy}{dx}$ given that $y = 2at$, $x = at^2$.

Solution: $\frac{dy}{dt} = 2a$ and $\frac{dx}{dt} = 2at$

$$\therefore \quad \frac{dy}{dx} = \frac{\left(\frac{dy}{dt}\right)}{\left(\frac{dx}{dt}\right)} = \frac{2a}{2at} = \frac{1}{t}$$

9. $\frac{d}{dx}\left(a^x\right) = a^x$ log a

$\frac{d}{dx}\left(a^x\right) = a^x \log a$ where 'a' is a constant.

Examples.

(a) $\frac{d}{dx}\left(2^x\right) = 2^x \log 2$

(b) $\frac{d}{dx}\left(8^x\right) = 8^x \log 8$

(c) $\frac{d\left(x \cdot 12^x\right)}{dx} = x\frac{d}{dx}\left(12^x\right) + \left(12^x\right)\frac{d(x)}{dx}$

$= x \cdot 12^x \log 12 + 12^x \cdot 1$

Successive Differentiation

The derivative of $\frac{dy}{dx}$ with respect to x is called the "second derivative of y" *w.r.t* 'x' and is denoted by $\frac{d^2y}{dx^2}$ (read as De two y by de x squared).

$$\therefore \quad \frac{d}{dx}\left(\frac{dy}{dx}\right) = \frac{d^2y}{dx^2}$$ and it is denoted by

$$D^2y,\ y_2,\ y'',\ \ddot{y},\ f''(x),\ \frac{d}{dx}\left[f'(x)\right].$$

Memorise the Following Formulae

1. $\frac{d}{dx}\left(x^{n}\right) = n \cdot x^{n-1}$

2. $\frac{d}{dx}\left(\sqrt{x}\right) = \frac{1}{2\sqrt{x}}$

3. $\frac{d}{dx}\left(\frac{1}{x}\right) = \frac{-1}{x^{2}}$

4. $\frac{d}{dx}\left(e^{x}\right) = e^{x}$

5. $\frac{d}{dx}(\log x) = \frac{1}{x}$

6. $\frac{d}{dx}$ (Constant) = Zero

7. $\frac{d}{dx}(k \cdot f(x)) = k\frac{d}{dx}[f(x)]$ Where k is a constant.

8. Product formula:

$$\frac{d}{dx}(uv) = u\frac{dv}{dx} + v\frac{du}{dx}$$

9. Quotient formula:

$$\frac{d}{dx}\left(\frac{u}{v}\right) = \frac{v\frac{du}{dx} - u\frac{dv}{dx}}{v^{2}}$$

10. $\frac{d}{dx}(uvw) = uv\frac{dw}{dx} + uw\frac{dv}{dx} + vw\frac{du}{dx}$

Note: $\frac{\partial z}{\partial y}$ Means the partial derivative of 'z' with respect to 'y' and it is denoted by Z_y.

Solved Examples

Example 1. Differentiate *w.r.t* '*x*': $y = \dfrac{1}{5 + 3x^2}$

Solution:

$$y = \frac{1}{5 + 3x^2} \quad 11$$

$$\frac{dy}{dx} = \frac{d}{dx}\left(\frac{1}{5 + 3x^2}\right)$$

$$= \frac{\left(5 + 3x^2\right)0 - 1 \cdot \frac{d}{dx}\left(5 + 3x^2\right)}{\left(5 + 3x^2\right)^2}$$

$$= \frac{0 - (0 + 6x)}{\left(5 + 3x^2\right)^2} = \frac{-6x}{\left(5 + 3x^2\right)^2}$$

Example 2. If $u = x^2 + xy + y^2$. find $\dfrac{\partial u}{\partial x}$ and $\dfrac{\partial u}{\partial y}$.

Solution: $u = x^2 + xy + y^2$

$$\frac{\partial u}{\partial x} = 2x + y + 0 = 2x + y$$

$$\frac{\partial u}{\partial y} = x + 2y$$

Example 3. If $y = \dfrac{3x + \log x}{4 - 3x}$ find $\dfrac{dy}{dx}$.

Solution: $\dfrac{dy}{dx} = \dfrac{d}{dx}\left(\dfrac{3x + \log x}{4 - 3x}\right)$

$$= \frac{(4 - 3x)\left(3 + \frac{1}{x}\right) - (3x + \log x)(-3)}{(4 - 3x)^2}$$

$$= \frac{12 - \cancel{9x} + \frac{4}{x} - 3 + \cancel{9x} + 3\log x}{(4 - 3x)^2}$$

$$= \frac{9 + \frac{4}{x} + 3 \log x}{(4 - 3x)^2}$$

Example 4. Differentiate with respect to '*x*':

$$\frac{x^2 - 3x + 4}{x^2 + x + 2}$$

Solution:

$$\frac{dy}{dx} = \frac{d}{dx}\left(\frac{x^2 - 3x + 4}{x^2 + x + 2}\right)$$

$$= \frac{(x^2 + x + 2)(2x - 3) - (x^2 - 3x + 4)(2x + 1)}{(x^2 + x + 2)^2}$$

$$= \frac{\cancel{2x^3} + 2x^2 + 4x - 3x^2 - \cancel{3x} - 6 - \cancel{2x^3} + 6x^2 - 8x - x^2 + \cancel{3x} - 4}{(x^2 + x + 2)^2}$$

$$= \frac{4x^2 - 4x - 10}{(x^2 + x + 2)^2}$$

Example 5. If $y = 8x^4 - 5x^3 + \frac{1}{2}x^2 - 10$ find $\frac{d^2y}{dx^2}$.

Solution:

$$\frac{dy}{dx} = \frac{d}{dx}\left(8x^4 - 5x^3 + \frac{1}{2}x^2 - 10\right)$$

$$= \frac{d}{dx}\left(8x^4\right) + \frac{d}{dx}\left(-5x^3\right) + \frac{d}{dx}\left(\frac{1}{2}x^2\right) + \frac{d}{dx}(-10)$$

$$= 8\frac{d}{dx}\left(x^4\right) + (-5)\frac{d}{dx}\left(x^3\right) + \frac{1}{2}\frac{d}{dx}\left(x^2\right) + 0$$

$$= 8 \cdot 4x^3 - 5 \cdot 3x^2 + \frac{\cancel{2}}{\cancel{2}} \cdot x + 0$$

$$\therefore \frac{dy}{dx} = 32x^3 - 15x^2 + x$$

$$\frac{d}{dx}\left(\frac{dy}{dx}\right) = \frac{d^2y}{dx^2} = \frac{d}{dx}\left(32x^3 - 15x^2 + x\right)$$

$$= 96x^2 - 30x + 1$$

Hence $\frac{d^2y}{dx^2} = 96x^2 - 30x + 1$

Example 6. Find $\frac{dy}{dx}$ if $y = x^4 - x^3 + 2x^2 - 5$.

Solution:

$$\frac{dy}{dx} = \frac{d}{dx}\left(x^4 - x^3 + 2x^2 - 5\right)$$

$$= \frac{d}{dx}\left(x^4\right) + \frac{d}{dx}\left(-x^3\right) + \frac{d}{dx}\left(2x^2\right) + \frac{d}{dx}(-5)$$

$$= 4x^3 - 3x^2 + 4x + 0$$

$$= 4x^3 - 3x^2 + 4x$$

Example 7. For $f(x) = \frac{1}{x}$ find $f''(x)$.

Solution:

$$f(x) = \frac{1}{x}$$

$$f'(x) = \frac{d}{dx}[f(x)] = \frac{d}{dx}\left(\frac{1}{x}\right)$$

$$\therefore\ f'(x) = \frac{-1}{x^2}$$

$$f''(x) = \frac{d}{dx}[f'(x)] = \frac{d}{dx}\left(\frac{-1}{x^2}\right)$$

$$= \frac{d}{dx}\left(-x^{-2}\right)$$

$$= +2x^{-3}$$

$$\therefore\ f''(x) = \frac{2}{x^3}$$

Example 8. If $y = \frac{x+5}{x^2 - 2x + 3}$ evaluate $\frac{dy}{dx}$.

Solution: $\frac{dy}{dx} = \frac{d}{dx}\left(\frac{x+5}{x^2 - 2x + 3}\right)$

$$= \frac{\left(x^2 - 2x + 3\right)(1) - (x+5)(2x-2)}{\left(x^2 - 2x + 3\right)^2}$$

$$= \frac{x^2 - 2x + 3 - 2x^2 - 10x + 2x + 10}{\left(x^2 - 2x + 3\right)^2}$$

$$= \frac{-x^2 - 10x + 13}{\left(x^2 - 2x + 3\right)^2}$$

Example 9. If $y = \dfrac{-1}{\left(1-\sqrt{x}\right)^2}$ find $\dfrac{dy}{dx}$.

Solution:

$$y = \frac{-1}{\left(1-\sqrt{x}\right)^2}$$

$$\frac{dy}{dx} = \frac{\left(1-\sqrt{x}\right)^2 (0) - (-1)\, 2\left(1-\sqrt{x}\right)\left(-\frac{1}{2\sqrt{x}}\right)}{\left[\left(1-\sqrt{x}\right)^2\right]^2}$$

$$= \frac{+\,2\left(1-\sqrt{x}\right)\left(-\frac{1}{2\sqrt{x}}\right)}{\left(1-\sqrt{x}\right)^4}$$

$$= \frac{-\frac{2}{2\sqrt{x}}}{\left(1-\sqrt{x}\right)^3}$$

$$\therefore \frac{dy}{dx} = \frac{-1}{\sqrt{x}\left(1-\sqrt{x}\right)^3}$$

Example 10. Find $\dfrac{dy}{dx}$ if $y = 4x^5 - 16x^3 + 4x^2 - 4x + 3$.

Solution:

$$y = 4x^5 - 16x^3 + 4x^2 - 4x + 3$$

$$\frac{dy}{dx} = 20x^4 - 48x^2 + 8x - 4$$

Example 11. Differentiate with respect to 'x' $\dfrac{4x^2 - 7x}{6x + 5}$.

Solution: $y = \dfrac{4x^2 - 7x}{6x + 5}$

$$\frac{dy}{dx} = \frac{(6x + 5)(8x - 7) - (4x^2 - 7x)(6)}{(6x + 5)^2}$$

$$= \frac{48x^2 + 40x - \cancel{42x} - 35 - 24x^2 + \cancel{42x}}{(6x + 5)^2}$$

$$\therefore \frac{dy}{dx} = \frac{24x^2 + 40x - 35}{(6x + 5)^2}$$

Example 12. If $y = \dfrac{5x^3 + 3x^2 + 8}{x^4}$ evaluate $\dfrac{dy}{dx}$.

Solution: $\dfrac{dy}{dx} = \dfrac{d}{dx}\left(5/x + 3/x^2 + 8/x^4\right)$

$$\therefore \frac{dy}{dx} = \frac{-5}{x^2} - \frac{6}{x^3} - \frac{32}{x^5}$$

Example 13. If $y = x^3\left(x^2 + 2x - 1\right)$ find $\dfrac{dy}{dx}$.

Solution: $\dfrac{dy}{dx} = \dfrac{d}{dx}\left[x^5 + 2x^4 - x^3\right]$

$$\therefore \frac{dy}{dx} = 5x^4 + 8x^3 - 3x^2$$

Example 14. Differentiate *w.r.t* 'x' $y = 4\sqrt{x} + \dfrac{3}{x} + 4e^x$.

Solution: $\dfrac{dy}{dx} = \dfrac{d}{dx}\left[4\sqrt{x} + \dfrac{3}{x} + 4e^x\right]$

$$= 4 \cdot \frac{1}{2\sqrt{x}} + 3\left(\frac{-1}{x^2}\right) + 4e^x$$

$$\therefore \frac{dy}{dx} = \frac{2}{\sqrt{x}} - \frac{3}{x^2} + 4e^x$$

Example 15. If $u = (x - y)(x + y)$ find $\frac{\partial u}{\partial x}$ and $\frac{\partial u}{\partial y}$.

Solution: $\frac{\partial u}{\partial x} = (x - y)(1) + (x + y)(1)$

$= x - y + x + y = 2x$

$$\frac{\partial u}{\partial y} = (x - y)(1) + (x + y)(-1)$$

$= x - y - x - y = -2y$

Example 16. Find $\frac{dy}{dx}$ if $y = \frac{2x+1}{(1-x)^2}$.

Solution: $\frac{dy}{dx} = \frac{d}{dx}\left(\frac{2x+1}{(1-x)^2}\right)$

$$= \frac{(1-x)^2(2) - (2x+1)(-2(1-x))}{(1-x)^4}$$

$$= \frac{2(1-x)^2 + 2(2x+1)(1-x)}{(1-x)^4}$$

$$= \frac{(1-x)[2(1-x) + 2(2x+1)]}{(1-x)^4}$$

$$= \frac{2 - 2x + 4x + 2}{(1-x)^3}$$

$$\therefore \frac{dy}{dx} = \frac{4+2x}{(1-x)^3}$$

Example 17. Find $\frac{dy}{dx}$ if $y = \left(x + \frac{1}{x}\right)\log x$.

Solution: $y = x \log x + \frac{\log x}{x}$

$$\frac{dy}{dx} = \frac{x}{x} + \log x + \frac{1}{x}\left(\frac{1}{x}\right) + \log x\left(\frac{-1}{x^2}\right)$$

$$= 1 + \log x + \frac{1}{x^2} - \frac{\log x}{x^2}$$

$$\therefore \frac{dy}{dx} = 1 + \frac{1}{x^2} + \log x - \frac{\log x}{x^2}$$

Example 18. If $u = 2x^2 + 3xy + 4y^2$ find $\frac{\partial u}{\partial x}$ and $\frac{\partial u}{\partial y}$.

Solution: $\frac{\partial u}{\partial x} = 4x + 3y$

$$\frac{\partial u}{\partial y} = 3x + 8y.$$

Example 19. Find $\frac{dy}{dx}$ if (a) $y = \frac{e^x + 1}{e^x - 1}$ (b) $y = \sqrt{x} - 4x^{3/2} + \frac{5}{x}$.

Solution:

(a) $y = \frac{e^x + 1}{e^x - 1}$

$$\frac{dy}{dx} = \frac{(e^x - 1)(e^x) - (e^x + 1)(e^x)}{(e^x - 1)^2}$$

$$= \frac{e^{2x} - e^x - e^{2x} - e^x}{(e^x - 1)^2}$$

$$\therefore \frac{dy}{dx} = \frac{-2e^x}{(e^x - 1)^2}$$

(b) $y = \sqrt{x} - 4x^{3/2} + \frac{5}{x}$

$$\frac{dy}{dx} = \frac{1}{2\sqrt{x}} - 4\,\frac{3}{2}\,x^{3/2-1} + 5\left(\frac{-1}{x^2}\right)$$

$$\therefore \frac{dy}{dx} = \frac{1}{2\sqrt{x}} - 6x^{1/2} - \frac{5}{x^2}$$

Example 20. Differentiate *w.r.t* '*x*' $y = x^2 e^x$.

Solution: $\dfrac{dy}{dx} = x^2 e^x + e^x\, 2x$

$= x^2 e^x + 2xe^x$

Example 21. $y = x\left(x^2 + 1\right)$ find $\dfrac{dy}{dx}$.

Solution: $\dfrac{dy}{dx} = \dfrac{d}{dx}\left[x^3 + x\right]$

$= 3x^2 + 1$

Example 22 Differentiate the following with respect to '*x*' $y = \left(3x^2 + 4x - 5\right)^3$.

Solution $\dfrac{dy}{dx} = 3\left(3x^2 + 4x - 5\right)^2 (6x + 4)$

$\therefore \dfrac{dy}{dx} = (18x + 12)\left(3x^2 + 4x - 5\right)^2$

Example 23. Differentiate with respect ot '*x*' $y = 4x^2 + 3x + 12$.

Solution: $\dfrac{dy}{dx} = 8x + 3$

Example 24. If $y = \sqrt{x}\left(x^2 + 2\right)$ find $\dfrac{dy}{dx}$.

Solution: $\dfrac{dy}{dx} = \left(\sqrt{x}\right)(2x) + \left(x^2 + 2\right)\left(\dfrac{1}{2\sqrt{x}}\right)$

$\therefore \dfrac{dy}{dx} = 2x\sqrt{x} + \dfrac{x^2 + 2}{2\sqrt{x}}$

Example 25. Find $\dfrac{dy}{dx}$ if $y = (3x - 5)\left(x^2 + 2x\right)$.

Solution: $\dfrac{dy}{dx} = (3x - 5)(2x + 2) + \left(x^2 + 2x\right)(3)$

$= 6x^2 - 10x + 6x - 10 + 3x^2 + 6x$

$\therefore \dfrac{dy}{dx} = 9x^2 + 2x - 10$

Example 26. Find $\dfrac{dy}{dx}$ if $y = \left(x^2 + 1\right)^4$.

Solution: $\frac{dy}{dx} = 4\left(x^2 + 1\right)^3 (2x)$

$\therefore \frac{dy}{dx} = 8x\left(x^2 + 1\right)^3$

Example 27. Find $\frac{dy}{dx}$ if $y = \frac{x + 1}{x}$.

Solution: $\frac{dy}{dx} = \frac{x(0 + 1) - (x + 1)(1)}{x^2}$

$= \frac{\cancel{x} - \cancel{x} - 1}{x^2} = \frac{-1}{x^2}$

Example 28. If $y = \frac{7x^5 + 6x^3 - 5x}{x^4}$ find $\frac{dy}{dx}$.

Solution: $\frac{dy}{dx} = \frac{d}{dx}\left(7x + \frac{6}{x} + \frac{5}{x^3}\right)$

$\therefore \frac{dy}{dx} = \left(7 - \frac{6}{x^2} + \frac{5}{x^3}\right)$

Example 29. Find $\frac{dy}{dx}$ if $x = at^2$, $y = 2at$.

Solution: $\frac{dx}{dt} = 2\,at$, $\frac{dy}{dt} = 2a$

$\therefore \frac{dy}{dx} = \frac{\left(\frac{dy}{dt}\right)}{\left(\frac{dx}{dt}\right)} = \frac{\cancel{2a}}{\cancel{2a}t} = \frac{1}{t}$

Example 30. If $y = 3x^4 - 4x^2 + 18x$ find $\frac{d^2y}{dx^2}$.

Solution: $\frac{dy}{dx} = 12x^3 - 8x + 18$

$\frac{d}{dx}\left(\frac{dy}{dx}\right) = \frac{d}{dx}\left(12x^3 - 8x + 18\right)$

$\therefore \frac{d^2y}{dx^2} = 36x^2 - 8$

Example 31. Differentiate $\frac{3x-1}{2x+3}$ with respect to 'x'.

Solution:

$$\frac{dy}{dx} = \frac{d}{dx}\left(\frac{3x-1}{2x+3}\right)$$

$$= \frac{(2x+3)(3)-(3x-1)(2)}{(2x+3)^2}$$

$$= \frac{\cancel{6x}+9-\cancel{6x}+2}{(2x+3)^2}$$

$$\therefore \frac{dy}{dx} = \frac{11}{(2x+3)^2}$$

Example 32. Find $\frac{dy}{dx}$ if $y=\left(x^2-3\right)^4$.

Solution:

$$\frac{dy}{dx} = 4\left(x^2-3\right)^3(2x)$$

$$\therefore \frac{dy}{dx} = 8x\left(x^2-3\right)^3$$

Example 33. If $y=\frac{2x-1}{3x+7}$ find $\frac{dy}{dx}$.

Solution:

$$\frac{dy}{dx} = \frac{d}{dx}\left(\frac{2x-1}{3x+7}\right)$$

$$= \frac{(3x+7)(2)-(2x-1)(3)}{(3x+7)^2}$$

$$= \frac{\cancel{6x}+14-\cancel{6x}+3}{(3x+7)^2}$$

$$\therefore \frac{dy}{dx} = \frac{17}{(3x+7)^2}$$

FUNCTIONS

Definition:

A function is mathematical rule that assigns to each input value one and only one output value. The equation $y=f(x)$ denotes a functional relationship between the variables x and y. A verbal translation of this equation is 'y equals $f(x)$'. When we say that 'y is a function of x', we mean that the value of this variable y depends upon and is uniquely determined by the value of the variable x, here x is the input variable (called the independent variable) and y is the output variable (called the dependent variable).

Odd and Even Functions

A function $f(x)$ is called an odd function if $f(-x) = -f(x)$

Example 1.

$$f(x) = x^3 + 3x$$
$$= f(-x)\,(-x)^3 + 3\,(-x)$$
$$= -x^3 - 3x$$
$$= -(x^3 + 3x)$$
$$= -f(x)$$

$\therefore f(x)$ is an odd function.

A function $f(x)$ is called an even function if $f(-x) = f(x)$

Example 2.

$$f(x) = x^4 + 2x^2 + 5$$
$$f(-x) = (-x)^4 + 2\,(-x)^2 + 5$$
$$= x^4 + 2x^2 + 5$$
$$= f(x)$$

$\therefore f(x)$ is an even function.

Solved Examples

Example 1. If $f(x) = (4 - 2x)(x + 4)$ find $f(1)$ and $f(-3)$.

Solution: $f(x) = (4 - 2x)(x + 4)$

$$f(1) = (4 - 2)(1 + 4) = 2\,(5) = 10$$
$$f(-3) = (4 + 6)(1) = 10$$

Example 2. If $f(x) = 2x^2 - 4x + 4$ find x such that $2f(x) = f(2x)$.

Solution: $2f(x) = 2\,[2x^2 - 4x + 4] = 4x^2 - 8x + 8$

$$f(2x) = 2\,(2x)^2 - 4\,(2x) + 4 = 8x^2 - 8x + 4$$

if $\quad 2f(x) = f(2x)$

then $4x^2 - \cancel{8x} + 8 = 8x^2 - \cancel{8x} + 4$

$$4x^2 - 4 = 0$$
$$4x^2 = 4$$
$$x^2 = 1$$
$$\therefore \quad x = +1$$

Example 3. If $f(x) = x^{-1}$ find $f(x + 1) + 2f(x) + \dfrac{1}{f(x)}$.

Solution: $f(x + 1) + 2f(x) + \dfrac{1}{f(x)}$

$$= (x+1)^{-1} + 2x^{-1} + \frac{1}{x^{-1}}$$

$$= \frac{1}{x+1} + \frac{2}{x} + x$$

$$= \frac{x + 2(x+1) + x(x+1)x}{(x+1)x}$$

$$= \frac{x + 2x + 2 + x^3 + x^2}{x(x+1)}$$

$$= \frac{x^3 + x^2 + 3x + 2}{x(x+1)}$$

Example 4. If $f(x) = 2x^2 - 5x + 4$ for what values of 'x' is $2f(x) = f(2x)$?

Solution: $2f(x) = 2(2x^2 - 5x + 4) = 4x^2 - 10x + 8$

$f(2x) = 2(2x)^2 - 5(2x) + 4$

$= 2(4x^2) - 10x + 4$

$\therefore\ f(2x) = 8x^2 - 10x + 4$

if $2f(x) = f(2x)$

then $4x^2 - 10x + 8 = 8x^2 - 10x + 4$

$8x^2 - 10x + 4 - 4x^2 + 10x - 8 = 0$

$4x^2 - 4 = 0$

$4x^2 = 4$

$x^2 = 1$

$\therefore\ x = \pm 1$

Example 5. If $f(x) = 3x^2 - 2x + 5$ find $f(-3), f(2)$.

Solution: $f(x) = 3x^2 - 2x + 5$

$f(-3) = f(-3) = 3(-3)^2 - 2(-3) + 5$

$= 3(9) + 6 + 5$

$\therefore\ f(-3) = 27 + 11 = 38$

$f(2) = 3(2)^2 - 2(2) + 5$

$= 3(4) - 4 + 5$

$\therefore\ f(2) = 12 + 1 = 13$

Example 6. If $f(x) = x^4 + \frac{1}{x^4}$ show that $f(-2) = f(2)$.

Solution: $f(-2) = (-2)^4 + \frac{1}{(-2)^4}$

$$= 16 + \frac{1}{16}$$

$$\therefore\ f(-2) = \frac{(16)^2 + 1}{16} = \frac{256 + 1}{16} = \frac{257}{16}$$

$$f(2) = (2)^4 + \frac{1}{(2)^4}$$

$$= 16 + \frac{1}{16}$$

$$\therefore\ f(2) = \frac{256 + 1}{16} = \frac{257}{16}$$

$$\therefore\ f(-2) = f(2)$$

Example 7. If $f(x) = \dfrac{x}{1 + x^2}$ find $f(3), f(-3)$.

Solution: $f(3) = f(x) = \dfrac{3}{1 + 3^2}$

$$= \frac{3}{1 + 9} = \frac{3}{10}$$

$$f(-3) = \frac{-3}{1 + (-3)^2}$$

$$= \frac{-3}{1 + 9} = \frac{-3}{10}$$

Example 8. If $f(x) = x^3 - 3x^2 - 4x - 1$ then find $f(x)$.

Solution: $f(x) = x^3 - 3x^2 - 4x - 1$

if $x = 3$, $f(3) = (3)^3 - 3\,(3)^2 - 4\,(3) - 1$

$= 27 - 3\,(9) - 12 - 1$

$= \cancel{27} - \cancel{27} - 13$

$\therefore\ f(3) = -13$

if $x = -3, f(-3) = (-3)^3 - 3(-3)^2 - 4(-3) - 1$

$= -27 - 3\,(9) + 12 - 1$

$= -27 - 27 + 11$

$= -54 + 11$

$\therefore\ f(-3) = -43$

LIMITS

Definition:

A function $f(x)$ is said to tend to a limit l as x tends to a, if for a given small positive number ε (Epsilon) there exists a corresponding positive number δ (delta) such that $|f(x) - l| < \varepsilon$ when $|x - a| < \delta$.

This means that the numerical difference between $f(x)$ and l can be made as small as we please by taking the numerical difference between x and a as very small. When the above condition is satisfied it is written as $\lim_{x \to a} f(x) = l$.

The above definition also means that $f(x)$ is very nearly equal to l when x is very nearly equal to a.

Evaluation of Limits

To evaluate limits, first put the value of x and see whether the expression takes an indeterminate form or not. If the expression does not take an indeterminate form, the value obtained itself is the limit otherwise, the expressions in numerator and denominator are factorised (or some other suitable method is applied) to reduce the expression so that it will not rake the indeterminate form when the limiting value of x is substituted.

Solved Examples

Example 1. $\lim_{x \to 1} \left(x^2 - 5x + 7\right)$.

Solution: $= \lim_{x \to 1} x^2 - 5 \lim_{x \to 1} x + \lim_{x \to 1} 7$

$= (1)^2 - 5(1) + 7$

$= 1 - 5 + 7 = 3$

Example 2. $\lim_{x \to 2} (2x + 1)(3x + 1)$.

Solution: $= \lim_{x \to 2} (2x + 1) \cdot \lim_{x \to 2} (3x + 1)$

$= \left(\lim_{x \to 2} 2x + \lim_{x \to 2} 1\right)\left(\lim_{x \to 2} 3x + \lim_{x \to 2} 1\right)$

$= (2(2) + 1) \cdot (3(2) + 1)$

$= (4 + 1) \cdot (6 + 1)$

$= (5) \cdot (7)$

$= 35$

Example 3. $\lim_{x \to 4} \dfrac{x - 3}{x + 3}$.

Solution: $= \dfrac{\lim\limits_{x \to 4} (x-3)}{\lim\limits_{x \to 4} (x+3)}$

$= \dfrac{(4-3)}{(4+3)}$

$= \left(\dfrac{1}{7}\right)$

Example 4. $\lim\limits_{x \to 3} \sqrt{25 - x^2}$.

Solution: $= \sqrt{\lim\limits_{x \to 3} \left(25 - x^2\right)}$

$= \sqrt{(25-9)}$

$= \sqrt{16}$

$= 4$

Example 5. $\lim\limits_{x \to 2} \dfrac{x^3 - 8}{x - 2}$.

Solution: $= \lim\limits_{x \to 2} \dfrac{x^3 - 2^3}{x - 2}$

$= 3 \cdot (2)^{3-1}$ $\qquad \left(\because \lim\limits_{x \to a} \dfrac{x^n - a^n}{x - a} = n \cdot a^{n-1}\right)$

$= 3 \cdot (2)^2$

$= 3 \cdot 4$

$= 12$

Example 6. $\lim\limits_{x \to 4} \dfrac{x^{3/2} - 8}{x - 4}$.

Solution: $= \lim\limits_{x \to 4} \dfrac{x^{3/2} - (4)^{3/2}}{x - 4}$

$= \dfrac{3}{2} \cdot (4)^{3/2-1}$ $\qquad \left(\because n = \dfrac{3}{2}, a = 4\right)$

$$= \frac{3}{2}(4)^{1/2}$$

$$= \frac{3}{2}\left(\sqrt{4}\right)$$

$$= \frac{3}{2}(2) \qquad = 3$$

Example 7. $= \lim_{x \to -2} \dfrac{x^5 + 32}{x + 2}$.

Solution: $\lim_{x \to -2} \dfrac{x^5 - (-32)}{x - (-2)}$

$$\lim_{x \to -2} \frac{x^5 - (-2)^5}{x - (-2)}$$

$$= (5) \cdot (-2)^{5-1}$$

$$= 5 \cdot (-2)^4$$

$$= 5 \cdot (16)$$

$$= 80$$

Example 8. Evaluate $\lim_{x \to 2} \dfrac{2x^2 - 3x - 2}{3x^2 - 2x - 8}$.

Solution: $= \lim_{x \to 2} \dfrac{(x-2)(2x + 1)}{(x-2)(3x + 4)}$

$$= \lim_{x \to 2} \frac{(2x + 1)}{(3x + 4)}$$

$$= \frac{2(2) + 1}{3(2) + 4} = \frac{5}{10}$$

$2x^2 - 3x - 2 = 0$
$2x^2 - 4x + x - 2 = 0$
$2x(x - 2) + 1(x + 2) = 0$
$(x - 2)(2x + 1) = 0$
$3x^2 - 2x - 8 = 0$
$3x^2 - 6x + 4x - 8 = 0$
$3x(x - 2)(x - 2) = 0$
$(x - 2)(3x + 4) = 0$

Example 9. Evaluate $\lim_{x \to 2} \dfrac{x^2 - 4}{\sqrt{x+2} - \sqrt{3x-2}}$.

Solution:

$$= \lim_{x \to 2} \frac{(x^2 - 4)}{\left(\sqrt{x+2} - \sqrt{3x-2}\right)} \times \frac{\left(\sqrt{x+2} + \sqrt{3x-2}\right)}{\left(\sqrt{x+2} + \sqrt{3x-2}\right)}$$

$$= \lim_{x \to 2} \frac{(x+2)(x-2)\left(\sqrt{x+2}+\sqrt{3x-2}\right)}{(x+2)-(3x-2)}$$

$$= \lim_{x \to 2} \frac{(x+2)\cancel{(x-2)}\left(\sqrt{x+2}+\sqrt{3x-2}\right)}{-2\cancel{(x-2)}}$$

$$= \lim_{x \to 2} \frac{(x+2)\left(\sqrt{x+2}+\sqrt{3x-2}\right)}{-2}$$

$$= \frac{(2+2)\cdot\left(\sqrt{4}+\sqrt{4}\right)}{-2}$$

$$= \frac{4(2+2)}{-2}$$

$$= \frac{16}{-2} = -8$$

Example 10. Evaluate $\lim_{x \to 2} \left[\frac{1}{x^2-3x+2} - \frac{12}{x^3-8}\right]$.

Solution:

$$= \lim_{x \to 2}\left[\frac{1}{x^2-3x+2}\right] - \lim_{x \to 2}\left[\frac{12}{x^3-8}\right]$$

$$= \lim_{x \to 2}\left[\frac{1}{(x-2)(x-1)}\right] - 12 \lim_{x \to 2}\left[\frac{1}{(x-8)\left(x^2+4+2x\right)}\right]$$

$$= \lim_{x \to 2}\left[\frac{\left(x^2+4+2x\right)-12x+12}{(x-2)(x-1)\left(x^2+4+2x\right)}\right]$$

$x^2 - 3x + 2 = 0$
$x^2 - 2x - x + 2 = 0$
$x(x-2) - (x-2) = 0$
$(x-2)(x-1) = 0$

$$= \lim_{x \to 2}\left[\frac{x^2-10x+16}{(x-2)(x-1)\left(x^2+4+2x\right)}\right]$$

$$= \lim_{x \to 2} \frac{\cancel{(x-2)}(x-8)}{\cancel{(x-2)}(x-1)\left(x^2+4+2x\right)}$$

$$= \lim_{x \to 2} \frac{x-8}{(x-1)\left(x^2+4+2x\right)}$$

$$= \frac{2-8}{(2-1)(4+4+4)}$$

$$= \frac{-6}{12} \frac{1}{2}$$

$$= -\frac{1}{2}$$

Example 11. Evaluate $\lim_{x \to 3} \frac{x^2 - 5x + 6}{x^2 - 9}$.

Solution:

$$= \frac{\lim_{x \to 3} \left(x^2 - 5x + 6\right)}{\lim_{x \to 3} \left(x^2 - 9\right)}$$

$$x^2 - 5x + 6 = 0$$
$$x^2 - 3x - 2x + 6 = 0$$
$$x(x-3) - 2(x-3) = 0$$
$$(x-3)(x-2) = 0$$

$$= \frac{\lim_{x \to 3} (x-2)(x-3)}{\lim_{x \to 3} (x+3)(x-3)}$$

$$= \frac{\lim_{x \to 3} (x-2)}{\lim_{x \to 3} (x+3)}$$

$$= \frac{(3-2)}{(3+3)}$$

$$= \frac{1}{6}$$

Example 12 Evaluate

(a) $\lim_{x \to -2} \frac{3x^2 + 5x - 2}{x^2 - 3x - 10}$.

Solution:

$$= \lim_{x \to -2} \frac{(x+2)(3x-1)}{(x+2)(x-5)}$$

$$= \lim_{x \to -2} \frac{(3x-1)}{(x-5)}$$

$$= \frac{3(-2)-1}{(-2-5)}$$

$$= \frac{-6-1}{-7} = \frac{-7}{-7} = 1$$

(b) $\lim_{x \to \infty} \frac{(x+5)(2x+5)}{(3x+1)(x-2)}$.

$$= \lim_{x \to \infty} \frac{(2x^2+15x+25)}{(3x^2-5x-2)}$$

$$= \lim_{x \to \infty} \frac{\cancel{x^2}}{\cancel{x^2}} \frac{\left(2+\frac{15}{x}+\frac{25}{x^2}\right)}{\left(3-\frac{5}{x}-\frac{2}{x^2}\right)}$$

$$= \lim_{x \to \infty} \frac{(2+0+0)}{(3-0-0)}$$

$$= \frac{2}{3}$$

Example 13. Evaluate $\lim_{x \to 2} \frac{(3x^2-4x+1)}{(2x^2+4x-1)}$.

Solution:

$$= \lim_{x \to 2} \left[\frac{3(4)-4(2)+1}{8+8-1}\right]$$

$$= \frac{12-8+1}{16-1} = \frac{\cancel{5}}{\cancel{15}}\frac{1}{3} = \frac{1}{3}$$

Example 14. Evaluate $\lim_{x \to 3} \frac{x^2+2x-15}{x^2-9}$

Solution:

$$= \lim_{x \to 3} \frac{x^2+5x-3x-15}{(x+3)(x-3)}$$

$$= \lim_{x \to 3} \frac{x(x+5)-3(x+5)}{(x+3)(x-3)}$$

$$= \lim_{x \to 3} \frac{\cancel{(x-3)}(x+5)}{(x+3)\cancel{(x-3)}}$$

$$= \lim_{x \to 3} \left(\frac{x+5}{x+3}\right)$$

$$= \frac{3+5}{3+3} = \frac{8}{6}$$

Example 15. Evaluate $\lim_{x \to 1} \frac{x^2 + x - 2}{x^2 - 3x + 2}$.

Solution:

$$= \lim_{x \to 1} \frac{x^2 + 2x - x - 2}{x^2 - 2x - x + 2}$$

$$= \lim_{x \to 1} \frac{x(x+2) - (x+2)}{x(x-2) - (x-2)}$$

$$= \lim_{x \to 1} \frac{(x+2)\cancel{(x-1)}}{(x-2)\cancel{(x-1)}}$$

$$= \frac{1+2}{1-2} = \frac{3}{-1} = -3$$

Example 16. Evaluate $\lim_{x \to 4} \frac{x^2 - 16}{x - 4}$.

Solution: $= 2 \cdot (4)^{2-1}$

$= 2 \cdot 4 = 8$

Example 17. Evaluate $\lim_{x \to \infty} \frac{x^2 + 50}{x^2 + 3x + 2}$.

Solution:

$$= \lim_{x \to \infty} \frac{\cancel{x^2}\left[1 + \frac{50}{x^2}\right]}{\cancel{x^2}\left[1 + \frac{3}{x} + \frac{2}{x^2}\right]}$$

$$= \lim_{x \to \infty} \frac{\left[1 + \frac{50}{x^2}\right]}{\left[1 + \frac{3}{x} + \frac{2}{x^2}\right]}$$

$$= \frac{\left[1 + \frac{50}{\infty}\right]}{\left[1 + \frac{3}{\infty} + \frac{2}{\infty}\right]} = \frac{1}{1} = 1$$

Example 18. $\lim_{x \to \infty} \frac{x^2 + 8}{x^2 + 5x + 7}$.

Solution:

$$= \lim_{x \to \infty} \frac{x^2\left[1 + \frac{8}{x^2}\right]}{x^2\left[1 + \frac{5}{x} + \frac{7}{x^2}\right]}$$

$$= \lim_{x \to \infty} \frac{\left[1 + \frac{8}{x^2}\right]}{\left[1 + \frac{5}{x} + \frac{7}{x^2}\right]}$$

$$= \frac{\left[1 + \frac{8}{\infty}\right]}{\left[1 + \frac{5}{\infty} + \frac{7}{\infty}\right]} = \frac{1}{1} = 1$$

Example 19. Evaluate $\lim_{y \to -2} \frac{y^4 - 16}{y^3 + 8}$.

Solution:

$$= \lim_{y \to -2} \frac{y^4\left[1 - \frac{16}{y^4}\right]}{y^3\left[1 + \frac{8}{y^3}\right]}$$

$$= \lim_{y \to -2} \frac{y^4[1] - 16}{y^3 + 8}$$

$$= \lim_{y \to -2} \frac{\left[y^2\right]^2 - \left[(-2)^2\right]^2}{y^3 - (-2)^3}$$

$$= \lim_{y \to -2} \frac{[y^2 - (-2)^2][y^2 + (-2)^2]}{[y-(-2)][y^2 + (-2)^2 + (-2y)]}$$

$$= \lim_{y \to -2} \frac{[y-(-2)][y+(-2)][y^2 + 4]}{[y+4][y^2 + 4 - 2y]}$$

$$= \lim_{y \to -2} \frac{\cancel{[y+2]}[y-2][y^2 + 4]}{\cancel{[y+2]}[y^2 + 4 - 2y]}$$

$$= \lim_{y \to -2} \frac{[-2-2][4+4]}{[4+4+4]}$$

$$= \frac{\cancel{(-4)}(8)}{\cancel{12}} \frac{1}{3} = \frac{-8}{3}$$

Example 20. Evaluate $\lim_{x \to \infty} \frac{(8x+3)^2}{(x^2 - 3)}$.

Solution:

$$= \lim_{x \to \infty} \frac{64x^2 + 9 + 48x}{x^2 - 3}$$

$$= \lim_{x \to \infty} \frac{64x^2 + 48x + 9}{x^2 - 3}$$

$$= \frac{\cancel{x^2}\left[64 + \frac{48}{x} + \frac{9}{x^2}\right]}{\cancel{x^2}\left[1 - \frac{3}{x^2}\right]}$$

$$= \frac{64}{1} = 64$$

Example 21. $\lim_{x \to 0} \frac{x}{\sqrt{x+a} - \sqrt{a}}$.

Solution:

$$= \lim_{x \to 0} \left(\frac{x}{\sqrt{x+a} - \sqrt{a}}\right)\left(\frac{\sqrt{x+a} + \sqrt{a}}{\sqrt{x+a} + \sqrt{a}}\right)$$

$$= \lim_{x \to 0} \frac{x\left(\sqrt{x+a}+\sqrt{a}\right)}{(x+a)-a}$$

$$= \lim_{x \to 0} \frac{\cancel{x}\left(\sqrt{x+a}+\sqrt{a}\right)}{\cancel{x}}$$

$$= \lim_{x \to 0} \sqrt{x+a}+\sqrt{a}$$

$$= \sqrt{0+a}+\sqrt{a}$$

$$= \sqrt{a}+\sqrt{a}$$

$$= 2\sqrt{a}$$

Example 22. Evaluate $\lim_{x \to 3} \frac{x^2-5x+6}{x^2-9}$

Solution:

$$= \lim_{x \to 3} \frac{\left(x^2-3x-2x+6\right)}{(x+3)(x-3)}$$

$$= \lim_{x \to 3} \frac{x(x-3)-2(x-3)}{(x+3)(x-3)}$$

$$= \lim_{x \to 3} \frac{\cancel{(x-3)}(x-2)}{(x+3)\cancel{(x-3)}}$$

$$= \lim_{x \to 3} \frac{x-2}{x+3} = \frac{3-2}{6} = \frac{1}{6}$$

Example 23. $\lim_{x \to \infty} \frac{x^2+3x+10}{x^2+5x+2}$.

Solution:

$$= \lim_{x \to \infty} \frac{\cancel{x^2}\left[1+\frac{3}{x}+\frac{10}{x^2}\right]}{\cancel{x^2}\left[1+\frac{5}{x}+\frac{2}{x^2}\right]}$$

$$= \frac{1+0+0}{1+0+0} = 1$$

Example 24. (a) $\lim_{x \to 3} \frac{x^3 - 27}{x - 3}$.

Solution:

$$= \lim_{x \to 3} \frac{x^3 - 3^3}{x - 3}$$

$= 3\,(3)^{3-1}$

$= 3\,(3^2)$

$= 3\,(9)$

$= 27$

(b) $\lim_{x \to 3} \frac{x^3 - 27}{x^2 - 9}$.

$$= \lim_{x \to 3} \frac{(x - 3)(x^2 + 9 + 3x)}{(x + 3)(x - 3)}$$

$$= \frac{9 + 9 + 9}{6}$$

$$= \frac{27}{6} = \frac{9}{2}$$

Example 25. $\lim_{x \to 3} \frac{x^2 - 9}{x^2 - 5x + 6}$.

Solution:

$$= \lim_{x \to 3} \frac{(x + 3)(x - 3)}{(x^2 - 3x - 2x + 6)}$$

$$= \lim_{x \to 3} \frac{(x + 3)(x - 3)}{(x - 3)(x - 2)}$$

$$= \lim_{x \to 3} \frac{(x + 3)}{(x - 2)}$$

$$= \frac{6}{1} = 6$$

Example 26. $\lim_{x \to 1} \frac{x^2 - 1}{x^2 - 3x + 2}$.

Solution:

$$= \lim_{x \to 1} \frac{(x+1)(x-1)}{\left(x^2 - 2x - x + 2\right)}$$

$$= \lim_{x \to 1} \frac{(x+1)\cancel{(x-1)}}{(x-2)\cancel{(x-1)}}$$

$$= \lim_{x \to 1} \left(\frac{x+1}{x-2}\right)$$

$$= \left(\frac{1+1}{1-2}\right) = \frac{2}{-1} = -2$$

Example 27. $\lim_{n \to \infty} \frac{(2n+1)^2}{n^2+1}$.

Solution:

$$= \lim_{n \to \infty} \frac{\left(4n^2 + 1 + 4n\right)}{n^2+1}$$

$$= \lim_{n \to \infty} \frac{\cancel{n^2}\left[4 + \frac{1}{n^2} + \frac{4}{n}\right]}{\cancel{n^2}\left[1 + \frac{1}{n^2}\right]}$$

$$= \frac{4+0+0}{1+0}$$

Example 28. $\lim_{x \to 0} \frac{(x+1) - \left(x^2+1\right)}{x}$.

Solution:

$$= \lim_{x \to 0} \frac{x \cancel{+1} - x^2 \cancel{-1}}{x}$$

$$= \lim_{x \to 0} \frac{\cancel{x}(1-x)}{\cancel{x}}$$

$$= \lim_{x \to 0} \frac{\cancel{x}(1-0)}{\cancel{x}}$$

$$= 1 - 0$$

$$= 1$$

Example 29. $\lim_{x \to 1} \dfrac{x^2 + x - 2}{x^2 - 3x + 2}$.

Solution:

$$= \lim_{x \to 1} \frac{\left(x^2 + 2x - x - 2\right)}{\left(x^2 - 2x - x + 2\right)}$$

$$= \lim_{x \to 1} \frac{(x + 2)(x - 1)}{(x - 2)(x - 1)}$$

$$= \frac{1 + 2}{1 - 2} = \frac{3}{-1} = -3$$

Example 30. $\lim_{x \to 0} \left(7x^2 - 5x + 1\right)$.

Solution: $= 7(0) - 5(0) + 1 = 1$

Example 31. $\lim_{y \to 4} \dfrac{y^2 - 7y + 12}{y^3 - 64} = \dfrac{1}{48}$.

Solution:

$$= \lim_{y \to 4} \frac{y^2 - 4y - 3y + 12}{(y - 4)\left(y^2 + 16 + 4y\right)}$$

$$= \lim_{y \to 4} \frac{(y - 4)(y - 3)}{(y - 4)\left(y^2 + 16 + 4y\right)}$$

$$= \frac{4 - 3}{16 + 16 + 16}$$

$$= \frac{1}{32 + 16}$$

$$= \frac{1}{48}$$

Example 32. $\lim_{x \to 1} \dfrac{4x^4 + 3x^2 - 1}{x^3 + 7}$.

Solution:

$$= \lim_{x \to 1} \frac{x^4\left[4 + \frac{3}{x^2} - \frac{1}{x^4}\right]}{x^3\left[1 + \frac{7}{x^3}\right]}$$

$$= \lim_{x \to 1} \frac{x\left[4 + \frac{3}{x^2} - \frac{1}{x^4}\right]}{1 + \frac{7}{x^3}}$$

$$= \frac{1[4 + 3 - 1]}{1 + 7} = \frac{6}{8} = \frac{3}{4}$$

MAXIMA AND MINIMA

Working rule for finding maximum and minimum values of a function:

1. Find $\frac{dy}{dx}$ for the given function $y = f(x)$.
2. Putting $\frac{dy}{dx} = 0$, calculate the values of 'x'.
3. Find $\frac{d^2y}{dx^2}$.
4. Put $x = a$ in $\frac{d^2y}{dx^2}$, if the result is negative, i.e., if $\frac{d^2y}{dx^2} < 0$, the function is maximum at $x = a$ and max. $y = f(a)$.

On the other hand, if the result is positive, i.e., if $\frac{d^2y}{dx^2} > 0$, the function is minimum at $x = a$ and min. $y = f(a)$.

If $\frac{d^2y}{dx^2} = 0$ then compute $\frac{d^3y}{dx^3}$.

If $\frac{d^3y}{dx^3} \neq 0$ for a particular value $x = a$ then the function is neither a maximum nor a minimum function at that point.

If $\frac{d^3y}{dx^3} = 0$, then find $\frac{d^4y}{dx^4}$. If this derivative is negative (–ve) at that point then there is a maximum and if it is positive (+ve) then there is a minimum. In case $\frac{d^4y}{dx^4} = 0$, then compute $\frac{d^5y}{dx^5}$ if $\frac{d^5y}{dx^5} \neq 0$ for a particular point $x = a$, then the function is neither a maximum nor a minimum function at that point.

If $\frac{d^5y}{dx^5} = 0$ then find $\frac{d^6y}{dx^6}$, if this derivative is negative at that point then there is a maximum and if it is positive then there is a minimum. In case $\frac{d^6y}{dx^6} = 0$, then compute $\frac{d^7y}{dx^7}$ and proceed as above till we get a definite answer.

Solved Examples

Example 1. Find the maximum and minimum values of the expression $x^3 - 3x^2 - 9x + 27$.

Solution: Take $y = x^3 - 3x^2 - 9x + 27$

$$\frac{dy}{dx} = 3x^2 - 6x - 9$$

if $\frac{dy}{dx} = 0$, then $3x^2 - 6x - 9 = 0$

$$3x^2 - 9x + 3x - 9 = 0$$

$$3x(x - 3) + 3(x - 3) = 0$$

$$(x - 3)(3x + 3) = 0$$

$x - 3 = 0$ or $3x + 3 = 0$

$x = 3$ $\quad$ $3x = -3$

$x = -1$

now $\frac{d^2y}{dx^2} = 6x - 6.$

at $x = 3$, $\frac{d^2y}{dx^2} = 18 - 6 = 12 > 0$

$\therefore$ $x = 3$ gives a minimum point of 'y'.

Hence minimum value of 'y'

i.e., $y = x^3 - 3x^2 - 9x + 27$

$= \cancel{27} - \cancel{27} - \cancel{27} + \cancel{27}$

$\therefore$ min, $y = 0$

at $x = -1, \dfrac{d^2y}{dx^2} = -6 - 6 = -12 < 0$

$\therefore x = -1$ gives a maximum point of 'y' hence maximum value of 'y'.

That is $y = x^3 - 3x^2 - 9x + 27$

$= -1 - 3 + 9 + 27$

$= -4 + 9 + 27$

$= 5 + 27$

$= 32$

Example 2. Find the maximum and minimum values of the function $\frac{2}{3}x^3 + \frac{1}{2}x^2 - 6x + 8$.

Solution: Take $y = \frac{2}{3}x^3 + \frac{1}{2}x^2 - 6x + 8$

$$\frac{dy}{dx} = 2x^2 + x - 6$$

if $\dfrac{dy}{dx} = 0, 2x^2 + x - 6 = 0$

$2x^2 + 4x - 3x - 6 = 0$

$2x(x + 2) - 3(x + 2) = 0$

$(x + 2)(2x - 3) = 0$

$x + 2 = 0$ or $2x - 3 = 0$

$x = -2$ $\quad 2x = 3$

$\therefore x = \dfrac{3}{2}$

-12 → $+4$, -3

Now $\dfrac{d^2y}{dx^2} = 4x + 1$

at $x = -2, \dfrac{d^2y}{dx^2} = -8 + 1 = -7 < 0$

$\therefore f(x)$ is maximum at $x = -2$

$\therefore$ Maximum value is

$$y = \frac{2}{3}x^3 + \frac{1}{2}x^2 - 6x + 8$$

$$= \frac{2}{3}(-2)^3 + \frac{1}{2}(-2)^2 - 6(-2) + 8$$

$$= \frac{-16}{3} + \frac{4}{2} + 12 + 8$$

$$= \frac{-32 + 12 + 120}{6}$$

$$= \frac{100}{6}$$

at $x = \frac{3}{2}, \frac{d^2y}{dx^2} = 6 + 1 = 7 > 0$

$\therefore f(x)$ is minimum at $x = \frac{3}{2}$

$\therefore$ minimum value is

$$y = \frac{2}{3}\left(\frac{27}{8}\right) + \frac{1}{2}\left(\frac{9}{4}\right) - 6\left(\frac{3}{2}\right) + 8$$

$$= \frac{9}{4} + \frac{9}{8} - \frac{18}{2} + 8$$

$$= \frac{18 + 9 - 72 + 84}{8}$$

$$= \frac{27 + 64 - 72}{8}$$

$$= \frac{91 - 72}{8}$$

$$= \frac{19}{8}$$

4	4,	8,	2
2	1,	2,	2
	1,	1,	1

Example 3. Find the maximum and minimum values of the function $x^4 + 2x^3 - 3x^2 - 4x + 4$.

***Solution*:** $y = x^4 + 2x^3 - 3x^2 - 4x + 4$

$$\frac{dy}{dx} = 4x^3 + 6x^2 - 6x - 4$$

if $\frac{dy}{dx} = 0,\ 4x^3 + 6x^2 - 6x - 4 = 0$

by inspection method,

put $x = 1,\ 4 + 6 - 6 - 4 = 0$

$\therefore\ x = 1$

Using synthetic division method,

$x = 1$	4	6	−6	−4
	0	4	10	4
	4	10	4	0

$$4x^2 + 10x + 4 = 0$$

$\div 2,\quad 2x^2 + 5x + 2 = 0$

4

4 1

$$2x^2 + 4x + x + 2 = 0$$

$$2x(x + 2) + (x + 2) = 0$$

$$(x + 2)(2x + 1) = 0$$

$$x = -2,\ x = -\frac{1}{2}$$

$$\therefore\ x = 1, -2, -\frac{1}{2}$$

at $x = 1, \dfrac{d^2y}{dx^2} = 12x^2 + 12x - 6$

$= 12 + 12 - 6$

$= 18 > 0$

$\therefore f(x)$ is minimum at $x = 1$

$\therefore$ Minimum value of y is

$y = x^4 + 2x^3 - 3x^2 - 4x + 4$

$= \cancel{1} + \cancel{2} - \cancel{3} - \cancel{4} + \cancel{4}$

$= 0$

at $x = -2, \dfrac{d^2y}{dx^2} = 12(4) + 12(-2) - 6$

$= 48 - 24 - 6$

$= 48 - 30 = 18 > 0.$

$\therefore f(x)$ is minimum at $x = -2$

$\therefore$ Minimum value of y is

$y = x^4 + 2x^3 - 3x^2 - 4x + 4$

$= \cancel{16} - \cancel{16} - \cancel{12} + \cancel{8} - \cancel{4}$

$= 0$

at $x = \dfrac{-1}{2}, \dfrac{d^2y}{dx^2} = 12\left(\dfrac{1}{4}\right) + 12\left(\dfrac{-1}{2}\right) - 6$

$= 3 - 6 - 6$

$= -9 < 0$

$\therefore f(x)$ is maximum at $x = \dfrac{-1}{2}$

$\therefore$ Maximum value of 'y' is

$y = x^4 + 2x^3 - 3x^2 - 4x + 4$

$$= \frac{1}{16} + \left(\frac{-2}{8}\right) - \frac{3}{4} + \frac{4}{2} + 4$$

$$= \frac{1}{16} - \frac{2}{8} + \frac{4}{2} - \frac{3}{4} + 4$$

4	16,	8,	2,	4
2	4,	2,	2,	1
2	2,	1,	1,	1
	1,	1,	1,	1

$$= \frac{1 - 4 + 32 - 12 + 64}{16}$$

$$= \frac{-3 + 20 + 64}{16}$$

$$= \frac{-3 + 84}{16}$$

$$= \frac{81}{16}$$

Example 4. Find the maximum and minimum values of the function $8x^5 - 15x^4 + 10x^2$.

Solution: $y = 8x^5 - 15x^4 + 10x^2$

$$\frac{dy}{dx} = 40x^4 - 60x^3 + 20x$$

If $f'(x) = 0,\ 40x^4 - 60x^3 + 20x = 0$

$\div 20,\ 20,\ 2x^4 - 3x^3 + x = 0$

Put $x = 1, 2 - 3 + 1 = 0$

$\therefore\ x = 1$

$x = 1$

2	−3	0	1
0	2	−1	−1
2	−1	−1	0

$2x^2 - x - 1 = 0$

$2x^2 - 2x + x - 1 = 0$

$2x(x - 1) + (x - 1) = 0$

$(x - 1)(2x + 1) = 0$

$x - 1 = 0$ or $2x + 1 = 0$

$x = +1$ $\qquad x = -\frac{1}{2}$

$$\therefore\ x = 1, +1, -\frac{1}{2}, 0$$

Now $\dfrac{d^2y}{dx^2} = 8x^3 - 9x^2 + 1$

at $x = 1,\ \dfrac{d^2y}{dx^2} = 8 - 9 + 1 = 0$

$$\frac{d^3y}{dx^3} = 24x^2 - 18x$$

at $x = 1, \frac{d^3y}{dx^3} = 24 - 18 = 6 \neq 0$

$\therefore$ the $f(x)$ is neither maximum nor minimum at $x = 1$.

at $x = 0, \frac{d^2y}{dx^2} = 1 > 0$

$\therefore f(x)$ is minimum at $x = 0$

$\therefore$ minimum value of y is

$y = 0$

at $x = -\frac{1}{2}, \frac{d^2y}{dx^2} = -45 < 0$

$\therefore f(x)$ is maximum at $x = -\frac{1}{2}$

$\therefore$ maximum value of y is

$y = \frac{21}{16}$

Exercises

I. Question carrying THREE marks each (Short Answers):

1. Find $\frac{dy}{dx}$ if $y = x^2 + 2x + 9e^x - \log x$.

   [*Ans:* $\frac{dy}{dx} = 2x + 2 + 9e^x - \frac{1}{x}$]

2. Find $\frac{dy}{dx}$ if $y = \frac{xe^x}{1 + x^2}$

   [*Ans:* $\frac{dy}{dx} = \frac{(1 + x^2)(e^x + xe^x) - xe^x \cdot 2x}{(1 + x^2)^2}$]

3. Find $\frac{dy}{dx}$ if $x^2 + y^2 + 2xy + 8x = 0$.

   [*Ans:* $\frac{dy}{dx} = \frac{-(x + y + 4)}{(x + y)}$]

4. If $e^x + e^y = e^{x+y}$ show that $\frac{dy}{dx} = e^{y-x}$.

5. Find $\frac{dy}{dx}$ if $y = x^4 e^{3x}$.

[*Ans:* $\frac{dy}{dx} = x^3 e^{3x}[3x + 4]$]

6. Find $\frac{dy}{dx}$ and $\left(\frac{dy}{dx}\right)_{x=3}$ for $y = x^3$.

[*Ans:* $\frac{dy}{dx} = 3x^2$ *and* $\left(\frac{dy}{dx}\right)_{x=3} = 27$]

7. If $x^m y^n = (x + y)^{m+n}$, prove that $\frac{dy}{dx} = \frac{y}{x}$.

8. Find $\frac{dy}{dx}$ if $y = (x)^{e^x}$

[*Ans:* $\frac{dy}{dx} = (x)^{e^x} \cdot e^x \cdot \log x + e^x \cdot (x)^{e^x - 1}$]

9. If $y = x^{x^{x \ldots \infty}}$ then find $\frac{dy}{dx}$.

[*Ans:* $\frac{dy}{dx} = \frac{y^2}{x(1 - y \log x)}$]

10. If $y = \sqrt{x + \sqrt{x + \sqrt{x + \ldots + \infty}}}$ then find $\frac{dy}{dx}$.

[*Ans:* $\frac{dy}{dx} = \frac{1}{1 - 2y}$]

11. find $\frac{dy}{dx}$ if $y = (x^7 + 2x + 4)(x^5 + 4x^3 + 6)$.

[*Ans:* $\frac{dy}{dx} = (x^7 + 2x + 4)(5x^4 + 12x^2) + (x^5 + 4x^3 + 6)(7x^6 + 2)$]

12. find $\frac{dy}{dx}$ if $y = a\,x^m + \frac{b}{x^n}$.

[*Ans:* $\frac{dy}{dx} = a\,m\,x^{m-1} - b\,n\,x^{-(n+1)}$]

13. Find $\frac{dy}{dx}$ if $y = \frac{ax^2 + bx + c}{x^3}$.

[*Ans:* $\frac{dy}{dx} = \frac{-a}{x^2} \frac{-2b}{x^3} \frac{-3c}{x^4}$]

14. Find $\frac{dy}{dx}$ if

(a) $x^2 + y^2 = a^2$

(b) $x^3 + y^3 = 3axy$

(c) $ax^2 + 2hxy + by^2 = 0$

[*Ans: (a)* $\frac{dy}{dx} = -\frac{x}{y}$ *(b)* $\frac{dy}{dx} = \frac{ay - x^2}{y^2 - ax}$ *(c)* $\frac{dy}{dx} = \frac{-(ax + y)}{(hx + by)}$]

15. Differentiate $\sqrt{1 - x^2}$ *w.r.t* '*x*'.

[*Ans:* $\frac{dy}{dx} = \frac{-x}{\sqrt{1 - x^2}}$]

16. Differentiate $e^{\sqrt{x}}$ *w.r.t* '*x*'.

[*Ans:* $\frac{dy}{dx} = \frac{e^{\sqrt{x}}}{2\sqrt{x}}$]

17. Differentiate $(x + 1)(x + 2)(2x - 1)$ *w.r.t* '*x*'.

[*Ans:* $\frac{dy}{dx} = (x + 1)(x + 2)(2x - 1)\left[\frac{1}{x + 1} + \frac{1}{x + 2} + \frac{2}{2x - 1}\right]$]

18. Differentiate $\sqrt{x^2 + 2x + 3}$ *w.r.t* '*x*'.

[*Ans:* $\frac{dy}{dx} = \frac{x + 1}{\sqrt{x^2 + 2x + 3}}$]

19. Find $\frac{dy}{dx}$ if $x = at^2$ and $y = 2at$.

[*Ans:* $\frac{dy}{dx} = \frac{1}{t}$]

20. If $f(x) = 4x^4 - 3x^2 + 5$ is an even function or an odd function.

[***Ans: An even function***]

21. If $f(x) = 3x - 2$, find $f(4)$, $f(k + 1)$ and $f(k^2)$.

[***Ans:*** $f(4) = 10, f(k + 1) = 3k + 1$ ***and*** $f(k^2) = 3k^2 - 2$]

22. If $f(x) = (x - 1)(x - 2)(x - 3)$, find $f(4)$ and $f(-1)$.

[*Ans:* $f(4) = 6$ *and* $f(-1) = -24$.]

23. If $f(x) = 3^x$, find $f(0), f(1)$ and $f\left(\frac{1}{2}\right)$.

[*Ans:* $f(0) = 1,\ f(1) = 3,$ *and* $f\left(\frac{1}{2}\right) = \sqrt{3}$]

24. Find $f(2)$ if $f(x) = \dfrac{x}{1 + x^2}$ [*Ans:* $f(2) = \frac{2}{5}$]

25. If $f(x) = 3x + 22$, find $\dfrac{f(x+h) - f(x)}{h}$. [*Ans:* 3]

26. Find whether the function $f(x) = x^5 - 3x^2 + 2x + 7$ is even or odd function.

[*Ans: Odd function.*]

27. Evaluate:

$$\lim_{x \to \infty} \frac{(x+1)(2x+3)}{x^2}.$$ [*Ans:* 2]

28. Evaluate:

$$\lim_{x \to 0} \frac{\sqrt{1+x} - 1}{x}.$$ [*Ans:* $\frac{1}{2}$]

29. Evaluate:

$$\lim_{x \to \infty} \frac{6x^2 + 7x + 3}{4x^2 + 3x - 7}.$$ [*Ans:* $\frac{3}{2}$]

30. Evaluate:

$$\lim_{x \to 2} \frac{x^2 + 5x - 14}{x^2 - 4}.$$ [*Ans:* $\frac{9}{4}$]

31. Evaluate:

$$\lim_{n \to \infty} \frac{(9n+7)(4n-1)}{n(6n+3)}.$$ [*Ans:* 6]

32. Find the maximum value of $7 - 3x - 2x^2$.

[*Ans: max. at* $x = \frac{-3}{4}$. $\therefore\ y_{max} = \frac{65}{8}$]

33. Find the minimum value of $4x^2 - 8x + 5$.

[*Ans: min. at* $x = 1$. $\therefore\ y_{min} = 1$]

34. Find the maximum value of $3x - x^2$.

[*Ans: max. at* $x = \frac{3}{2}$. $\therefore\ y_{max} = \frac{9}{4}$]

35. Find the minimum value of $2x^2 - 3x + 1$.

[*Ans: min. at* $x = \frac{3}{4}$. $\therefore y_{min} \frac{-1}{8}$]

II. Question carrying FIVE marks each (Essay Type Answer).

1. if $x^y = e^{x-y}$ show that $\frac{dy}{dx} = \frac{\log x}{(1 + \log x)^2}$.

2. Find $\frac{dy}{dx}$ if $y = x^{x^{x}}$. [*Ans:* $\frac{dy}{dx} = x^{x^{x}} \cdot x^{x \log x}\left[1 + \log x + \frac{1}{x \log x}\right]$]

3. If $x\sqrt{1 + y} + y\sqrt{1 + x} = 0$, then prove that $\frac{dy}{dx} = \frac{-1}{(1 + x)^2}$.

4. Find $\frac{dy}{dx}$, if (a) $y = x^2 e^x$ and (b) $y = x \log x$.

[*Ans: (a)* $\frac{dy}{dx} = e^x\left[2x + x^2\right]$, *(b)* $\frac{dy}{dx} = [1 + \log x]$]

5. If $y = (\log a)^{\log x} + (\log x)e^x$ then find $\frac{dy}{dx}$.

[*Ans:* $\frac{dy}{dx} = \frac{\log[\log a] \cdot (\log a)^{\log x} + xe^x \log x + e^x}{x}$]

6. Differentiate $\frac{(3x - 4)^3 (2 - x)}{(5x^2 + 1)}$ *w.r.t 'x'*.

[*Ans:* $\frac{dy}{dx} = \frac{2(11 + 34x + 5x^2 - 15x^3)(3x - 4)^2}{(5x^2 + 1)^2}$]

7. Differentiate $\frac{3x^2 + 6x + 7}{\sqrt{x + 3}}$ *w.r.t 'x'*. [*Ans:* $\frac{dy}{dx} = \frac{15x^2 + 54x + 43}{2(x + 3)^{3/2}}$]

8. Prove that $\frac{d}{dx}\left(x^x\right) = x^x(1 + \log x)$.

9. Differentiate $x^2 (x + 1)\left(x^3 + 3x + 1\right)$ *w.r.t 'x'*.

[*Ans:* $\frac{dy}{dx} = x^2 (x + 1)\left(x^3 + 3x + 1\right)\left[\frac{2}{x} + \frac{1}{x + 1} + \frac{3\left(x^2 + 1\right)}{x^3 + 3x + 1}\right]$]

10. Find $\frac{dy}{dx}$ if $x = \frac{3at}{1+t^3}$ and $y = \frac{3at^2}{1+t^3}$. [*Ans:* $\frac{dy}{dx} = \frac{t(2-t^3)}{(1-2t^3)}$]

11. Differentiate the following *w.r.t* '*x*' using quotient rule.

(a) $\frac{e^x - 1}{e^x + 1}$

(b) $\frac{\log x}{1 + e^x}$

(c) $\frac{x^2 + 3x + 1}{x^2 - x + 1}$

(d) $\frac{ax^2 + bx + c}{(a+b)(x^2 - 1)}$

(e) $\frac{e^x + \log x}{x}$

[*Ans: (a)* $\frac{dy}{dx} = \frac{2e^x}{(e^x+1)^2}$, *(b)* $\frac{dy}{dx} = \frac{(1+e^x) - xe^x \log x}{x(1+e^x)^2}$, *(c)* $\frac{dy}{dx} = \frac{-4x^2+4}{(x^2-x+1)^2}$,

(d) $\frac{dy}{dx} = \frac{(a+b)(x^2-1)(2ax+b) - (ax^2+bx+c)(a+b)(2x)}{(a+b)^2(x^2-1)^2}$,

(e) $\frac{dy}{dx} = \frac{xe^x + 1 - e^x - \log x}{x^2}$]

12. If $3x^2 - 4xy + 5y^2 = 1$, prove that $\frac{d^2y}{dx^2}$ when $x = 1$ and $y = 0$ is $\frac{11}{8}$.

13. If $y = \left(\frac{1}{x}\right)^x$ prove that $\frac{d^2y}{dx^2}$ at $x = 1$ is equal to zero.

14. Evaluate $\frac{\partial z}{\partial x}$ and $\frac{\partial z}{\partial y}$ when $z = \frac{x^2}{x - y + 1}$.

[*Ans:* $\frac{\partial z}{\partial x} = \frac{x^2 + 2x - 2xy}{(x-y+1)^2}$ *and* $\frac{\partial z}{\partial y} = \frac{x^2}{(x-y+1)^2}$]

15. Verify that for $z = \log\left(x^2 - y^2\right)$, $\dfrac{\partial^2 z}{\partial x\,\partial y} = \dfrac{\partial^2 z}{\partial y\,\partial x}$.

16. Differentiate $x^3 + 5x^2 - 7x + 2$ four times *w.r.t* 'x'.

[*Ans:* $\dfrac{d^4y}{dx^4} = 0$]

17. If $f(x) = x^2 + 2x + 3$ find $f(0)$, $f(-1)$ and $f(3)$.

[*Ans:* $f(0) = 3, f(-1) = 2$ *and* $f(3) = 18$]

18. If $f(x) = \dfrac{2^x - 2^{-x}}{2^x + 2^{-x}}$ find $f(x)$ when $x = \dfrac{1}{2}$.

[*Ans:* $f\left(\dfrac{1}{2}\right) = \dfrac{1}{3}$]

19. If $f(x) = \log x$, show that $f(xy) = f(x) + f(y)$.

20. if $f(x) = x^2 + 4$, find (a) $f(2 + h) - f(2)$ (b) $f(x + 1) - f(x - 1)$.

[*Ans: (a)* $h^2 + 4h$, *(b)* $4x$]

21. If $f(x) = 2x^2 - 5x + 4$, for what values of x is $2f(x) = f(2x)$?

[*Ans:* $x = \pm 1$]

22. Determine

    (a) $f(0)$, (b) $f(-2)$, (c) $f(3)$, (d) $f(a + b)$ in each of the following:

    (i) $f(x) = 25$

    (ii) $f(x) = \dfrac{3x}{2} - 1$

    (iii) $f(x) = \sqrt{x^2}$

[*Ans: (i) (a) = (b) = (c) = (d) = 25*]

[*Ans: (ii) (a) –1 (b) –4 (c) –3.5 (d)* $\dfrac{3a + 3b - 2}{2}$]

[*Ans: (iii) (a) Zero (b) 2 (c) 3 (d)* $(a + b)$]

23. If $f(x) = 3x^2 - 2x + 5$, find $f(2)$, $f(-3)$, $f(0)$ and $f\left(\dfrac{1}{2}\right)$.

[*Ans:* $f(2) = 13, f(-3) = 38, f(0) = 5$ *and* $f\left(\dfrac{1}{2}\right) = 4.75$]

24. Find $f(x) + f\left(\dfrac{1}{x}\right)$, if $f(x) = x^3 - \dfrac{1}{x^3}$.

[*Ans: Zero*]

25. Find for what values of x, $f(x) = 0$, if $f(x) = x^2 - 3x + 2$.

[*Ans:* $x = 1$, $x = 2$]

26. Evaluate: $\lim_{x \to -2} \frac{2x^2 + 5x - 2}{3x^2 + 5x + 2}$.

[*Ans:* -1]

27. Evaluate: $\lim_{x \to -1} \frac{x^2 - 7x - 8}{x^2 + 8x + 7}$.

[*Ans:* $\frac{-3}{2}$]

28. Evaluate: $\lim_{x \to 2} \frac{x^2 - 4}{\sqrt{x+2} - \sqrt{3x-2}}$.

[*Ans:* -8]

29. Evaluate: $\lim_{x \to a} \frac{x^{-3} - a^{-3}}{x^{-5} - a^{-5}}$.

[*Ans:* $\frac{3}{5}a^2$]

30. Evaluate: $\lim_{x \to 0} \frac{\sqrt{a+x} - \sqrt{a}}{x}$.

[*Ans:* $\frac{1}{2\sqrt{a}}$]

31. Evaluate: $\lim_{x \to \infty} \frac{(2x-1)(3x+5)(4x-1)}{x^2(8x+1)}$.

[*Ans:* 3]

32. Evaluate: $\lim_{x \to 3} \frac{x^4 - 81}{x^5 - 243}$.

[*Ans:* $\frac{4}{15}$]

33. Evaluate: $\lim_{x \to a} \frac{x^9 - a^9}{x^5 - a^5}$.

[*Ans:* $\frac{9}{5}a^4$]

34. Evaluate: $\lim_{x \to 2} \frac{x^{5/2} - \sqrt{32}}{x - 2}$.

[*Ans:* $5\sqrt{2}$]

35. Evaluate: $\lim_{x \to 1} \frac{x^{3/2} - 1}{x^{5/2} - 1}$.

[*Ans:* $\frac{3}{5}$]

36. Discuss the maxima and minima of:
    (a) $x^5 - 5x^4 + 5x^3 - 10$
    (b) $x^3 - 3x^2 - 9$

[*Ans: (a) Max. at x = 1, min. at x = 3 (b) Max. at x = 0, Min. at x = 2*]

37. Find the maximum and minimum values of the function.
    (a) $2x^3 - 3x^2 - 36x + 10$
    (b) $x^3 - 2x^2 + x + 5$

[*Ans: (a) Max. at x = –2,* $\therefore y_{max} = 54$
Min. at x = 3, **$\therefore y_{min} = -71$**

(b) Max. at x = **$\frac{1}{3}$, $\therefore y_{max} = \frac{139}{27}$**

Min. at x = 1, **$\therefore y_{min} = 5$]**

38. Show that $x^5 - 5x^4 + 5x^3 - 1$ has a max. at $x = 1$, a min. at $x = 3$ and neither max. nor min. at $x = 0$.

5

Integration

Contents

INTRODUCTION

Integration is the reverse process of differentiation. It is also called anti-derivative.

We define integral of $f(x)$, w·r·t 'x' to be the function $g(x)$. This is put in the notational form as $\int f(x)\,dx = g(x) + c$. The function $f(x)$ is called the integrand. Here arbitrary constant 'C' is called the constant of integration. The symbol $\int$ is derived from the first letter of the Greek word "Summa".

Standard Formulae:

1. $\int x^n\,dx = \dfrac{x^{n+1}}{n+1} + c$ Where 'C' is the integrating constant.

2. $\int \dfrac{1}{x}\,dx = \log_e x + c$

3. $\int e^{ax}\,dx = \dfrac{e^{ax}}{a} + c$

4. $\int a^x\,dx = \dfrac{a^x}{\log a} + c$

5. $\int k\,dx = kx + c$ Where K is a constant.

6. $\int e^{x}\,dx = e^{x} + c$

7. $\int 1 \cdot dx = x + c$

8. $\int (ax+b)^{n}\,dx = \frac{1}{a} \cdot \frac{(ax+b)^{n+1}}{(n+1)} + c$

9. $\int \frac{dx}{ax+b} = \frac{1}{a} \cdot \log(ax+b) + c$

10. $\int e^{ax+b} \cdot dx = \frac{1}{a} \cdot e^{ax+b} + c$

11. Integration by parts:

$$\int u \frac{dv}{dx}\,dx = uv - \int v \frac{du}{dx}\,dx + c$$

or

The generalised formula for integration of the product of two functions u and v called the Bernoulli's rule may be used for finding a_n and b_n.

i.e., $\int uv\,dx = uv^{1} - u'v^{2} + u''v^{3} - u'''v^{4} + \ldots$

Where dashes denote differentiation *w.r.t* 'x' and suffixes 1, 2, 3, ... denote integration *w.r.t* 'x'.

12. $$\int_{-a}^{a} f(x)\,dx = \begin{cases} 2\int_{0}^{a} f(x)\,dx & \text{if } f(x) \text{ is even.} \\ 0 & \text{if } f(x) \text{ is odd.} \end{cases}$$

13. $$\int_{a}^{b} f(x)dx = [g(x)+c]_{a}^{b}$$

$$= \{g(b)+c\} - \{g(a)+c\}$$
$$= g(b) - g(a)$$

14. To evaluate $\int \frac{f'(x)}{f(x)}\,dx$ where $f'(x)$ is the derivative of $f(x)$.

put $f(x) = t$, then $f'(x)\,dx = dt$.

Thus $\int \frac{f'(x)}{f(x)} dx = \int \frac{dt}{t} = \log t = \log f(x)$

15. To evaluate $\int [f(x)]^n f'(x)\,dx, n \neq -1$ put $f(x) = t$, then $f'(x)\,dx = dt$.

Thus $\int [f(x)]^n f'(x)\,dx = \int t^n dt = \frac{t^{n+1}}{n+1}$

$= \frac{[f(x)]^{n+1}}{n+1}$

16. To evaluate $\int f'(ax+b)dx$ put $(ax+b) = t$, then $a\,dx = dt, dx = \frac{dt}{a}$

Thus $\int f'(ax+b)\,dx = \int f'(t)\frac{dt}{a}$

$= \frac{1}{a}\int f'(t)\,dt$

$= \frac{1}{a}[f(t)]$

$= \frac{f(ax+b)}{a}$

Worked Examples

Example 1. Find $\int x\,dx$.

Solution: $\int x\,dx = \frac{x^{1+1}}{1+1} + c$

$= \frac{x^2}{2} + c$

Example 2. $\int x^3\,dx$.

Solution: $\int x^3\, dx = \dfrac{x^{3+1}}{3+1} + c$

$$= \frac{x^4}{4} + c$$

Example 3. $\int \dfrac{1}{x^2}\, dx$.

Solution: $\int \dfrac{1}{x^2}\, dx = \int x^{-2}\, dx$

$$= \frac{x^{-2+1}}{-2+1} + c$$

$$= \frac{x^{-1}}{-1} + c$$

$$= \frac{-1}{x} + c$$

Example 4. $\int \sqrt{x}\, dx$.

Solution: $\int \sqrt{x}\, dx = \int (x)^{1/2}\, dx$

$$= \frac{x^{1/2+1}}{\frac{1}{2}+1} + C$$

$$= \frac{x^{3/2}}{\frac{3}{2}} + c$$

$$= \frac{2x^{3/2}}{3} + c$$

Example 5. $\int \dfrac{1}{\sqrt[3]{x^2}}\, dx$.

Solution: $\int x^{-2/3}\, dx$

$$= \frac{x^{1/3}}{\frac{1}{3}} + c \qquad = 3x^{1/3} + c$$

Example 6. $\int 3x^4\, dx$.

Solution: $\int 3x^4\, dx = 3\int x^4\, dx$

$$= \frac{3x^5}{5} + c$$

Example 7. $\int \frac{2}{3x^6}\, dx$.

Solution: $\int \frac{2}{3x^6}\, dx$

$$= \frac{2}{3}\int x^{-6}\, dx$$

$$= \frac{2}{3}\left(\frac{x^{-5}}{-5}\right) + c$$

$$= \frac{-2x^{-5}}{15} + c$$

$$= \frac{-2}{15x^5} + c$$

Example 8. $\int \left(x^3 + 3x^2 + 7x - 13\right) dx$.

Solution:

$$= \int x^3\, dx + \int 3x^2\, dx + \int 7x\, dx - \int 13x\, dx$$

$$= \frac{x^4}{4} + \frac{3 \cdot x^3}{3} + \frac{7 \cdot x^2}{2} - 13x + c$$

Example 9. $\int \left(\sqrt{x} + \frac{1}{\sqrt{x}}\right) dx$.

Solution: $\int \left(\sqrt{x} + \frac{1}{\sqrt{x}}\right) dx = \int \left[(x)^{1/2} + (x)^{-1/2}\right] dx$

$$= \frac{x^{3/2}}{\frac{3}{2}} + \frac{x^{1/2}}{\frac{1}{2}} + c$$

$$= \frac{2x^{3/2}}{3} + 2x^{1/2} + c$$

Example 10. $\int\left(2x^3 - \frac{1}{3x^2}\right) dx$.

Solution: $\int\left(2x^3 - \frac{1}{3x^2}\right) dx$

$$= 2 \cdot \frac{x^4}{4} - \frac{x^{-1}}{3(-1)} + c$$

$$= \frac{x^4}{2} + \frac{1}{3x} + c$$

Example 11. $\int\left(e^x + x^e + e^e\right) dx$.

Solution: $\int\left(e^x + x^e + e^e\right) dx$

$$= \int e^x\, dx + \int x^e\, dx + \int e^e\, dx$$

$$= e^x + \frac{x^{e+1}}{e+1} + e^e \cdot x + c$$

$$= e^x + \frac{x^{e+1}}{e+1} + xe^e + c$$

Example 12. $\int x^0\, dx$.

Solution: $\int x^0\, dx$

$$= \frac{x^{0+1}}{0+1} + c$$

$$= \frac{x}{1} + c$$

$$= x + c$$

Example 13. $\int\left(3x^4 - 4x^2 + \frac{1}{x} + \frac{1}{x^2}\right) dx$.

Solution: $\int 3x^4\, dx - \int 4x^2\, dx + \int \frac{1}{x}\, dx + \int \frac{1}{x^2}\, dx$

$$= 3 \cdot \frac{x^5}{5} - 4 \cdot \frac{x^3}{3} + \log x + \left(\frac{-1}{x}\right) + c$$

$$= \frac{3x^5}{5} - \frac{4x^3}{3} + \log x - \frac{1}{x} + c$$

Example 14. $\int (x^2 + 1)(2x^3 - 3)\, dx$.

Solution: $\int (x^2 + 1)(2x^3 - 3)\, dx$

$$= \int (2x^5 + 2x^3 - 3x^2 - 3)\, dx$$

$$= 2\int x^5\, dx + 2\int x^3\, dx - 3\int x^2\, dx - 3\int dx$$

$$= 2 \cdot \frac{x^6}{6_3} + 2 \cdot \frac{x^4}{4_2} - 3 \cdot \frac{x^3}{3} - 3x + c$$

$$= \frac{x^6}{3} + \frac{x^4}{2} - x^3 - 3x + c$$

Example 15. $\int \left(x^3 - \frac{2}{x^3}\right)^2 dx$.

Solution: $\int \left(x^3 - \frac{2}{x^3}\right)^2 dx$ $\qquad (\because (a - b)^2 = a^2 + b^2 - 2ab)$

$$= \int \left(x^6 + \frac{4}{x^6} - 2(x^3)\left(\frac{2}{x^3}\right)\right) dx$$

$$= \int x^6\, dx + 4\int \frac{dx}{x^6} - 4\int dx$$

$$= \frac{x^7}{7} + 4\int x^{-6}\, dx - 4x + c$$

$$= \frac{x^7}{7} - \frac{4}{5 \cdot x^5} - 4x + c$$

Example 16. $\int \left(x + \frac{1}{x}\right)^3 dx$.

Solution: $\int \left[x^3 + \frac{1}{x^3} + 3x \cdot \frac{1}{x}\left(x + \frac{1}{x}\right)\right] dx$

$$= \int \left[x^3 + \frac{1}{x^3} + 3x + \frac{3}{x} \right] dx$$

$$= \frac{x^4}{4} - \frac{1}{2x^2} + \frac{3x^2}{2} + 3 \log x + c$$

Example 17. $\int \left(\frac{x^3 + 3x^2 + 4}{\sqrt{x}} \right) dx.$

Solution: $\int \frac{x^3}{\sqrt{x}} dx + 3\int \frac{x^2}{\sqrt{x}} dx + \int \frac{4}{\sqrt{x}} dx$ 11

$$= \int x^3 \cdot x^{-1/2} dx + 3\int x^2 \cdot x^{-1/2} dx + 4\int x^{-1/2} dx$$

$$= \int x^{5/2} dx + 3\int x^{3/2} dx + 4\int x^{-1/2} dx$$

$$= \frac{x^{7/2}}{\frac{7}{2}} + 3\left(\frac{x^{5/2}}{\frac{5}{2}} \right) + 4\left(\frac{x^{1/2}}{\frac{1}{2}} \right) + c$$

$$= \frac{2x^{7/2}}{2} + \frac{6x^{5/2}}{5} + 8x^{1/2} + c$$

Example 18. $\int \left(\frac{a}{x} + \frac{b}{x^2} + \frac{c}{x^3} + \frac{d}{x^4} \right) dx.$

Solution:

$$= \int \frac{a}{x} dx + \int \frac{b}{x^2} dx + \int \frac{c}{x^3} dx + \int \frac{d}{x^4} dx$$

$$= a\int \frac{dx}{x} + b\int x^{-2} dx + c\int x^{-3} dx + d\int x^{-4} dx$$

$$= a \log x + b\left(\frac{x^{-1}}{-1} \right) + c\left(\frac{x^{-2}}{-2} \right) + d\left(\frac{x^{-3}}{-3} \right) + c$$

$$= a \log x - \frac{b}{x} - \frac{cx^{-2}}{2} - \frac{d}{3x^3} + c$$

$$= a \log x - \frac{b}{x} - \frac{c}{2x^2} + \left(\frac{-d}{3x^3} \right) + c$$

Example 19. $\int \frac{1}{(2x+1)^3}\,dx$

Solution: $\int \frac{1}{(2x+1)^3}\,dx$

$$= \frac{(2x+1)^{-3+1}}{2(-3+1)} + c$$

$$= \frac{(2x+1)^{-2}}{2(-2)} + c$$

$$= \frac{-1}{4(2x+1)^2} + c$$

Example 20. $\int \frac{1}{\sqrt{4-3x}}\,dx$.

Solution: $\int \frac{1}{\sqrt{4-3x}}\,dx$

$$= \frac{(4-3x)^{-1/2+1}}{-3\left(-\frac{1}{2}+1\right)} + c$$

$$= \frac{(4-3x)^{1/2}}{-\frac{3}{2}} + c$$

$$= \frac{-2}{3}\sqrt{4-3x} + c$$

Example 21. $\int \frac{1}{1-8x}\,dx$.

Solution:

Put $1 - 8x = t$

$-8dx = dt$

$\therefore\ dx = \frac{-dt}{8}$

$$= \frac{-1}{8}\int \frac{1}{t}\,dt$$

$$= \frac{-1}{8}\log t + c$$

$$= \frac{-1}{8}\log(1-8x) + c$$

Example 22. $\int (2x^2 - 6x + 4)^{3/2} (2x - 3)\, dx$.

Solution:

$$= \int (t)^{3/2} \cdot \frac{dt}{2}$$

Put $2x^2 - 6x + 4 = t$

$(4x - 6)\, dx = dt$

$2\,(2x - 3)\, dx = dt$

$(2x - 3) dx = \frac{dt}{2}$

$$= \frac{1}{2} \int t^{3/2}\, dt$$

$$= \frac{1}{2} \left(\frac{t^{5/2}}{\frac{5}{2}} \right) + c$$

$$= \frac{1}{5} t^{5/2} + c$$

$$= \frac{1}{5} (2x^2 - 6x + 4)^{5/2} + c$$

Example 23. $\int \frac{x + 2}{\sqrt{x^2 + 4x + 3}}\, dx$.

Solution:

$$= \int \frac{\frac{dt}{2}}{\sqrt{t}}$$

Put $x^2 + 4x + 3 = t$

$2x + 4 = \frac{dt}{dx}$

$(x + 2) dx = \frac{dt}{2}$

$$= \frac{1}{2} \int (t)^{-1/2}\, dt$$

$$= \frac{1}{2} \left[\frac{t^{1/2}}{\frac{1}{2}} \right] + c$$

$$= \frac{1}{\not{2}} \left(\frac{\not{2}}{1} \right) \left(\sqrt{x^2 + 4x + 3} \right) + c$$

$$= \sqrt{x^2 + 4x + 3} + c$$

Example 24. $\int x\sqrt{1 - x^2}\, dx$.

Solution:

Put $1 - x^2 = t$

$-2x\, dx = dt$

$x\, dx = \frac{-dt}{2}$

$$= \int \sqrt{t} \left(\frac{-dt}{2} \right)$$

$$= \frac{-1}{2}\int \sqrt{t}\, dt$$

$$= \frac{-1}{2}\left(\frac{t^{3/2}}{\frac{3}{2}}\right) + c$$

$$= \frac{-1}{\cancel{2}} \times \frac{\cancel{2}}{3}(1 - x^2)^{3/2} + c$$

$$= \frac{-1}{3}(1 - x^2)^{3/2} + c$$

Example 25. $\int \frac{1}{x\sqrt{1 + \log x}}\, dx$.

Solution:

Put $1 + \log x = t$

$\frac{1}{x} \cdot dx = dt$

$$= \int \frac{1}{\sqrt{t}}\, dt$$

$$= \int (t)^{-1/2}\, dt$$

$$= \left(\frac{t^{1/2}}{\frac{1}{2}}\right) + c$$

$$= 2(1 + \log x)^{1/2} + c$$

Example 26. $\int \frac{1}{x \log x}\, dx$.

Solution:

Put $\log x = t$

$\frac{1}{x} \cdot dx = dt$

$$= \int \frac{1}{t}\, dt$$

$$= \log t + c$$

$$= \log(\log x) + c$$

Example 27. $\int \frac{x\, dx}{a^2 + x^2}$.

Solution:

Put $a^2 + x^2 = t$

$2x\, dx = dt$

$x\, dx = \frac{dt}{2}$

$$= \int \frac{\frac{dt}{2}}{t}$$

$$= \frac{1}{2}\int \frac{dt}{t}$$

$$= \frac{1}{2}\log t + c$$

$$= \frac{1}{2}\log\left(a^2 + x^2\right) + c$$

Example 28. $\int \frac{e^x}{1-e^x}\,dx$.

Solution:

$$= \int \frac{-dt}{t}$$

Put $1 - e^x = t$

$-e^x\,dx = dt$

$$= -\log t + c$$

$$= -\log\left(1 - e^x\right) + c$$

Example 29. $\int \frac{1}{1+e^x}\,dx$.

Solution:

$$= \int \frac{1 + e^x - e^x}{1+e^x}\,dx$$

(Add and subtract e^x in *Nr*.)

$$= \int \left(\frac{1+e^x}{1+e^x} - \frac{e^x}{1+e^x}\right)dx$$

$$= \int 1\,dx - \int \frac{e^x}{1+e^x}\,dx$$

$$= x - \int \frac{dt}{t}$$

$$= x - \log t + c$$

$$= x - \log\left(1 + e^x\right) + c$$

Example 30. $\int \frac{(\log x)^n}{x}\,dx$.

Solution:

$$= \int t^n \cdot dt$$

Put $\log x = t$

$\frac{1}{x}dx = dt$

$$= \frac{t^{n+1}}{n+1} + c$$

$$= \frac{(\log x)^{n+1}}{n+1} + c$$

Example 31. $\int x^3 e^{ax}\, dx$.

Solution:

$$= x^3 \cdot \left(\frac{e^{ax}}{a}\right) - 3x^2 \cdot \left(\frac{e^{ax}}{a^2}\right) + 6x \cdot \left(\frac{e^{ax}}{a^3}\right) - 6\frac{e^{ax}}{a^4} + c$$

$$= e^{ax}\left(\frac{x^3}{a} - \frac{3x^2}{a^2} + \frac{6x}{a^3} - \frac{6}{a^4}\right) + c$$

Example 32. $\int \log x\, dx$.

Solution:

$$= \int \log x \cdot 1\, dx$$

$$= \log x(x) - \frac{1}{x} \cdot \frac{x^2}{2} + c$$

$$= x \log x - \frac{x}{2} + c$$

or

$$\int \log x\, dx$$

$$= \int \log x \cdot 1\, dx$$

Put $u = \log x$, $u' = \frac{1}{x}$

$v' = 1\, dx$

$v = x$

$$= \log x(x) - \int (x)\left(\frac{1}{x}\right) dx$$

$$= x \log x - \int 1\, dx$$

$$= x \log x - x + c$$

Example 33. $\int x^n \log x\, dx$.

Solution:

Put $u = \log x$, $u' = \frac{1}{x}$

$v' = x^n$, $v = \frac{x^{n+1}}{n+1}$

$$= \log x\left(\frac{x^{n+1}}{n+1}\right) - \int \left(\frac{x^{n+1}}{n+1}\right)\left(\frac{1}{x}\right) dx$$

$$= \frac{x^{n+1} \log x}{n+1} - \frac{1}{n+1} \int x^n \, dx$$

$$= \frac{x^{n+1} \log x}{n+1} - \left(\frac{1}{n+1}\right)\left(\frac{x^{n+1}}{n+1}\right) + c$$

$$= \frac{x^{n+1} \log x}{n+1} - \frac{x^{n+1}}{(n+1)^2} + c$$

Example 34. $\int \frac{\left(ax^3 + bx^2 + cx + d\right)}{x} \, dx$.

***Solution*:**

$$= \int\left(ax^2 + bx + c + \frac{d}{x}\right) dx$$

$$= \int ax^2 \, dx + \int bx \, dx + \int c \, dx + \int \frac{d}{x} \, dx$$

$$= \frac{a \cdot x^3}{3} + \frac{b \cdot x^2}{2} + cx + d \log x + c$$

Example 35. $\int_a^b 1 \, dx$.

***Solution*:**

$$= [x]_a^b$$
$$= b - a$$

Example 36. $\int x\left(x^2 + 1\right)^3 \, dx$.

***Solution*:**

Put $x^2 + 1 = t$

$2x \, dx = dt$

$x \, dx = \frac{dt}{2}$

$$= \int (t)^3 \cdot \frac{dt}{2}$$

$$= \frac{1}{2}\int t^3 \cdot dt$$

$$= \frac{1}{2} \cdot \frac{t^4}{4} + c$$

$$= \frac{1}{2} \cdot \frac{\left(x^2 + 1\right)^4}{4} + c$$

Example 37. $\int x^2 e^x \, dx$.

Solution:

$$= x^2 \left(e^x\right) - 2x\left(e^x\right) + 2e^x + c$$

$$= x^2 e^x - 2xe^x + 2e^x + c$$

Example 38. $\int \left(2x^3 + 3x^2 - 4x + 5\right) dx$.

Solution:

$$= \int 2x^3 \, dx + \int 3x^2 \, dx - \int 4x \, dx + \int 5 \, dx$$

$$= 2 \cdot \frac{x^4}{4} + 3 \cdot \frac{x^3}{3} - 4 \cdot \frac{x^2}{2} + 5x + c$$

$$= \frac{x^4}{2} + x^3 - 2x^2 + 5x + c$$

Example 39. $\int \frac{ax^2 + bx + c}{\sqrt{x}} \, dx$.

Solution:

$$= \int \frac{ax^2}{\sqrt{x}} \, dx + \int \frac{bx}{\sqrt{x}} \, dx + \int \frac{c}{\sqrt{x}} \, dx + c'$$

$$= a\int x^{3/2} \, dx + b\int x^{1/2} \, dx + c\int x^{1/2} \, dx + c'$$

$$= a\left(\frac{x^{5/2}}{\frac{5}{2}}\right) + b\left(\frac{x^{3/2}}{\frac{3}{2}}\right) + c\left(\frac{x^{1/2}}{\frac{1}{2}}\right) + c'$$

$$= \frac{2a}{5}\left(x^{5/2}\right) + \frac{2b}{5}\left(x^{3/2}\right) + 2c\left(x^{1/2}\right) + c'$$

Example 40. $\int \left(x^5 + 5^x + \frac{5}{x}\right) dx$.

Solution:

$$= \int x^5 \, dx + \int 5^x \, dx + \int \frac{5}{x} \, dx$$

$$= \frac{x^6}{6} + \frac{5^x}{\log 5} + 5 \log x + c$$

Example 41. $\int (2x+3)^5 \cdot dx$.

Solution:

$$= \frac{(2x+3)^{5+1}}{2(5+1)} + c$$

$$= \frac{(2x+3)^6}{12} + c$$

Example 42. $\int \frac{dx}{3x+1}$.

Solution:

$$= \frac{1}{3}\log(3x+1) + c$$

Example 43. $\int e^{3x}\, dx$.

Solution:

$$= \frac{1}{3} \cdot e^{3x} + c$$

Example 44. $\int (x^2 - 2x + 3)^5 \cdot (x-1)\, dx$.

Solution:

$$= \int t^5 \cdot \frac{dt}{2}$$

Put $(x^2 - 2x + 3) = t$
$(2x - 2)\, dx = dt$
$2(x-1)\, dx = dt$
$(x-1)\, dx = \frac{dt}{2}$

$$= \frac{1}{2}\left(\frac{t^6}{6}\right) + c$$

$$= \frac{1}{12}\left(x^2 - 2x + 3\right)^6 + c$$

Example 45. $\int \left(x + \frac{1}{x}\right)^3 \cdot \left(1 - \frac{1}{x^2}\right) dx$.

Solution:

$$= \int t^3 \cdot dt$$

Put $x + \frac{1}{x} = t$

$\left(1 - \frac{1}{x^2}\right) dx = dt$

$$= \frac{t^4}{4} + c$$

$$= \frac{\left(x + \frac{1}{x}\right)^4}{4} + c$$

Example 46. $\int x\, e^x\, dx$.

Solution:

$$= x\, e^x - (1)e^x + c$$

$$= x\, e^x - e^x + c$$

Example 47. $\int_1^2 x^2\, dx$.

Solution:

$$= \left(\frac{x^3}{3}\right)_1^2$$

$$= \frac{(2)^3}{3} - \frac{(1)^3}{3}$$

$$= \frac{8}{3} - \frac{1}{3}$$

$$= \frac{7}{3}$$

Example 48. $\int_1^3 \left(3x^2 - 2x + 1\right) dx$.

Solution:

$$= \left[\frac{\cancel{3} \cdot x^3}{\cancel{3}} - \frac{\cancel{2} \cdot x^2}{\cancel{2}} + x\right]_1^3$$

$$= \left(x^3 - x^2 + x\right)_1^3$$

$$= \left\{\left[(3)^3 - (3)^2 + (3)\right] - \left[(1)^3 - (1)^2 + (1)\right]\right\}$$

$$= \{27 - 9 + 3 - 1 + \cancel{1} - \cancel{1}\}$$

$$= \{27 - 7\}$$

$$= 20$$

Example 49. $\int_0^1 (2x-1)^8\,dx$.

Solution:

$$= \left.\frac{(2x-1)^{8+1}}{2(8+1)}\right]_0^1$$

$$= \left.\frac{(2x-1)^9}{18}\right]_0^1$$

$$= \frac{1}{18}[(1)-(-1)]$$

$$= \frac{1}{18}[2]$$

$$= \frac{1}{9}$$

Example 50. $\int_1^4 \left(\sqrt{x} + \frac{1}{\sqrt{x}}\right) dx$

Solution:

$$= \int_1^4 \left(x^{1/2} + x^{-1/2}\right) dx$$

$$= \left[\frac{2x^{3/2}}{3} + 2x^{1/2}\right]_1^4$$

$$= \frac{2}{3}\cdot(2)^3 + 2\cdot(2) - \frac{2}{3} - 2$$

$$= \frac{20}{3}$$

Example 51. $\int_1^1 \left(x^2 + 2x + 3\right)^2 (x+1)\,dx$.

Solution:

$$= \int_1^1 (t)^2 \left(\frac{dt}{2}\right)$$

Put $x^2 + 2x + 3 = t$

$(2x+2)\,dx = dt$

$2\,(x+1)\,dx = dt$

$(x+1)\,dx = \frac{dt}{2}$

$$= \frac{1}{2}\left[\frac{t^3}{3}\right]_0^1$$

$$= \frac{1}{6}\left[t^3\right]_0^1$$

$$= \frac{1}{6}[1 - 0]$$

$$= \frac{1}{6}$$

Example 52. $\int x \log x \, dx$.

Solution:

$$u = \log x, \; u' = \frac{1}{x}$$

$$v' = x, \; v = \int v' = \int x \, dx = \frac{x^2}{2}$$

$$\therefore = uv - \int vu'$$

$$= \frac{x^2 \log x}{2} - \frac{1}{2}\int x^2 \frac{1}{x} \, dx$$

$$= \frac{x^2 \log x}{2} - \frac{1}{2}\int x \, dx$$

$$= \frac{x^2 \log x}{2} - \frac{1}{2} \cdot \frac{x^2}{2} + c$$

$$= \frac{x^2 \log x}{2} - \frac{x^2}{4} + c$$

Example 53. $\int_0^2 \left(x^3 + 2x^2 - x + 1\right) dx$

Solution: $\int_0^2 \left(x^3 + 2x^2 - x + 1\right) dx$

$$= \left(\frac{x^4}{4} + \frac{2 \cdot x^3}{3} - \frac{x^2}{2} + x\right)_0^2$$

$$= \left[\frac{(2)^4}{4} + \frac{2 \cdot (2)^3}{3} - \frac{(2)^2}{2} + 2\right]_0^2 - [0 + 0 + 0 + 0]$$

$$= \left[\frac{16}{4} + \frac{16}{3} - \frac{4}{2} + 2\right]$$

$$= 4 + \frac{16}{3} - \cancel{2} + \cancel{2}$$

$$= \frac{12 + 16}{3}$$

$$= \frac{28}{3}$$

Example 54. $\int \left(x^3 - x^2 + x\right) dx$.

***Solution*:** $\int \left[x^3 + \left(-x^2\right) + x\right] dx$

$$= \int x^3\, dx - \int x^2\, dx + \int x\, dx$$

$$= \frac{x^4}{4} - \frac{x^3}{3} + \frac{x^2}{2} + c$$

Example 55. $\int_{-1}^{1} \left(x^3 + 4x - 2\right) dx$.

***Solution*:**

$$= \left(\frac{x^4}{4} + \frac{4 \cdot x^2}{2} - 2x\right)_{-1}^{1}$$

$$= \left(\frac{x^4}{4} + 2x^2 - 2x\right)_{-1}^{1}$$

$$= \left(\frac{1}{4} + \cancel{2} - \cancel{2}\right) - \left(\frac{+1}{4} + 2 + 2\right)$$

$$= \frac{1}{4} - \frac{1}{4} - 4$$

$$= -4$$

Example 56. Evaluate (a) $\int\left(3^x + e^x + x^{-7/2}\right) dx$ (b) $\int\left(x + \frac{1}{x}\right)^2 dx$.

Solution:

(a) $\int 3^x\, dx + \int e^x\, dx + \int x^{-7/2}\, dx$

$$= \frac{3^x}{\log 3} + e^x + \frac{x^{-7/2+1}}{\frac{-7}{2}+1} + c$$

$$= \frac{3^x}{\log 3} + e^x + \frac{x^{-5/2}}{\frac{-5}{2}} + c$$

$$= \frac{3^x}{\log 3} + e^x - \frac{2x^{-5/2}}{5} + c$$

$$= \frac{3^x}{\log 3} + e^x - \frac{2}{5x^{5/2}} + c$$

(b) $\int\left(x + \frac{1}{x}\right)^2 dx$

$$= \int\left(x^2 + \frac{1}{x^2} + 2(x)\left(\frac{1}{x}\right)\right) dx + c$$

$$= \int x^2\, dx + \int x^{-2}\, dx + \int 2\, dx + c$$

$$= \left[\frac{x^3}{3} + \frac{x^{-1}}{-1} + 2x\right] + c$$

$$= \frac{x^3}{3} - \frac{1}{x} + 2x + c$$

Example 57. $\int\left(3x^3 + 7x^2 + 1\right) dx$.

Solution: $\int\left(3x^3 + 7x^2 + 1\right) dx$

$$= \int 3x^3\, dx + \int 7x^2\, dx + \int 1\, dx$$

$$= 3 \cdot \frac{x^4}{4} + 7 \cdot \frac{x^3}{3} + x + c$$

Example 58. $\int_1^2 \left(x^2 + \frac{1}{x^2} - x \right) dx$.

***Solution*:**

$$= \left(\frac{x^3}{3} + \frac{x^{-1}}{-1} - \frac{x^2}{2} \right)_1^2$$

$$= \left(\frac{(2)^3}{3} - \frac{1}{2} - \frac{4}{2} \right) - \left(\frac{1}{3} - \frac{1}{1} - \frac{1}{2} \right)$$

$$= \left(\frac{8}{3} - \frac{1}{2} - 2 \right) - \left(\frac{1}{3} - 1 - \frac{1}{2} \right)$$

$$= \left(\frac{16 - 3 - 12}{6} \right) - \left(\frac{2 - 6 - 3}{6} \right)$$

$$= \frac{1 + 7}{6}$$

$$= \frac{\not{8}\ 4}{\not{6}\ 3}$$

$$= \frac{4}{3}$$

Example 59. $\int \left(x^n \log n \right) dx$.

***Solution*:**

$$= \log n \int x^n \, dx$$

$$= \log n \cdot \frac{x^{n+1}}{n+1} + c$$

$$= \left(\frac{x^{n+1}}{n+1} \right) \log n + c$$

Example 60. $\int \left(\frac{x^2 + 2x - 1}{\sqrt{x}} \right) dx$.

***Solution*:**

$$= \int \frac{x^2}{\sqrt{x}} dx + \int \frac{2x}{\sqrt{x}} dx - \int \frac{1}{\sqrt{x}} dx$$

$$= \int x^2 \cdot x^{-1/2}\, dx + \int 2x \cdot x^{-1/2}\, dx - \int x^{-1/2}\, dx$$

$$= \int x^{2-1/2}\, dx + 2\int x^{1-1/2}\, dx - \int x^{-1/2}\, dx$$

$$= \int x^{3/2}\, dx + 2\int x^{1/2}\, dx - \int x^{-1/2}\, dx$$

$$= \frac{x^{3/2+1}}{\frac{3}{2}+1} + 2 \cdot \frac{x^{1/2+1}}{\frac{1}{2}+1} - \frac{x^{-1/2+1}}{\frac{-1}{2}+1} + c$$

$$= \frac{2x^{5/2}}{5} + \frac{4x^{3/2}}{3} - \frac{x^{1/2}}{\frac{1}{2}} + c$$

$$= \frac{2x^{5/2}}{5} + \frac{4x^{3/2}}{3} - 2x^{1/2} + c$$

Example 61. $\int_1^2 \left(6x^{-5} + 2x^4 + 4\right) dx$.

Solution:

$$= \left(\frac{6x^{-4}}{-4} + \frac{2 \cdot x^5}{5} + 4x\right)_1^2$$

$$= \left[\frac{6(2)^{-4}}{-4} + \frac{2\,(2)^5}{5} + 4(2)\right] - \left[\frac{-6(1)^{-4}}{4} + \frac{2}{5} + 4\right]$$

$$= \left[\frac{-6}{4(16)} + \frac{32}{5} + 8\right] - \left[\frac{-6}{4} + \frac{2}{5} + 4\right]$$

$$= \left[\frac{-6}{64} + \frac{32}{5} + 8\right] - \left[\frac{-6}{4} + \frac{2}{5} + 4\right]$$

$$= \left[\frac{-30 + 32(64) + 8(320)}{320}\right] - \left[\frac{-30 + 8 + 80}{20}\right]$$

$$= \left[\frac{-30 + 2048 + 2560}{320}\right] - \left[\frac{58}{20}\right]$$

$$= \frac{4578}{320} - \frac{58}{20}$$

$$= \frac{4578 - 928}{320}$$

$$= \frac{3650}{320}$$

$$= \frac{365}{32}$$

Example 62. $\int \left(x^3 + 3x^2 - 13\right) dx$.

Solution:

$$= \frac{x^4}{4} + 3 \cdot \frac{x^3}{3} - 13x + c$$

$$= \frac{x^4}{4} + x^3 - 13x + c$$

Example 63. $\int_{-1}^{2} \left(x^3 + 3x^2 + x\right) dx$.

Solution:

$$= \left(\frac{x^4}{4} + \frac{3 \cdot x^3}{3} + \frac{x^2}{2}\right)_{-1}^{2}$$

$$= \left(\frac{x^4}{4} + x^3 + \frac{x^2}{2}\right)_{-1}^{2}$$

$$= \left(\frac{16}{4} + 8 + \frac{4}{2}\right) - \left(\frac{1}{4} - 1 + \frac{1}{2}\right)$$

$$= \left(\frac{16 + 32 + 8}{4}\right) - \left(\frac{1 - 4 + 2}{4}\right)$$

$$= \left(\frac{24 + 32}{4}\right) - \left(\frac{-1}{4}\right)$$

$$= \left(\frac{24 + 32 + 1}{4}\right)$$

$$= \left(\frac{24 + 33}{4}\right)$$

$$= \frac{57}{4}$$

Example 64. $\int (4x^3 + 3x^2 + 12)\, dx$.

Solution:

$$= \int 4x^3\, dx + \int 3x^2\, dx + \int 12\, dx$$

$$= 4 \cdot \frac{x^4}{4} + 3 \cdot \frac{x^3}{3} + 12x + c$$

$$= x^4 + x^3 + 12x + c$$

Example 65. $\int_1^5 (5x^4 + 4x^3 + 3x^2)\, dx$.

Solution:

$$= \left(5 \cdot \frac{x^5}{5} + 4 \cdot \frac{x^4}{4} + 3 \cdot \frac{x^3}{3}\right)_1^5$$

$$= \left(x^5 + x^4 + x^3\right)_1^5$$

$$= \left[(5)^5 + (5)^4 + (5)^3\right] - (1 + 1 + 1)$$

$$= [3125 + 625 + 125] - (3)$$

$$= 3875 - 3$$

$$= 3872$$

Example 66. $\int \left(4x^3 - \frac{3}{x} + \sqrt{x}\right) dx$.

Solution:

$$= \int 4x^3\, dx - \int \frac{3}{x}\, dx + \int \sqrt{x}\, dx$$

$$= 4 \cdot \frac{x^4}{4} - 3 \log x + \frac{x^{1/2+1}}{\frac{1}{2}+1} + c$$

$$= x^4 - 3 \log x + \frac{2}{3} x^{3/2} + c$$

Example 67. $\int (x^3 + 4x - 5)\, dx$.

Solution: $= \frac{x^4}{4} + 4 \cdot \frac{x^2}{2} - 5x + c$

$$= \frac{x^4}{4} + 2x^2 - 5x + c$$

Example 68. (a) $\int (1 + e^{-x})\, dx$, (b) $\int x \log x\, dx$.

Solution:

(a) $\int (1 + e^{-x})\, dx$

$$= \int 1\, dx + \int e^{-x}\, dx$$

$$= x + \frac{e^{-x}}{-1} + c$$

$$= x - e^{-x} + c$$

(b) $\int x \log x\, dx$ Refer question No. (52).

Example 69. $\int (x^2 + 1)(2x^3 - 3)\, dx$.

Solution:

$$= \int (2x^5 + 2x^3 - 3x^2 - 3)\, dx$$

$$= \int 2x^5\, dx + \int 2x^3\, dx - \int 3x^2\, dx - \int 3\, dx$$

$$= 2\int x^5\, dx + 2\int x^3\, dx - 3\int x^2\, dx - 3\int 1\, dx$$

$$= 2 \cdot \frac{x^6}{6} + 2 \cdot \frac{x^4}{4} - \cancel{3} \cdot \frac{x^3}{\cancel{3}} - 3x + c$$

$$= \frac{x^6}{3} + \frac{x^4}{2} - x^3 - 3x + c$$

Example 70. $\int_2^3 x^2\, dx$.

Solution:

$$= \left(\frac{x^3}{3}\right)_2^3$$

$$= \frac{(3)^3}{3} - \frac{(2)^3}{3}$$

$$= \frac{27}{3} - \frac{8}{3}$$

$$= \frac{19}{3}$$

Example 71. $\int (2x - 1)^{-1} \, dx$.

Solution:

Put $2x - 1 = t$

$2dx = dt$

$\therefore \; dx = \frac{dt}{2}$

$$= \int \frac{1}{(2x-1)} \, dx$$

$$= \int \frac{1}{t} \cdot \frac{dt}{2}$$

$$= \frac{1}{2} \int \frac{1}{t} \, dt$$

$$= \frac{1}{2} \log t + c$$

$$= \frac{1}{2} \log(2x - 1) + c$$

Example 72. $\int \left(2x^3 - \frac{4}{x} + \sqrt{x}\right) dx$.

Solution:

$$= \int 2x^3 \, dx - \int \frac{4}{x} \, dx + \int \sqrt{x} \, dx$$

$$= 2 \cdot \frac{x^4}{4} - 4 \log x + \frac{x^{1/2+1}}{\frac{1}{2}+1} + c$$

$$= \frac{x^4}{4} - 4 \log x + \frac{x^{3/2}}{\frac{3}{2}} + c$$

$$= \frac{x^4}{4} - 4 \log x + \frac{2x^{3/2}}{3} + c$$

Example 73. $\int_0^2 \left(4x^3 - 5x^2 + 6x + 9\right) dx$.

Solution:

$$= \left[4 \cdot \frac{x^4}{4} - 5 \cdot \frac{x^3}{3} + 6 \cdot \frac{x^2}{2} + 9x \right]_0^2$$

$$= \left[x^4 - \frac{5x^3}{3} + 3x^2 + 9x \right]_0^2$$

$$= \left(16 - \frac{40}{3} + 12 + 18\right) - (0)$$

$$= \left(16 + 30 - \frac{40}{3}\right)$$

$$= \left(46 - \frac{40}{3}\right)$$

$$= \left(\frac{138 - 40}{3}\right)$$

$$= \frac{98}{3}$$

Example 74. $\int \left(x^4 - 5x^3 + 16\right) dx$.

Solution:

$$= \frac{x^5}{5} - \frac{5 \cdot x^4}{4} + 16x + c$$

Example 75. $\int_1^5 \left(3x^3 - 4x^2 + 5x + 3\right) dx$.

Solution:

$$= \left(3 \cdot \frac{x^4}{4} - 4 \cdot \frac{x^3}{3} + 5 \cdot \frac{x^2}{2} + 3x\right)_1^5$$

$$= \left(\frac{3(5)^4}{4} - \frac{4(5)^3}{3} + \frac{5(25)}{2} + 15\right) - \left(\frac{3}{4} - \frac{4}{3} + \frac{5}{2} + 3\right)$$

$$= \left(\frac{3(625)}{4} - \frac{4(125)}{3} + \frac{125}{2} + 15\right) - \left(\frac{3}{4} - \frac{4}{3} + \frac{5}{2} + 3\right)$$

$$= \left(\frac{5625 - 2000 + 750 + 180}{12}\right) - \left(\frac{9 - 16 + 30 + 36}{12}\right)$$

$$= \frac{4555}{12} - \frac{59}{12}$$

$$= \frac{4496}{12}$$

$$= \frac{1124}{3}$$

Exercises

I. Questions carrying THREE marks each (Short Answers):

1. Integrate *w.r.t* '*x*':

(a) $\left(\sqrt{x} + \frac{1}{\sqrt{x}}\right)^2$, (b) $\frac{(x+1)(x+2)}{x^2}$, (c) $\left(\frac{ax^2 + bx + c}{x}\right)$,

(d) $\left(3x^4 - \frac{5}{x} + \sqrt[3]{x}\right)$, (e) $\left(\frac{8x^3 + 1}{2x + 1}\right)$.

[*Ans: (a)* $\frac{x^2}{2} + 2x + \log x + c$, *(b)* $x + 3\log x - \frac{2}{x} + c$,

(c) $\frac{ax^2}{2} + bx + c\log x + \underset{\text{Constant}}{\underset{\downarrow}{k}}$, *(d)* $\frac{3x^5}{5} - 5\log x + \frac{3x^{4/3}}{4} + c$,

(e) $\frac{4}{3}x^3 - x^2 + x + c$]

2. Integrate *w.r.t* '*x*':

(a) $(2x + 1)(3x - 2)$, (b) $x^7 - 4x^3 + x$, (c) $(ax + b)(x + d)$,

(d) $\frac{1}{(2x-1)^3}$, (e) $\frac{a}{\sqrt{(ax+b)^3}}$.

[*Ans: (a)* $2x^3 - \frac{x^2}{2} - 2x + c$, *(b)* $\frac{x^8}{8} - x^4 + \frac{x^2}{2} + c$,

(c) $\frac{ax^3}{3} + (ad + b)\frac{x^2}{2} + bdx + c$, *(d)* $\frac{-1}{4(2x-1)^2} + c$, *(e)* $-2(ax + b)^{-1/2} + c$]

3. Integrate *w.r.t* '*x*':

(a) $x^3 + 3x^2 - \frac{1}{x^2} + \frac{1}{x} + 1$, (b) $\frac{1}{(2x+1)^4}$, (c) $(x+1)(x+2)(x+3)$.

[*Ans: (a)* $\frac{x^4}{4} + x^3 + \frac{1}{x} + \log x + x + c$, *(b)* $\frac{-1}{6(2x+1)^3} + c$,

(c) $\frac{x^4}{4} + 2x^3 + \frac{11x^2}{2} + 6x + c$]

4. Integrate $\log x$ *w.r.t* '*x*'. **[*Ans:* $x \log x - x + c$]**

5. Evaluate:

(a) $\int \left(e^x + x^e + e^e\right) dx$, (b) $\int \left(e^x - \frac{1}{x} + \frac{1}{x^2}\right) dx$, (c) $\int \frac{1}{(3x+5)} dx$,

(d) $\int \frac{1}{1-8x} dx$, (e) $\int \left(e^x + e^{-x}\right) dx$, (f) $\int \frac{x}{x+4} dx$, (g) $\int \frac{1}{(1-x)\sqrt{1-x}} dx$

[*Ans: (a)* $e^x + \frac{x^{e+1}}{e+1} + e^e x + c$, *(b)* $e^x - \log_e x - \frac{1}{x} + c$,

(c) $\frac{1}{3}\log_e(3x+5) + c$, *(d)* $\frac{-1}{8}\log_e(1-8x) + c$, *(e)* $e^x - e^{-x} + c$,

(f) $x - 4\log(x+4) + c$, *(g)* $\frac{2}{\sqrt{1-x}} + c$]

II. Questions carrying FIVE marks each (Essay Type Answers):

1. Integrate *w.r.t* '*x*':

(a) $x^2\sqrt{a^3 + x^3}$, (b) $\frac{(x + \log x)^3(1+x)}{x}$, (c) $\frac{e^x - 1}{e^x + 1}$

(d) $\frac{8x+6}{\sqrt{2x^2 + 3x + 4}}$, (e) $\frac{x+3}{x^2 + 6x + 8}$, (f) $(x+1)\sqrt{2x^2 + 4x + 5}$,

(g) $\frac{e^x + e^{-x}}{e^x - e^{-x}}$, (h) a^{4x+5}

[*Ans: (a)* $\frac{2}{9}\left(a^3 + x^3\right)^{3/2} + c$, *(b)* $\frac{(x + \log x)^4}{4} + c$, *(c)* $2\log\left(e^x + 1\right) - x + c$,

(d) $4\sqrt{2x^2 + 3x + 4} + c$, *(e)* $\frac{1}{2}\log\left(x^2 + 6x + 8\right) + c$,

(f) $\frac{1}{6}\left(2x^2 + 4x + 5\right)^{3/2} + c$, *(g)* $\log\left(e^x - e^{-x}\right) + c$, *(h)* $\frac{a^{4x+5}}{4\log a}$]

2. Evaluate:

(a) $\int \frac{1}{x\sqrt{1+\log_e x}}\,dx$, (b) $\int \frac{1}{x \log_e x}\,dx$, (c) $\int \frac{x\,dx}{a^2+x^2}$

(d) $\int \frac{e^x}{1-e^x}\,dx$, (e) $\int \frac{1}{1+e^x}\,dx$, (f) $\int \left(\frac{1-e^x}{1+e^x}\right) dx$, (g) $\int x e^{x^2}\,dx$,

(h) $\int \frac{e^x}{e^x+1}\,dx$, (i) $\int \left(\frac{2ax+b}{ax^2+bx+c}\right) dx$

[*Ans: (a)* $2\sqrt{1+\log_e x}+c$, *(b)* $\log_e(\log_e x)+c$, *(c)* $\frac{1}{2}\log_e(a^2+x^2)+c$,

***(d)* $-\log_e(1-e^x)+c$, *(e)* $x-\log_e(1+e^x)+c$,**

***(f)* $-\log(1+e^{-x})-\log(1+e^x)+c$, *(g)* $\frac{e^{x^2}}{2}+c$, *(h)* $\log_e(e^x+1)+c$,**

***(i)* $\log_e(ax^2+bx+c)+c$]**

3. Evaluate : $\int_0^2 (5x^4+2x^3+4x^2+5x+10)\,dx$. **[*Ans:* $\frac{242}{3}$]**

4. Evaluate (a) $\int_0^1 \sqrt{x}(1+x^2)\,dx$, (b) $\int_0^4 \left(\sqrt{x}+\frac{1}{\sqrt{x}}\right) dx$.

[*Ans: (a)* $\frac{20}{21}$, *(b)* $\frac{28}{3}$]

5. Evaluate:

(a) $\int_0^{10} (x^2-x+1)\,dx$, (b) $\int_0^2 (3x^2+1)(1-x)\,dx$, (c) $\int_0^1 x^2(1-x)^{5/2}\,dx$

(d) $\int_0^1 x(x+1)(x-1)\,dx$, (e) $\int_0^1 (x+1)(x-2)\,dx$

[*Ans: (a)* $\frac{880}{3}$, *(b)* $\frac{-19}{4}$, *(c)* $\frac{16}{693}$, *(d)* $\frac{-1}{4}$, *(e)* $\frac{-4}{3}$]

6. Evaluate:

$$\text{(a)} \int_1^2 \log_e x \, dx, \text{ (b)} \int_0^1 \frac{x}{x+1} dx, \text{ (c)} \int_{-1}^1 x^2 e^{-x} dx$$

[*Ans: (a)* **2 log$_e$ 2 – 1**, *(b)* **1 – log$_e$ 2**, *(c)* $\frac{e^2 - 5}{e}$]

7. Evaluate:

$$\text{(a)} \int_{-1}^1 \left(x^3 + 3x\right) dx, \text{ (b)} \int_0^1 x(x-1)^n dx, \text{ (c)} \int_{-1}^1 \left(x^2 + 1\right) dx,$$

$$\text{(d)} \int_{-1}^1 \left(x^2 - 3x + 2\right) dx, \text{ (e)} \int_4^9 \left(\sqrt{x} + \frac{1}{\sqrt{x}}\right) dx.$$

[*Ans: (a)* **0**, *(b)* $\frac{(-1)^{n+1}}{(n+1)(n+2)}$, *(c)* $\frac{8}{3}$, *(d)* $\frac{14}{3}$, *(e)* $\frac{44}{3}$]

6

Applications to Economics

Contents

- Revenue Functions, Cost Functions, Profit Function, Elasticity Demand Function, Break Even Point.

REVENUE FUNCTIONS

The most fundamental way of computing Total revenue (TR) from selling a product (or service) is Total revenue (TR) = R = Price × Quantity sold = pq. Where p is the price per unit and q is the quantity sold or quantity demanded.

For any demand function $p = f(q)$

$$\text{Total revenue } (TR) = R = q \times p$$
$$= q \times f(q)$$

$$\text{Average revenue } (AR) = \frac{\text{Total Revenue}}{\text{Total Quantity}}$$

$$\therefore \; AR = \frac{(TR)}{q}$$

Total revenue = Average Revenue × Total Quantity.

Marginal revenue (MR) is the additional revenue derived from selling one more unit of a product or service.

$$\text{Hence } MR = \frac{d(TR)}{dq} = \frac{dR}{dq}$$

Where $\frac{dR}{dq}$ = derivative of Total revenue with reference to quantity demanded.

We can also determine the level of output q, where Total revenue will be maximum since we know that TR is maximum when Marginal revenue = 0. Thus, by equating Marginal revenue function to zero, we find the output level q and then substituting this value of q in the given demand function. Finally, the price of product and level of output gives maximum level of Total revenue.

COST FUNCTIONS

Total cost (TC) is a function of total quantity of output when q, the quantity of output produced.

i.e., Total cost = $TC = C = f(q)$.

Average cost is the per unit of output. It is obtained by dividing the total cost by the total quantity produced.

$$\text{i.e., Average cost} = AC = \frac{\text{Total cost}}{\text{Quantity produced}}$$

$$\therefore\ AC = \frac{C}{q} \text{ or } \frac{f(q)}{q}$$

Where $C = f(q)$ is the total cost function and q is the total output.

Marginal cost is the derivative of the total cost function with respect to the quantity q.

$$\text{i.e., Marginal cost} = MC = \frac{d(TC)}{dq} = \frac{dC}{dq}$$

In the short run analysis,

Short run Total cost = Total variable cost + Total fixed cost.

$$= f(q) + F \quad (\because \ F = \text{Fixed cost})$$

$$\text{Short run average cost} = \frac{\text{Short run Total cost}}{q}$$

$$= \frac{(f(q) + F)}{q}$$

$$\text{Short run average fixed cost} = \frac{F}{q}$$

$$\text{Short run average variable cost} = \frac{f(q)}{q}$$

$$\text{Short run average marginal cost} = \frac{d(TC)}{dq}$$

$$= \frac{d\,f(q)}{dq}$$

Normally, the Total cost functions have different shapes under different circumstances. We assume Average and Marginal cost to have *U*-shape curves therefore, the relation is

(a) When Average cost curve falls, Marginal cost curve is below the Average cost curve.

i.e., $MC < AC \Rightarrow \dfrac{d(TC)}{dq} < \dfrac{TC}{q}$

(b) When Average cost curve rises Marginal cost curve is above the Average cost curve.

i.e., $MC > AC \Rightarrow \dfrac{d(TC)}{dq} > \dfrac{TC}{q}$

(c) When Average cost is at the minimum point, Average cost curve and the Marginal cost curve Intersect each other.

i.e., $MC = AC \Rightarrow \dfrac{d(TC)}{dq} = \dfrac{TC}{q}$

PROFIT FUNCTION

Profit is the difference between Total Revenue and Total cost.

i.e., Profit = Total Revenue – Total cost

$P(q) = TR - TC.$

Marginal profit is given by:

Marginal Profit = Marginal Revenue – Marginal cost.

i.e., $P(q) = \dfrac{d(TR)}{dq} - \dfrac{d(TC)}{dq}$

ELASTICITY OF DEMAND FUNCTION

The Average price elasticity of demand is the proportionate response of quantity demanded to the change in price. Let dp be small change in price p and dq be a small change in the quantity demanded q.

i.e., Average price elasticity of demand $= \frac{P}{q}\frac{dq}{dP}$ and it is denoted by e_d.

As demand curve slopes downward $\frac{dq}{dP}$ should be negative, so we put

$$e_d = \frac{-P}{q}\frac{dq}{dP}.$$

BREAK EVEN POINT

Break even point is that, output at which there is no profit and no loss. i.e., Total Revenue = Total cost.

List of formulae:

1. Total Revenue $= TR = R = pq$.
   Where, p is the price and q is quantity demanded.

2. Average Revenue $= AR = \frac{R}{q} = \frac{\text{Total Revenue}}{\text{Total quantity}}$

3. Marginal Revenue $= MR = \frac{d(TR)}{dq} = \frac{dR}{dq}$

   $$= \frac{d}{dq}(\text{Total Revenue})$$

4. Marginal Revenue $= MR = P + q \cdot \frac{dP}{dq}$

   $$= P\left(1 - \frac{1}{e_d}\right)$$

   where, e_d is elasticity of demand.

5. Total cost $TC = C = f(q)$.

6. Average cost $= AC = \frac{C}{q} = \frac{\text{Total cost}}{\text{Quantity Produced}}$

7. Marginal cost $= MC = \frac{d}{dq}(TC) = \frac{dC}{dq}$

   $$= \frac{d}{dq}(\text{Total cost})$$

8. Profit = P or $N = TR - TC = R - C$ = Total Revenue – Total cost

9. Marginal profit $= MP = P = \dfrac{d(TR)}{dq} - \dfrac{d(TC)}{dq}$

$$= \frac{d}{dq}(\text{Total Revenue}) - \frac{d}{dq}(\text{Total cost})$$

10. Marginal Average cost $= MAC = \dfrac{d}{dq}\left(\dfrac{C}{q}\right)$

$$= \frac{d}{dq}(\text{Average cost})$$

11. For maximum or minimum average cost:

Take $\dfrac{d}{dq}\left(\dfrac{C}{q}\right) = 0$

If $\dfrac{d^2}{dq^2}\left(\dfrac{C}{q}\right) > 0$ then we get minimum average cost.

and If $\dfrac{d^2}{dq^2}\left(\dfrac{C}{q}\right) < 0$ then we get maximum average cost.

12. For maximum or minimum profit:

Take $\dfrac{dP}{dq} = 0$ or $\dfrac{dN}{dq} = 0$

If $\dfrac{d^2P}{dq^2} < 0$, then we get maximum profit

and If $\dfrac{d^2P}{dq^2} > 0$, then we get minimum profit.

13. Elasticity of demand function $= e_d = \dfrac{-P}{q}\dfrac{dq}{dP}$.

14. Break even point: Total revenue = Total cost

15. Average cost = Average variable cost + Average fixed cost.

Example. $C = \frac{x^3}{10} - 9x^2 + 85x + 17.$

$$AC = \frac{x^2}{10} - 9x + 85 + \frac{17}{x}$$

$$A.V.C = \frac{x^2}{10} - 9x + 85$$

$$A.F.C = \frac{17}{x}$$

Note: In place of small letters we can use capital letters.

Example $\frac{d}{dq}\left(\frac{C}{q}\right) = \frac{d}{dQ}\left(\frac{C}{Q}\right)$

Worked Examples

Example 1 The total revenue R and the total cost C function of a firm are given by $R = 30x - x^2$ and $C = 20 + 6x$.

(a) Find the marginal revenue and marginal cost.

(b) Find the rate of change in R when $x = 2$.

(c) Show that the rate at which cost increases with increase in output is constant.

(d) Find the average revenue and average cost.

(e) Find the output at which marginal revenue is equal to marginal cost.

Solution Total Revenue function = $R = 30x - x^2$.

Total cost function = $C = 20 + 6x$.

(a) Marginal revenue (MR) = $\frac{dR}{dx} = 30 - 2x$

Marginal cost (MC) = $\frac{dc}{dq} = 6$

(b) Rate of change in R when $x = 2$ is $\frac{dR}{dx}$ at $x = 2$.

i.e., $\left[\frac{dR}{dx}\right]_{x=2} = 30 - 4 = 26$

(c) The rate at which cost increases with increase in output 'x' is $\frac{dc}{dq} = 6$ which is a constant.

(d) Average revenue $= \dfrac{\text{Total Revenue}}{\text{Output}} = \dfrac{R}{x}$.

$$= \frac{30x - x^2}{x} = 30 - x$$

Average cost $= \dfrac{\text{Total cost}}{\text{Output}} = \dfrac{C}{x} = \dfrac{20 + 6x}{x} = \dfrac{20}{x} + 6$

(e) Marginal revenue = Marginal cost.

$30 - 2x = 6$

$2x = 30 - 6$

$2x = 24$

$\therefore x = 12$

Example 2. The total cost function of output is given by $C = \dfrac{2}{3}x + \dfrac{35}{2}$. Find (a) cost when output is 4 units. (b) average cost of output of 10 units. (c) marginal cost when output is 3 units.

Solution: The cost function $C = \dfrac{2}{3}x + \dfrac{35}{2}$

(a) Total cost at $x = 4$ is $C = \dfrac{2}{3} \times 4 + \dfrac{35}{2}$

$$C = \frac{121}{6} = 20.17$$

(b) Average cost $= \dfrac{\text{Total cost}}{\text{Output}} = \dfrac{C}{x}$

$$= \frac{\frac{2}{3}x + \frac{35}{2}}{x}$$

$$= \frac{2}{3} + \frac{35}{2x}$$

at $x = 10$, $= \dfrac{2}{3} + \dfrac{35}{2 \times 10}$

$$= \frac{2}{3} + \frac{7}{4}$$

$$= \frac{29}{12} = 2.42$$

(c) Marginal cost $= \dfrac{dc}{dx}$

$$= \frac{d}{dx}\left[\frac{2}{3}x + \frac{35}{2}\right]$$

$$= \frac{2}{3} + 0 = \frac{2}{3}$$

Example 3. **Find the average cost and marginal cost. If total cost of an article is given by $C = 5x^2 + 2x + 3$. Where C is total cost and x is quantity.**

***Solution*:** Average cost $= \dfrac{\text{Total cost}}{\text{quantity}} = \dfrac{C}{x}$

$$= \frac{5x^2 + 2x + 3}{x}$$

$$= 5x + 2 + \frac{35}{2}$$

Marginal cost $= \dfrac{dC}{dx}$

$$= \frac{d}{dx}\left(5x^2 + 2x + 3\right)$$

$$= 10x + 2$$

Example 4. **The average cost for a commodity is given by $AC = x + 4 + \dfrac{4}{x}$. Find the total and marginal cost functions. Also, find output when average cost = marginal cost.**

***Solution*:** Average cost, $AC = x + 4 + \dfrac{4}{x}$

Total cost $= TC = AC \times x = x\left(x + 4 + \dfrac{4}{x}\right)$

$$= x^2 + 4x + 4$$

$\therefore$ Marginal cost $= MC = \dfrac{d}{dx}(TC)$

$$= 2x + 4$$

Average cost = Marginal cost

$$x + 4 + \frac{4}{x} = 2x + 4$$

$$\frac{x^2 + 4x + 4}{x} = 2x + 4$$

$$x^2 + 4x + 4 = 2x^2 + 4x$$

$$x^2 - 2x^2 + \cancel{4x} - \cancel{4x} + 4 = 0$$

$$-x^2 + 4 = 0$$

$$x^2 = 4$$

$$x = \pm 2$$

$\therefore$ $x = 2$ Neglecting negative sign.

Example 5. The total cost C is making x units of a product is $C = 0.00003x^3 - 0.045x^2 + 8x + 25000$. Find the marginal cost at 1000 units output.

Solution: Given $C = 0.00003x^3 - 0.045x^2 + 8x + 25000$.

$$\text{Marginal cost} = \frac{dc}{dx}$$

$$\text{But } \frac{dc}{dx} = 0.00009x^2 - 0.090x + 8$$

$$\text{at } x = 1000, \frac{dc}{dx} = \cancel{90} - \cancel{90} + 8 = 8$$

$\therefore$ Marginal cost at $x = 1000$ is 8.

Example 6. If the cost of making x units is C Rupees where $c = 10 + 9x^{2/3}$, at what number of units with marginal cost be Rs 0.90?

Solution: $c = 10 + 9x^{2/3}$

$$\therefore \text{Marginal cost} = MC = \frac{dc}{dx}$$

$$= {}^{3}\cancel{9} \times \frac{2}{\cancel{3}} x^{2/3-1}$$

$$= 6x^{-1/3}$$

$$= \frac{6}{x^{1/3}}$$

When marginal cost is Rs 0.90,

$$\text{We have } 0.90 = \frac{6}{x^{1/3}}$$

$$0.90x^{1/3} = 6$$

$$x^{1/3} = \frac{6}{0.90} = \frac{6 \times 10\not{0}}{9\not{0}}$$

$$x^{1/3} = \frac{\not{60}\ 20}{\not{9}\ 3} = \frac{20}{3}$$

$$x = \left(\frac{20}{3}\right)^3$$

$$\therefore\ x = \frac{8000}{27} = 296.30$$

***Example** 7.* **Find Total revenue, Marginal revenue and elasticity of demand at $p = 3$ if the demand function is given by $D = 30 - 4p - p^2$.**

***Solution*:** Total revenue $TR = PD$

$= P(30 - 4p - p^2)$

$= 30p - 4p^2 - p^3$

at $p = 3$, $TR = 90 - 36 - 27$

$= 27$

elasticity of demand $= e_d = \frac{-P}{D} \cdot \frac{dD}{dP}$

$$= \frac{-P}{D}(0 - 4 - 2P)$$

$$= \frac{4P + 2P^2}{D}$$

$$= \frac{4P + 2P^2}{30 - 4P - P^2}$$

at $p = 3$, $= \frac{12 + 18}{30 - 12 - 9}$

$$= \frac{\not{30}\ 10}{\not{9}\ 3} = \frac{10}{3}$$

Marginal revenue $= P\left(1 - \frac{1}{e_d}\right)$

at $p = 3$, $MR = 3\left[1 - \frac{1}{\left(\frac{10}{3}\right)}\right]$

$$= 3\left[1 - \frac{3}{10}\right]$$

$$= 3\left[\frac{7}{10}\right]$$

$$= \frac{21}{10}$$

$\therefore$ at $p = 3$, $MR = 2.1$

Example 8. Find the elasticity of demand, e_d for the demand function $x = 50 - 4p^2$ at $p = 1$ and at $p = 3$

Solution: $e_d = \frac{-P}{x}\frac{dx}{dp}$

since $x = 50 - 4P^2$

$$\frac{dx}{dp} = 0 - 8P = -8P$$

Hence $e_d = \frac{-P}{x}(-8P) = \frac{8P^2}{x}$

$$e_d = \frac{8P^2}{50 - 4P^2}$$

at $P = 1$, $e_d = \frac{8}{50 - 4}$

$\therefore$ at $P = 1$, $e_d = \frac{8 \; 4}{46 \; 23} = \frac{4}{23}$

and at $P = 3$, $e_d = \frac{72}{50 - 36}$

$$= \frac{72 \; 36}{14 \; 7}$$

$\therefore$ *at* $P = 3$, $e_d = \frac{36}{7}$

Example 9. The function of a firm is given by $C = 900x - 30x^2 + x^3$, where C is the cost and x the output. Find the output when average cost is equal to marginal cost.

Solution: $C = 900x - 30x^2 + x^3$, $MC = \frac{dc}{dx} = 900 - 60x + 3x^2$

Given: $AC = MC$

$$\frac{900x - 30x^2 + x^3}{x} = 900 - 60x + 3x^2$$

$\cancel{900} - 30x + x^2 = \cancel{900} - 60x + 3x^2$
$2x^2 - 30x = 0$
$2x(x - 15) = 0$
$2x = 0$ or $x - 15 = 0$
$\therefore x = 0$ or $x = 15$

Example 10. Find maximum profit. If $R = 30x - x^2$ and $C = 20 + 4x$.

Solution: Given $R = 30x - x^2$

$$\therefore\ MR = \frac{dR}{dx} = 30 - 2x$$

and given: $C = = 20 + 4x$

$$\therefore\ MC = \frac{dC}{dx} = 4$$

For maximum profit,

$MR = MC$
$30 - 2x = 4$
$-2x = 4 - 40$
$-2x = -26$

$$x = \frac{26}{2}$$

$\therefore\ x = 13$

Maximum profit $= R - C$

At $x = 13$, Max. profit $= [(30x - x^2) - (20 + 4x)]$
$= (30x - x^2 - 20 - 4x)\ x = 13$
$= (26x - x^2 - 20)\ x = 13$
$= [26(13) - 169 - 20]$
$= 338 - 169 - 20$
$= 149$

Example 11. Find the elasticity of demand for the function $p = 100 - x - x^2$.

Solution: $e_d = \frac{-P}{x} \cdot \frac{1}{\left(\frac{dP}{dx}\right)}$

$$= \frac{-P}{x} \cdot \left(\frac{1}{\frac{d}{dx}\left(100 - x - x^2\right)}\right)$$

$$= \frac{-P}{x} \cdot \left[\frac{1}{-1 - 2x}\right]$$

$$= \frac{+P}{x(1+2x)}$$

$$= \frac{P}{x+2x^2}$$

$$\therefore e_d = \frac{100 - x - x^2}{x + 2x^2}$$

Example 12. Find the minimum average cost if the cost function is given by $T = 36q - 10q^2 + 2q^3$. Also obtain $M.C$ at $q = \frac{5}{2}$.

Solution: $A.C = \frac{T}{q}$

$$= \frac{q(36 - 10q + 2q^2)}{q}$$

$$\therefore A.C = 36 - 10q + 2q^2$$

for maximum or minimum $\frac{d}{dq}\left(\frac{T}{q}\right) = 0$

$$\frac{d}{dq}\left(36 - 10q + 2q^2\right) = 0$$

$$-10 + 4q = 0$$

$$4q = 10$$

$$\therefore q = \frac{10}{4} = \frac{5}{2}$$

$$\frac{d^2}{dq^2}\left(\frac{T}{q}\right) = \frac{d}{dq}(-10 + 4q)$$

$$= 4 > 0$$

$\therefore$ It is minimum at $q = \frac{5}{2}$

$\therefore$ Minimum $AC = 36 - 10q + 2q^2$

$$= 36 - 10\left(\frac{5}{2}\right) + 2\left(\frac{25}{4}\right)$$

$$= 36 - 25 + \frac{25}{2}$$

$$= 11 + \frac{25}{2}$$

$$= \frac{22 + 25}{2} = \frac{47}{2}$$

$\therefore$ Minimum $AC = 23.5$

Marginal cost = = $MC = \frac{dT}{dq}$

$$= \frac{d}{dq}\left(36q - 10q^2 + 2q^3\right)$$

$\therefore$ $M.C = 36 - 20q + 6q^2$

at $q = \frac{5}{2}$, $M.C = 36 - 20\left(\frac{5}{2}\right) + \overset{3}{\not{6}}\left(\frac{25}{\not{4}2}\right)$

$$= 36 - 10(5) + \frac{75}{2}$$

$$= 36 - 50 + \frac{75}{2}$$

$$= -14 + \frac{75}{2}$$

$$= \frac{-28 + 75}{2}$$

$$= \frac{47}{2} = 23.5$$

Example 13. A firm has the revenue function $R = 100q - q^2$ and the cost function $T = q^3 - \frac{57}{2}q^2$ find maximum profit.

Solution: *Pr ofit* $\Rightarrow N = R - T$

$$= \left(100q - q^2\right) - \left(q^3 - \frac{57}{2}q^2\right)$$

$$= 100q - q^2 - q^3 + \frac{57}{2}q^2$$

$$= 100q - q^3 + \frac{55}{2}q^2$$

For maximum or minimum profit $\Rightarrow \frac{dN}{dq} = 0$

$$100 - 3q^2 + 55q = 0$$
$$3q^2 - 55q - 100 = 0$$
$$3q^2 - 60q + 5q - 100 = 0$$
$$3q(q - 20) + 5(q - 20) = 0$$
$$(q - 20)(3q + 5) = 0$$

2	300
2	150
3	75
5	25
	5

$$\therefore q = 20 \text{ or } 3q = -5$$
$$\therefore q = \frac{-5}{3}$$

$$\text{If } \frac{d^2N}{dq^2} = \frac{d}{dq}\left(100 - 3q^2 + 55q\right)$$
$$\text{at } q = 20 \qquad = -6q + 55$$
$$= -120 + 55 = -65 < 0$$

$\therefore$ at $q = 20$ it gives maximum profit

$$\therefore \text{Maximum profit} = 100q - q^3 + \frac{55}{2}q^2$$
$$\text{at } q = 20, = 100(20) - (20)^3 + \frac{55}{2}(20)^2$$
$$= 2000 - 8000 + 55\,(200)$$
$$= -6000 + 11000$$
$$= 5000$$

Example 14. Show that if $T = aq^b + k\ (a > 0, b > 0, k \geq 0)$ then $MC = abq^{b-1}$ and Marginal average cost $= a(b-1)q^{b-2} - \dfrac{k}{q^2}$.

Solution: Marginal cost $= MC = \dfrac{dT}{dq}$

$$= abq^{b-1} + 0$$
$$\therefore MC = abq^{b-1}$$

$$\text{Marginal average cost} = \frac{d}{dq}\left(\frac{T}{q}\right)$$
$$= \frac{d}{dq}\left[\frac{aq^b + k}{q}\right]$$
$$= \frac{d}{dq}\left[aq^{b-1} + \frac{k}{q}\right]$$
$$\therefore M.A.C = a(b-1)q^{b-2} - \frac{k}{q^2}$$

Example 15. Find Average cost and Marginal cost, if $T = 1000 + 100q - 10q^2 + q^3$.

Solution: $A.C = \frac{T}{q} = \frac{1000 + 100q - 10q^2 + q^3}{q}$

$$= \frac{1000}{q} + 100 - 10q + q^2$$

$$MC = \frac{dT}{dq} = \frac{d}{dq}\left(1000 + 100q - 10q^2 + q^3\right)$$

$$\therefore \quad M.C = 100 - 20q + 3q^2$$

Example 16. Show that if $T = ae^{bq}$, then Marginal cost is equal to Average cost provided $q = b^{-1}$.

Solution: $A.C = \frac{T}{q}$

$$\therefore A.C = \frac{ae^{bq}}{q}$$

$$M.C = \frac{d}{dq}(T)$$

$$= \frac{d}{dq}\left(ae^{bq}\right)$$

$$\therefore M.C = abe^{bq}$$

$$MC = AC$$

$$abe^{bq} = \frac{abe^{bq}}{q}$$

$$b = \frac{1}{q}$$

$$qb = 1$$

$$q = \frac{1}{b}$$

$$\therefore q = b^{-1}$$

Example 17. The relation between a country's import M and income y is given by $M = 10 + \frac{1}{6} \cdot y^2$ what is the marginal propensity to import?

Solution: Marginal propensity to import

$$= \frac{dM}{dy}$$

$$= \frac{d}{dy}\left(10 + \frac{y^2}{6}\right)$$

$$= 0 + \frac{2y}{6\,3}$$

$$= \frac{y}{3}$$

Example 18. A long run is that period of time over which all resources can be varied. A firm's total costs '*C*' are related to total out put '*x*' over a five years period by the function $C = ax^3 - bx^2 + cx$. Where *a, b, c* are positive constants.

(a) What is the Average cost function?

(b) At what output is Average cost a minimum?

(c) What is the Marginal cost function?

(d) At what output does Average cost equal Marginal cost?

Solution:

(a) *A.C* function $= \frac{C}{x}$

$$= \frac{ax^3 - bx^2 + cx}{x}$$

$$= ax^2 - bx + c$$

(b) For minimum or maximum $\frac{d}{dx}\left(\frac{C}{x}\right) = 0$

$$\frac{d}{dx}\left(ax^2 - bx + c\right) = 0$$

$$2ax - b = 0$$

$$\therefore\ x = \frac{b}{2a}$$

If $\frac{d^2}{dx^2}\left(\frac{C}{x}\right) = \frac{d}{dx}(2ax - b)$

$$= 2a > 0.$$

$\therefore$ It gives minimum average cost at $x = \frac{b}{2a}$

(c) Marginal cost function $= \frac{dc}{dx}$

$$= \frac{d}{dx}\left(ax^3 - bx^2 + cx\right)$$

$\therefore M.C = 3ax^2 - 2bx + c$

(d) If $AC = MC$

$ax^2 - bx + c = 3ax^2 - 2bx + c$

$3ax^2 - 2bx - ax^2 + bx = 0$

$2ax^2 - bx = 0$

$x(2ax - b) = 0$

$\therefore x = 0$ or $2ax = b$

$$x = \frac{b}{2a}$$

$\therefore$ at $x = 0$ or $x = \dfrac{b}{2a}$ average cost equal to marginal cost.

Exercises

I Questions carrying THREE marks each (Short Answers):

1. The total cost C of making x litres of a product in $C = 10 + 30\sqrt{x}$. Find the marginal cost of 225 litres of output. [*Ans: M.C = 1*]
2. The short run total cost function is given by $C = Q^3 - 3Q^2 + 15Q + 27$, find the average cost and marginal cost.

   [*Ans:* $AC = Q^2 - 3Q + 15 + \dfrac{27}{Q}$, $MC = 3Q^2 - 6Q + 15$]
3. The total cost function is $C = 2Q - 2Q^2 + Q^3$. Find AC function and MC function. [*Ans:* $AC = 2 - 2Q + Q^2$, $MC = 2 - 4Q + 3Q^2$]
4. The total cost function is $C = 4Q - Q^2 + 2Q^3$. Find AC function and MC function. [*Ans:* $AC = 4 - Q + 2Q^2$, $MC = 4 - 2Q + 6Q^2$]
5. Given the following total cost function, find the average and marginal cost, $C = 0.0003Q^2 + 6.75Q - 10485$.

   [*Ans:* $AC = 0.0003Q + 6.75 - \dfrac{10485}{Q}$, $MC = 0.0006Q + 6.75$]
6. The total cost function is $C = 2Q - 2Q^2 + Q^3$ where C = Total cost, Q = Total quantity produced. Find (a) Average cost function. (b) Marginal cost function. [*Ans:* $AC = 2 - 2Q + Q^2$, $MC = 2 - 4Q + 3Q^2$]
7. If the total cost function is $C = \dfrac{Q^3}{3} - 3Q^2 + 9Q$. Find average cost function and marginal cost function.

   [*Ans:* $AC = \dfrac{Q^2}{3} - 3Q + 9$, $MC = Q^2 - 6Q + 9$]

8. If the total cost function is $C = 4Q - Q^2 + 2Q^3$. Find average cost function and marginal cost function. **[Ans: $AC = 4 - Q + 2Q^2$, $MC = 4 - 2Q + 6Q^2$]**

9. The average cost 'C' of producing and marketing per unit sold is given by $C = 0.02 + 10^{-6} S$. The total cost $C = Sc$. What is the marginal cost of producing and marketing? **[Ans: $MC = 0.02 + 2(10^{-6})S$]**

10. If the demand curve is given by $P = (10 - 2D)(20 - D^2)$. Find the marginal revenue. **[Ans: $M.R = 200 - 80D - 30D^2 + 8D^3$]**

11. The price P is given in terms of the demand D, by $P = 100 - 2D - D^2$.find the elasticity of the demand *w.r.t* price, at $D = 2$. **[Ans: at $D = 2$, $e_d = \frac{23}{3}$]**

12. Find the elasticity of demand for the demand function $D = 50 - 5p - p^2$ at $p = 2$. **[Ans: at $P = 2$, $e_d = \frac{1}{2}$]**

II Questions carrying FIVE marks each (Essay Type Answers):

1. The total cost of x units is y rupees, where $y = \frac{x^3}{300} - 0.2x^2 - 27.5x$ and all units made are sold at Rs 10 per unit. At what two points does marginal cost equal marginal revenue? **[Ans: $x = 84.42$ and $x = -44.42$]**

2. If total cost function is $C = 5000 + 1000Q - 500Q^2 + \frac{2}{3}Q^3$ find (a) Average cost function (b) Marginal cost (c) The output at which marginal cost (MC) would be equal to the average variable cost (AVC).

   [Ans: (a) $AC = \frac{5000}{Q} + 1000 - 500Q + \frac{2Q^2}{3}$,

   (b) $MC = 1000 - 1000Q + 2Q^2$

   (c) $TVC = 1000Q - 500Q^2 + \frac{2}{3}Q^3$

   $AVC = 1000 - 500Q + \frac{2}{3}Q^2$

   $MC = AVC \Rightarrow Q = 375$]

3. The total revenue function R of a certain commodity is given by $R = 60x + 2x^2$, where x is the output. Find (a) Marginal revenue, (b) Marginal revenue when output is 10. **[Ans: (a) $MR = 60 + 4x$, (b) at $x = 10$, $MR = 100$]**

4. Find out the total revenue and marginal revenue from the demand curve $P = 10 - 3x - 2x^2$. **[Ans: $TR = 10x - 3x^2 - 2x^3$, $MR = 10 - 6x - 6x^2$]**

5. A monopolist's cost function is $C = q^2 + 3q$ and average revenue function $P = 23 - 3q$.

   (a) Find the output level for which marginal cost is equal to the average cost.

   (b) Find the profit at that point. **[Ans: (a) $q = 5$, (b) Profit = 30]**

6. If the demand function for commodity is given by $P = 30 - x$ and cost function $C = 20 + 4x$, where x is the output find the output that will maximise the profit.

[*Ans:* $x = 13$]

7. Find the elasticity of demand for the demand function:
(a) $x = 20 - p^2$ (b) $x = 40 + p - p^2$.

[*Ans: (a)* $e_d = \frac{-2P^2}{20 - P^2} = \frac{2P^2}{P^2 - 20}$, *(b)* $e_d = \frac{P(2P-1)}{P^2 - P - 40}$]

8. If $q = 32 - 4p - p^2$, find elasticity when $p = 3$. [*Ans:* $e_d = 2.727$]

9. A firm produces x units of output at a total cost $C = \frac{x^3}{10} - 9x^2 + 85x + 17$, find the Average cost, Average variable cost and Average fixed cost. Find the value of these at the level of output of 10 units.

[*Ans:* $AC = \frac{x^2}{10} - 9x + 85 + \frac{17}{x}$, $A.V.C = \frac{x^2}{10} - 9x + 85$, $A.F.C = \frac{17}{x}$,
at $x = 10$ *units,* $AC = 6.7$, $A.V.C = 5$, $A.F.C = 1.7$]

10. If the total cost function is $C = 8q^3 - 12q^2 + 20q$, where q = level of output, find the level of output at which the average cost will be minimized?

[*Ans:* $q = \frac{3}{4} = 0.75$]

Part II

Statistics

R.A. Fisher
(1890-1962)
Father of Statistics

Statistics is a branch of applied mathematics which specializes in data.

Part II

Statistics

Internal Marks: 5
Examination Marks: 45
Total Marks: 50

R.A. Fisher
(1890–1962)
Father of Statistics

Statistics is a branch of applied mathematics which specialises in data

–R.A. Fisher

1

Introduction

Contents

The term 'Statistics' has been derived from the Latin word 'STATUS' which means political states. Initially statistics was concerned with economical, political and demographical particulars of a country. In the 17th century Prof. Gottfried Achenwall recognised statistics, a specialized branch of human knowledge. Slowly development of theory of probability added importance for statistics. The development of modern statistics started after industrial revolution. Now statistics has become a universally applicable science. In any field where ever information can be measured. Statistics acts as a required tool for further action.

1.1 DIMENSION OF STATISTICS

The word 'Statistics' is used in two senses.

1. In plural form, it stands for a numerical facts.

   Example In a population of 80,00,00,000 in a country 30,00,00,000 lives below poverty, 40,00,00,000 lives moderate life and 1,00,000 lives a luxurious life. This gives the statistics or the numerical facts of that country.

2. In singular form, it stands for the science of collection, organization analysis and interpretation of numerical facts.

1.2 DEFINITION OF STATISTICS IN TERMS OF NUMERICAL DATA

There are so many definitions, but this definition covers all aspects of statistics as a numerical facts.

By statistics we mean aggregate of facts affected to a marked extent by multiplicity of causes, numerically expressed, enumerated or estimated according to reasonable standards of accuracy, collected in a systematic manner for a pre-determined purpose and placed in relation to each other.

–Prof Horce Secrist

Definition of Statistics as a Science

The most suitable definition for statistics as science (methods) is given by Croxton and Cowden:

Croxton and Cowden Statistics or statistical methods may be defined as the collection, presentation, analysis and interpretation of numerical data.

1.3 FUNCTIONS OF STATISTICS

The functions of statistics are numerous, that is why statistics is used in almost all fields. The important functions of statistics are as follows:

1. *It simplifies complex data* Data presented in raw form are unwieldy and complex, these complex mass of data are made simple and understandable with the help of statistical methods. This is done with the help of diagrams and averages etc.
2. *It facilitates comparison* With the help of statistical methods it is easy to compare between data of two different products, periods, places etc.
3. *It facilitates the study relationship between different facts* Statistics facilities in investigating the existing relationship between two or more phenomena. For example, we can study the relationship between demand and supply, price level and demand etc.
4. *It enhances an individual's knowledge* Statistics helps one to analyse the reason for variation in data. statistics sharpens the faculty of rational thinking and reasoning thus enlarges individuals knowledge:

   For example, When the production data of company for past two or more years are presented. One may be able to do the comparative study and will be able to observe the variations in data, so one starts to analyse the reason for the variations.
5. *It facilitates in formulating policies* Vital statistics and demography studies helps government, non-government bodies to frame various policies.

6. *It tests the hypothesis* Statistics are useful in the field of research. It helps to study the effects of an experiment and its significance.
7. *It facilitates forecasting* Statistical methods are not only helpful in estimating the present but also in forecasting the future. Time series, regression analysis are used in forecasting.

1.4 SCOPE OF STATISTICS

The application of statistical techniques is so wide spread, it is used in almost all fields. Statistics is considered to be a distinct branch of study applicable in many other branch of study.

Statistics in Business

In the field of business, statistics plays a vital role from its germinal stages to its each stage of progress. It helps a new business entrepreneur to take decision regarding whether to start a new business. This is done with the help of statistical market survey. As he innovated his business environment he starts a production, here statistical quality control, helps him in assuring the quality of his product. The company can also calculate the cost of production and selling price. The company makes a comparative study about its performance with other companies. Then the company uses statistics tool in decision making process to improve the business and to make profit. Thus statistics has vital role to play in the field of business.

Statistics in Economics

Statistical data are a powerful aid in economical analysis. Prof. A. Marshall states statistics are the straw out of which I, like every other economist have to make bricks.

Statistics are useful in the study of demand-supply analysis. Statistics are the very foundation stone of the theory of exchange. It is applicable in almost all areas of economics. Thus statistics are inevitable tools for economics.

Statistics in Banking and Finance

Banks use statistics in maintaining the data regarding customer deposits, loan, it uses statistical techniques to calculate the interest. Banks uses statistical methods to keep track of its performance. Stock exchanges, financial institutions use statistical methods for future projection, to planning policies etc.

1.5 LIMITATIONS OF STATISTICS

As no branch of science is perfect in itself statistics also has its own limitation in certain aspects. The limitations can be stated as follows.

1. Statistics deals with quantitative data only.

Qualitative facts are converted into quantitative measures and conclusions are made on that basis.

For example, Intelligence of students are calculated with the help of marks scored by them in examinations.

2. Statistics is true on an average only.
3. It does not deal with an individual but with the masses.
4. Statistics can be misused if it is handled by an in experienced person.
5. Statistical results are correct only in a general sense.

1.6 DISTRUST OF STATISTICS

The misuse of Statistical tools by biased people has lead to the distrust of Statistics. Statistics is distrusted as the figures put forward for arguments may be inaccurate or incomplete, which leads to wrong conclusion.

Here are a few interesting comments on Statistics.

1. "An ounce of truth will produce tons of Statistics".
2. "Statistics can prove anything"
3. "Figures do not lie liars figure".
4. "There are three types of lies,—lies, damned lies and Statistics, wicked in the order of their naming, and so on.

These misinterpretations happen as figures are innocent, easily understood and convincing. The facts supported by figures will have more psychological appeal, hence the wrong figures given by people for their own selfish need, causes distrust of Statistics.

Hence Statistics is a tool which if rightly used may prove extremely useful and if misused, might be disastrous. It is not the tool that has to be blamed but the person who misuses the tool is to be blamed, hence it is not Statistics that has to be blamed. Thus the distrust in Statistics can be minimized by skillful handling of the tool by an expert.

Few illustrations regarding the misinterpretation of Statistical data:

1. A Statistical report: City A has 80 tmc of water resources. City B has 60 tmc of water resource. Hence City A has more water resources. This conclusion is fallacious because the vital information regarding area of land in which this water is used is not taken into account.
2. A report: "The number of traffic accidents is lower on foggy weather than in clear weather days. Hence it is safer to drive in fog".

This conclusion is wrong since we are not given the proportion of the number of accidents to the number of persons travelling in the two cases.

Exercises

1. Define statistics.
2. Mention a few functions of statistics.
3. State the scope of statistics in business field.
4. State the limitations of statistics.
5. Briefly explain the fact 'Distrust of statistics'.

2

Organization of Data

Contents

In statistics, there may arise problems of differing natures, which may require different methods and techniques to find the right solution. Irrespective of the methods and the techniques adopted, the steps leading to the solution will be more or less same in almost all cases. They are the following.

1. Formulation of the problem
2. Experimentation
3. Tabulation
4. Statistical inference.

In this chapter we deal only with collection, classification and tabulation of data, which serves as the basic steps to solve a given problem.

2.1 COLLECTION OF DATA

Collection of data means the methods that are to be employed for getting the required information for the given problem. As the saying goes: if the foundation is strong, the building is also strong. Similarly in statistics the collection of data plays a vital role. The data are collected by a statistician or investigator based on the formulation of the problem, that is, it depends upon the nature and scope of statistical enquiry.

Depending upon the nature of the problem a statistician selects a primary or a secondary data.

2.2 PRIMARY AND SECONDARY DATA

The data collected directly by an investigator for studying a particular problem are called 'Primary Data'. The data collected through censuses and surveys provide primary data for the investigator.

Secondary data are those which have already been collected and analysed by some other agency. The chief source of secondary data are published or unpublished sources.

2.3 COLLECTION OF PRIMARY DATA AND SECONDARY DATA

(a) Primary Data:

We have seen that data which is collected directly by an investigator is known as primary data. The investigator collects the data by employing one of these methods.

1. Direct personal investigation
2. Indirect oral investigation
3. Information through correspondents
4. Schedules
5. Questionnaire

The best method to collect all the details necessary for the given problem under study is through questionnaire.

Questionnaire

Questionnaire is a list of questions, which is filled by the informants/respondents.

Qualities of a good Questionnaire

Depending upon the scope and objective of the given problem the investigator has to draft the questionnaire with utmost care so that all the relevant and essential information can be collected without difficulties. While drafting a questionnaire the investigator has to borne the following points.

1. The size of the questionnaire should be as small as possible.
2. It should be simple, clear and unambiguous.
3. Questions should be arranged in a logical sequences.
4. Personal and sensitive questions should be avoided.
5. Questions should be framed to facilitate the respondent by giving them the questions such as

   (a) Multiple choice, (b) Tick-options, (c) True or false etc.
6. Questions also can be framed to give an opportunity to express the independent ideas of the respondents.
7. Questions should be framed in conformity with the mental capacity, educational background and social status of the respondents.

8. It is important to include cross-check questions to check the varsity of the information given by the respondents.
9. To get accurate data, it is advisable to include questions to evaluate the socio-cultural status of the respondents.

(b) Secondary Data:

The data taken from other sources are called 'secondary data'.

The main source of secondary data

Published source | Unpublished source

Published sources

Census reported reports published by certain government, non-government organisation and international organisation for: To create the AIDS awareness programme, certain NGO worked and displayed the statistical report of area and number of people affected by it. This published source becomes secondary data for other social workers.

Unpublished sources

Certain prime companies, collect data for their business requirement. These information are private and not published. Certain organisations have unpublished data.

2.4 CLASSIFICATION OF DATA

Classification is a systematic grouping of the units according to their common characteristics.

For example: Given below is data of 25 observation of income of employees in an organization.

250, 1500, 500, 1000, 750, 750, 250, 1000, 1500, 250, 250, 500, 500, 750, 1000, 250, 1500, 750, 250, 500, 1000, 1000, 750, 250, 250

The given data can be classified as follows:

Income (Rs)	*Number of Persons*
250	8
500	4
750	5
1000	5
1250	0
1500	3

2.5 FUNCTIONS OF CLASSIFICATION

1. It reduces the bulk of the data that is, it condenses the data.
2. Simplifies the data.
3. Facilitate comparison.
4. Facilitate in further statistical analysis.
5. To find out mutual relationship between certain measures and their effects.
6. It provides a basis for tabulation.

2.6 TYPES OF CLASSIFICATION

There are four modes of classification of the data. They are:

1. Quantitative Classification (Variable)
2. Qualitative Classification (Attribute)
3. Spatial Classification (Geographical)
4. Temporal Classification (Chronological)

2.7 DIFFERENT KINDS OF CLASSIFICATION

1. Classification according to attributes.
2. Simple or two fold (dichotomous) classification.
3. Multiple classification.
4. Quantitative classification.

2.8 FORMATION OF STATISTICAL SERIES

In the problem of organising the data pertaining to a quantitative phenomenon, the following four series are considered:

1. Individual series
2. Discrete series
3. Grouped series
4. Continuous series

1. Individual Series The data as observed by an investigator is known as 'individual series'. For example: Marks stored by 10 students:

100, 90, 72, 59, 40, 20, 23, 75, 61, 43

2. Discrete Series When the data collected are large in number and they are distinct then we can form a frequency distribution (the number of units associated with each value of the variable). While constructing the discrete frequency distribution, the occurrence of the variable is counted with the help of vertical bars known as tally marks

For example: The following data represents size of shoes which were sold in a showroom on a particular day:

1, 3, 10, 9, 9, 7, 4, 5, 3, 2, 7, 8, 6, 9, 10, 6, 4, 3, 2, 8, 3, 10, 7, 3, 2, 10, 7, 3, 4, 5

Size	*Tally mark*	*Number of Shoes*
1	1	1
2	111	3
3	~~1111~~ 1	6
4	111	3
5	11	2
6	11	2
7	1111	4
8	11	2
9	111	3
10	1111	4

3. Grouped Series If the data are large in number, then the discrete series can be further reduced down by grouping them with the help of class intervals. That is, collecting data which will fall within two small limits.

For example: The following are the percentage marks of 50 students in an examination:

39, 39, 43, 62, 30, 47, 33, 46, 17, 38, 10, 36, 29, 40, 32, 24, 57, 42, 15, 30, 39, 53, 47, 64, 31, 07, 37, 47, 27, 43, 36, 39, 36, 25, 34, 30, 50, 76, 42, 43, 56, 40, 29, 22, 30, 65, 32, 36, 39, 70.

The data can be classified as follows: Here as the minimum value is 10 and the maximum value is 76. We form class interval as follows.

Class intervals (marks)	*Tally marks*	*Number of Students*
10–20	1111	4
20–30	~~1111~~	5
30–40	~~1111~~ ~~1111~~ ~~1111~~ ~~1111~~ 1	21
40–50	~~1111~~ ~~1111~~ 1	11
50–60	1111	4
60–70	111	3
70–80	11	2
	Total	50

4. Continuous Series There are certain variables which will not have a distinct integer value. For example: weight, height, etc. In those cases the data will be of continuous type and grouping of those continuous data is known as 'continuous series'.

For example: The following are the heights in centimeters of 40 students:

155.0, 163.3, 155.0, 169.9, 163.9, 148.2, 151.5, 128.9, 146.1, 145.2, 156.6, 159.9, 158.0, 153.4, 160.0, 149.3, 154.2, 164.7, 159.0, 160.8, 157.1, 176.0, 164.4, 170.0, 152.9, 169.3, 159.5, 162.4, 173.6, 154.8, 158.3, 172.5, 158.0, 154.0, 172.2, 148.0, 152.4, 156.6, 157.7, 165.3.

Heights	*Tally marks*	*Number of Students*
145–150	~~1111~~	5
150–155	~~1111~~ 11	7
155–160	~~1111~~ ~~1111~~ 11	12
160–165	~~1111~~ 11	7
165–170	111	3
170–175	1111	4
175–180	11	2
	Total	40

In case of both grouped series and continuous series the class width depends on number of classes. To find the width of the classes we use the formula:

$$i = \frac{L - S}{K}$$

Where i is the class width

L is the largest value of the given data

S is the smallest value of the given data

K is the number of classes

2.9 FREQUENCY DISTRIBUTION

A quantitative characteristic which varies from unit to unit is a variable. For example: height, weight. The number of units associated with each value of the variable is the frequency of the value. A systematic presentation of values taken by a variable and the corresponding frequencies is called frequency distribution of that variable.

A frequency distribution in which class intervals are considered is known as continuous frequency distribution. If class intervals are not considered it is a discrete frequency distribution.

For example: Let us consider for a set of data:

20, 30, 31, 35, 36, 32, 31, 33, 37, 39, 40, 41, 34, 35, 29, 25, 27, 30, 25, 23

Class interval	*Tally mark*	*Number of observation*
20–25	11	2
25–30	~~1111~~	5
30–35	~~1111~~ 11	7
35–40	1111	4
40–45	11	2

Let us see the various terms that are involved in the frequency distribution. If the range is vast in a frequency distribution then it is divided into mutually exclusive sub ranges called *class intervals*. Each class interval is specified by two limits which are called *class limits*.

The difference between the class limits of a class intervals is known as the *width* of the class interval. The mid point or the central value of a class interval is called *class mid value* or *class mark*.

In a frequency distribution, the class intervals may be inclusive or exclusive. If a class interval is such that the lower as well as the upper class limits are included in the same class interval it is inclusive *class interval*. If a class interval is such that the lower limit is included in the first class interval where as the upper class limit is included in the successive class interval it is *exclusive class* interval.

For example: 10, 25, 20, 12, 30, 46, 19, 15, 17, 20, 25.

Inclusive class interval		*Exclusive class interval*	
10–19	~~1111~~	10–20	~~1111~~
20–29	1111	20–30	1111
30–39	1	30–40	1
40–49	1	40–50	1

In inclusive class interval '20' will be included in 20–29 and in exclusive class interval '20' will be included in 20–30 instead of 10–20 class interval.

Thus classification helps simplifying and condensing the bulk data and also forms the basis for tabulation. Thus it helps for further statistical analysing.

Solved Examples

1. Classify the given paragraph in terms of the number of letters in each word by taking the class interval of 1–3, 4–6, 7–9, etc.

The extraordinary general meeting may be convened by the Board of Directors on the written requisition by members holding not less than one tenth of the total voting rights in the company, where it has no share capital or one tenth of the paid up capital of the company. The Board of Directors must arrange to convene a meeting to consider the matters referred to in the requisition with a period of twenty one days. In any case the meeting must be convened by the Directors on a date not later than forty five days from the date of deposit of the requisition.

(B.U., April, 1995)

Solution Frequency distribution of number of letters.

Class intervals	*Tally marks*	*Number of words*
1–3	~~1111~~ ~~1111~~ ~~1111~~ ~~1111~~ ~~1111~~ ~~1111~~ ~~1111~~ ~~1111~~ ~~1111~~ ~~1111~~	50
4–6	~~1111~~ ~~1111~~ ~~1111~~ ~~1111~~ ~~1111~~ 1	26
7–9	~~1111~~ ~~1111~~ ~~1111~~ ~~1111~~ 111	23
10–12	111	3
13–15	1	1
		Total 103

2. A psychologist estimates the I.Q of 28 students. The values are as follows:

103, 86, 94, 97, 100, 113, 102, 76, 95, 101, 99, 83, 94, 84, 78, 102, 105, 114, 88, 84, 90, 100, 96, 98, 78, 96, 79.

Form a frequency distribution with class intervals of width 5 each.

Solution

Class interval	*Tally marks*	*Frequency*
75–80	1111	4
80–85	11	2
85–90	111	3
90–95	111	3
95–100	~~1111~~ 11	7
100–105	~~1111~~ 1	6
105–110	1	1
110–115	11	2
		Total 28

3. A Geneticist observes a hybrid variety of beans. He counts the number of seeds in each of 50 bean pods. The data is as follows.

Number of seeds per bean pod:

4, 2, 3, 5, 3, 2, 1, 3, 1, 4, 5, 2, 2, 4, 7, 5, 3, 2, 3, 1, 5, 6, 4, 3, 4, 2, 5, 3, 5, 9, 2, 1, 3, 3, 2, 1, 4, 5, 2, 2, 4, 3, 2, 8, 4, 5, 6, 3, 2, 3.

Form a suitably frequency table.

Solution

Number of seeds	*Tally marks*	*Number of beans*
1	~~1111~~	5
2	~~1111~~ ~~1111~~ 11	12
3	~~1111~~ ~~1111~~ 11	12
4	~~1111~~ 111	8
5	~~1111~~ 111	8
6	11	2
7	1	1
8	1	1
9	1	1
	Total	50

4. The following data regarding weight of newborn babies are obtaining from the office records of a hospital obtain a frequency distribution.

Weight (kgs): 4.3, 3.3, 3.5, 4.0, 2.4, 3.0, 3.1, 2.7, 2.8, 2.7, 3.0, 3.3, 3.7, 2.2, 3.0, 2.5, 2.9, 3.0, 4.3, 3.0, 3.2, 3.3, 3.6, 2.8, 3.9, 3.1, 3.3, 3.7, 2.6, 1.2, 4.8, 2.7

Solution

Weight	*Tally mark*	*Number of Babies*
1–2	1	1
2–3	~~1111~~ ~~1111~~	10
3–4	~~1111~~ ~~1111~~ ~~1111~~ 11	17
4–5	1111	4
	Total	32

5. The following are the weights of 30 students. Draw up a frequency distribution with class intervals 40–44, 45–49, 50–54, ... kgs.

Weights (kgs): 51, 47, 50, 54, 62, 52, 42, 49, 52, 49, 44, 50, 53, 58, 46, 50, 51, 53, 48, 50, 55, 52, 55, 58, 63, 54, 52, 49, 50, 58.

Solution

Class interval	*Tally mark*	*Number of Students*
40–44	11	2
45–49	~~1111~~ 1	6
50–54	~~1111~~ ~~1111~~ ~~1111~~	15
55–59	~~1111~~	5
60–64	11	2
	Total	30

2.10 TABULATION OF DATA

Tabulation is the process of presenting data collected through survey experiment or records in rows and columns. So that it can be more easily understood and can be used for further statistical analysis.

2.11 OBJECTIVE OF TABULATION

1. To summarize the data systematically and present it in a minimum space.
2. To classify the data in a simple form.
3. To facilitate the data for a comparative study.
4. To facilitate analysis of data.

2.12 CLASSIFICATION AND TABULATION

The process of classification and tabulation is interacted. Classification is meant for arranging the data into groups where each group has the number of items attached to it. Whereas tabulation is the logical and systematic arrangement of data in rows and columns. Classification lay a basis for tabulation and tabulations are used for further statistical analysis.

2.13 THE PARTS OF A TABLE

A statistical table is an orderly arrangement of data in columns and rows. Following are the different parts of table.

1. *Table number*: Table number is an address of the table used in a report. This identification number is generally given at the top of each table for convenience sake.
2. *Title*: As the table number gives the address of the table, the title speaks of its contents.
3. *Caption and stubs*: Captions refers to the heading of the information shown at the top of the vertical column where as stubs refers to the heading of the information shown at the extreme left of the horizontal rows.
4. *Body*: The body is the most important part of the table which describes the entire data of the information that are presented by the way of caption and stubs.
5. *Head note*: It explains some of the important facts relating to the statistical units, which is written as a supplementary note to the title. It is preferably placed in brackets.
6. *Foot note*: Foot notes, given at the bottom of the table clarifies or explains the marked parts of the table. Foot notes are usually indicated by asterisk. (*)

7. *Source note*: It is relevant in case of quoting of secondary data. Source note generally gives the information of the source from were the data is collected.

Title

For example

Results of Degree students University of Bangalore, 2000 as follows:

(*Head note*: Marks in percentage)

Stubs   Caption

Degree / *Year*	*B.SC*		*B.BM*		*B.HM*	
	Pass	*Fail*	*Pass*	*Fail*	*Pass*	*Fail*
I Yrs	60*	40	75*	25	61*	39
II Yrs	50*	50	45*	65	45*	65
III Yrs	70*	30	76*	24	67*	33

body

Foot note: * given data

Source note: Bangalore University: Examination department

2.14 TYPES OF TABLES

1. One-way table (simple table)

The simple table provides the information of a single characteristics of the data.

Example (According to Age group).

Number of students participating in the zonal meet' 2000

Age (in years)	*Number of students*
10–13	20
13–15	25
15–17	15
Total	60

2. Two-way table

They provide the information about two characteristics of the data.

Example (According to Age and Sex of students).
Number of students participating in the zonal meet 2000.

Sex / *Age in years*	*Number of students*		
	Male	*Female*	*Total*
10–13	9	11	20
13–15	15	10	25
15–17	10	5	15
Total	34	26	60

3. Three-way table

These tables provide the information about three characteristics of the data.

Example (According to Age, Sex and events).
Number of students participating in the zonal meet 2000.

Sex/ events / *Age in years*	*Number of students*						
	Male			*Female*			
	Track event	*Field event*	*Total*	*Track event*	*Field event*	*Total*	*Grand total*
10–13	5	4	9	6	5	11	20
13–15	10	5	15	4	6	10	25
15–17	7	3	10	3	2	5	15
Total	22	12	34	13	13	26	60

4. Manifold tables

These tables provide information about a number of characteristics of the same data. They are also called as complex tables.

Generally three-way tables are used as manifold tables are more complicated for presentation and for understanding.

For e.g.: (According to age, sex, events and academic groups).
Number of students participating at zonal meet 2000.

Sex/ events / *Age/ group*		*Number of students*						
		Male			*Female*			
		Track event	*Field event*	*Total*	*Track event*	*Field event*	*Total*	*Grand total*
10–13	Arts	2	2	4	3	2	5	9
	Science	3	2	5	3	3	6	11

Contd.

Age/group	*Sex/events*	*Number of students*						
		Male			*Female*			
		Track event	*Field event*	*Total*	*Track event*	*Field event*	*Total*	*Grand total*
13–15	Arts	8	3	11	2	5	7	18
	Science	2	2	4	2	1	3	7
15–17	Arts	4	3	7	3	2	5	12
	Science	3	0	3	0	0	0	3
Total		22	12	34	13	13	26	60

Worked Examples

1. Draw up a blank table in which could be entered the number of work people employed in four industries at two dates. Distinguish male workers from female workers and subdivide the female worker in to single, married and widowed. ***(B.U., November, 1994)***

Solution

Number of employees in four different industries in two given dates.

Table—ABC

Date/Ind.	*Sex*	*Male*	*Female*				*Grand Total*
			Single	*Married*	*Widow*	*Total*	
Date 1	A						
	B						
	C						
	D						
Date 2	A						
	B						
	C						
	D						
Total							

2. In 1975 out of 1750 workers of a factory 1200 were members of a trade union. The number of women employed were 200 out of which 175 did not belong to any trade union. In 1980 the number of union workers increased to 1580 of which 1290 were men. In the other hand the number of non-union workers fell to 208 of which 180 were men. In 1985, there were 1800 employees who belonged to the trade union and 50 who did not belong to any trade union of all the employee in 1985, 300 were women of them only 8 did not belong to the trade union. Tabulate the above information giving suitable title, table number etc. ***(B.U., April, 1997)***

Solution

Year 1975	Total workers:	1750
	Trade union members:	1200
	∴ Non-trade union members:	550
	Number of women:	200
	Non-trade union women	175
	∴ Trade union women members	25
	Men trade union members:	1200 – 25 = 1175
	Men non-trade union members	550 – 175 = 375
Year 1980		
	Number union members:	1580
	Men members:	1290
	∴ Women union members:	1580 – 1290 = 290
	Number of non-union members	208
	Men	180
	∴ Women non-union member	208 – 180 = 28
Year 1985		
	Trade union members:	1800
	Non-trade union members:	50
	Number of women workers:	300
	Non-union women members	8
	∴ Union women members	300 – 8 = 292
	Men union members:	1800 – 292 = 1508
	Men non-union members	50 – 8 = 42

As we know all the data now, we will tabulate this data.

Title:

Table No. A1 Head note: Actual fig.

Year/sex / *Member*	*1975*			*1980*			*1985*		
	Men	*Women*	*Total*	*Men*	*Women*	*Total*	*Men*	*Women*	*Total*
Trade union member	1175	25	1200*	1290*	290	1580*	1508	292	180*
Non-Trade union member	375	175*	550	180*	28	208*	42	8*	50*
Total	1550	200*	1750*	1470	318	1788	1550	300*	1850

Foot note: * given date

Source:

3. Prepare a blank table to show the exports of three companies A, B, and C to the five countries, the U.K, the U.S.A, Turkey, France and Germany in each of the years 1995 and 1996.

(B.U., November, 1997)

Title:

Table No.

Year/Company \ Countries		UK	USA	Turkey	France	Germany	Grand Total
1995	A						
	B						
1996	A						
	B						
Total							

4. In a sample study about coffee habit in two towns, the following information was received:
Town A: Females were 40% Total coffee drinkers were 46% and male non-coffee drinkers were 20%.

Town B: Males were 55%. Male non-coffee drinkers were 30% and female coffee drinkers were 15%.

Represent the above data in a tabular form. ***(B.U., Oct./Nov., 2001)***

Solution

Let us assume the totals of town A and town B as 100.

Town A: Total = 100

Females (40%) = 40

Males = Total – Females

Males = 100 – 40 = 60

Coffee drinkers = 46

∴ Non-coffee drinkers = 100 – 46 = 54

Given that male non-coffee drinkers is 20% = 12

∴ Male coffee drinkers = Total males – non-coffee drinkers

= 60 – 12 = 48

Female coffee drinkers = Total coffee drinkers – male drinkers

= 46 – 40 = 6

Female non-coffee drinkers = Total female – female drinkers

= 40 – 6 = 34

Town B: Let the total population be 100 given males is 55%

Male = 55 person

∴ No. of female = total – No. of males

= 100 – 55 = 45

Given: 30% males non-coffee drinkers

$$= \frac{30}{100} \times 55 \simeq 17$$

Male coffee drinkers = total male – non-coffee drinkers

= 55 – 17 = 38.

Give that 15% of females are coffee drinkers

$$= \frac{15}{100} \times 45$$
$$= 7$$

No. of female non-coffee drinkers = 45 – 7 ≃ 38.

Now let us tabulate the above data.

Drinking habits of people in two towns A, B

Head note: figures are % of persons

Drinking habits \ Sex	Town A			Town B			Grand Total
	Male	Female	Total	Male	Female	Total	
Coffee drinkers	40	6	46*	38	7	45*	91
Non-coffee drinkers	20*	34	54	17*	38	55	109
	60	40*	100	55*	45	100	200

* given data.

5. In a sample study about the smoking habits in two towns following data were derived.

	Town A	*Town B*
Males in total population	52%	54%
Smokers	26%	28%
Male smokers	18%	20%

Tabulate the above data **(*B.U., April, 1998*)**

Solution

Town A:

Let the total population be 100

Males 52%

Total no of male is 52

∴ Total No. of females = Total – No. of Males

= 100 – 52 = 48

Total smokers is 26% = 26

$$\text{Male smokers is } 18\% = \frac{18}{100} \times 52$$
$$= 9$$

∴ Female smokers = Total smokers – Male smokers

= 26 – 9

= 17

Town B:

Let the total population be 100

Male is 54%

∴ No. of males is 54

Females = Total – No. of males

= 100 – 54

= 46

Total smokers is 28%

No. of smokers = 28

Given male smoker is 20%

$$\text{No. of male smokers} = \frac{20}{100} \times 54$$

= 11

∴ Female smokers = 28 – 11

= 17

Now let us tabulate the data:

Smoking habits in two towns

Sl:

Smokers habits \ Sex	Town A			Town B			Grand Total
	Male	Female	Total	Male	Female	Total	
Smokers	9*	17	26*	11*	17	28*	54
Non-smokers	52 – 9 = 43	48 – 17 = 31	74	54 – 11 = 43	46 – 17 = 29	72	146
Total	52*	48	100	54*	46	100	200

* given data.

6. In the house of Lok Sabha there were 540 members present. During the discussion on a non confidence motion put to vote, 350 voted against the motion. Ruling party members were 325. 180 members belonging to the opposition voted for it. All the members are belonging to either of the two groups and there were no absentees. Tabulate the information.

Voters choice \ Party	Ruling party	Opposite party	Grand total
For	10	180*	540 – 350 = 190
Against	315	350 – 315 = 35	350*
Total	325*	215	540*

* given data.

7. Convert the given ratios into corresponding numbers and prepare a complete table for the following information.

In the year 2000, the total strength of three colleges X, Y, Z in a city was in the ratio of 3:2:5. The strength of the college Y was 1500. The proportion of girls and boys in all the colleges was in the ratio 2:3. The faculty was distribution of girls and boys in the faculty of Arts and Science was in the ratio 1:2. Tabulate the information.

Solution

Strength of College Y is 1500*

$\therefore$ Strength of College X is $3 \times 1500 = 4500$

and Strength of College Z is $5 \times 1500 = 7500$

Calculation:

		College Y	College X	College Z
Boys		$\frac{3}{5} \times 1500 = 900$	$\frac{3}{5} \times 4500 = 2700$	$\frac{3}{5} \times 7500 = 4500$
Girls		$\frac{2}{5} \times 1500 = 600$	$\frac{2}{5} \times 4500 = 1800$	$\frac{2}{5} \times 7500 = 3000$
Boys:	Science	$\frac{2}{3} \times 900 = 600$	$\frac{2}{3} \times 2700 = 1800$	$\frac{2}{3} \times 4500 = 3000$
	Arts	$\frac{1}{3} \times 900 = 300$	$\frac{1}{3} \times 2700 = 900$	$\frac{1}{3} \times 4500 = 1500$
Girls:	Science	$\frac{2}{3} \times 600 = 400$	$\frac{2}{3} \times 1800 = 1200$	$\frac{2}{3} \times 3000 = 2000$
	Arts	$\frac{1}{3} \times 600 = 200$	$\frac{1}{3} \times 1800 = 600$	$\frac{1}{3} \times 3000 = 1000$

Let's tabulate the data.

Distribution of students in three colleges into science and arts faculty.

Sex / *Faculties*	*College Y*			*College X*			*College Z*		
	Boys	*Girls*	*Total*	*Boys*	*Girls*	*Total*	*Boys*	*Girls*	*Total*
Arts	300	200	500	900	600	1500	1500	1000	2500
Science	600	400	1000	1800	1200	3000	3000	2000	5000
Total	900	600	1500*	2700	1800	4500	4500	3000	7500

* given data.

Exercises

1. What are primary and secondary data?
2. What is meant by questionnaire?
3. What do you mean by classification?
4. What are the functions of classification?
5. What do you mean by class intervals?

6. Define class mark, class width.
7. What do you mean by tabulation?
8. What are the function of tabulation?
9. What are the different parts of statistical tables?
10. Explain any three parts of statistical tables.
11. In a sample study about the literacy rate in two cities following data was derived.

	Town A	*Town B*
Female population	60%	55%
Literacy rate	53%	72%
Female literates	60%	70%

Tabulate the above data.

12. In April 2000 examination of Bangalore University 5000 appeared for the BBM examination of which 70% were from Bangalore city college and the rest from rural college. Of the total students appearing for the examination 40% were girls and the remain boys. The results of the examination revealed a total pass of 50%. The ratio of boys and girls passing in the examination was 2:3.
13. In a sample study about the drug addicts in two cities following data was derived.

	Town A	*Town B*
College students	35%	25%
School drug addicts	20%	15%

(total population consists of school going and college students)

Tabulate the data.

14. Present the following information in the form of a table. In the year 1980 there were 200 banks out of which 50 were private banks. In this fifty banks 10 had ATM facility and non of the nationalized banks had this facility. In the year 2000 there were an increase of 50 more banks in the private sector and total number of banks is 275. 20 nationalized bank had ATM facility and 75 private sector has this facility.

3

Diagrammatical and Graphical Representation of Data

Contents

Key Points

1. For drawing an ogive or histogram of the class interval is of inclusive type convert them to exclusive type.
2. While drawing an histogram check whether the class intervals are at regular intervals.

INTRODUCTION

Diagrammatic and graphical presentation of data enables one to grasp the data at a glance. Thus the data that are presented in tabular form, when represented with the aid of diagram or graphs makes it clear and vivid. These types of representation has the following advantages and limitations.

3.1 ADVANTAGES

1. These forms of representation gives a bird eye view of complex data.
2. They have long lasting impression.
3. These representation enables even a common man to understand facts easily.
4. They save time and labour.
5. They facilitate comparison.

3.2 LIMITATIONS

1. Graphs and diagrams are visual aids, they cannot be considered as alternative for numerical data.
2. Diagrams are not subjected to further mathematical analysis.
3. Diagrammatical or graphical misrepresentations can mislead observers very easily.

3.3 TYPES OF DIAGRAM

There are various forms of presenting the data diagrammatically. The following are the common methods of these representations.

I. One dimensional

1. Simple bar diagram
2. Multiple bar diagram
3. Sub-divided or component bar diagram.
4. Percentage bar diagram.
5. Duo-directional bar diagram.
6. Paired bar diagram.

II. Two dimensional

1. Rectangles
2. Circles.
3. Pie diagram.

III. Three dimensional

1. Cubes
2. Cylinders
3. Spheres.

IV. Non-dimensional

1. Pictogram
2. Cartograms.

One Dimensional Diagram

1. Bar diagrams

A bar diagram represents the magnitude of a single factor according to time periods, places, items etc.

Example 1 The following table represents the No. of deaths that occurred in earth quake for the past five years.

Years:	1997	1998	1999	2000	2001
No. of deaths	4000	7000	25000	12000	35000

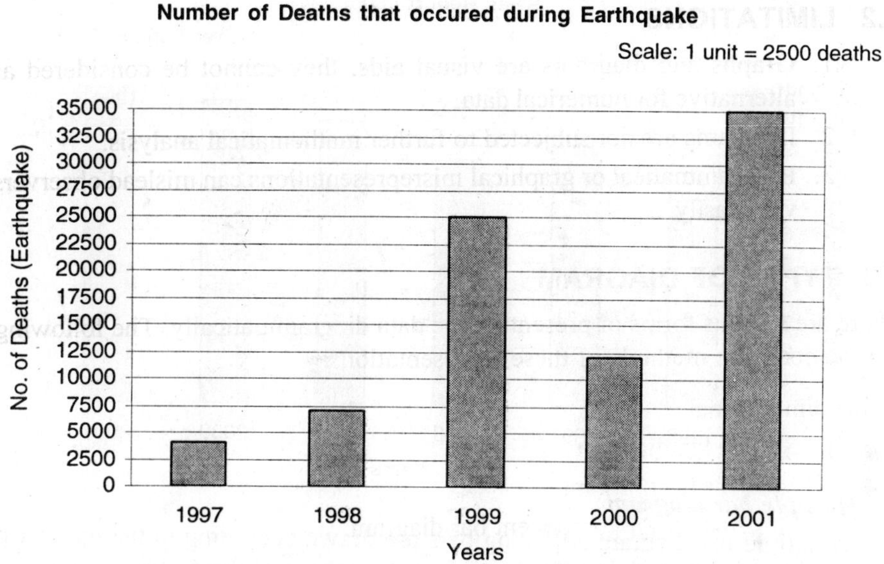

Example 2 The following table represents the number of road accidents in Bangalore.

Years:	1995	1996	1997	1998	1999	2000
No. of accidents	2200	3300	4400	4400	5400	7300

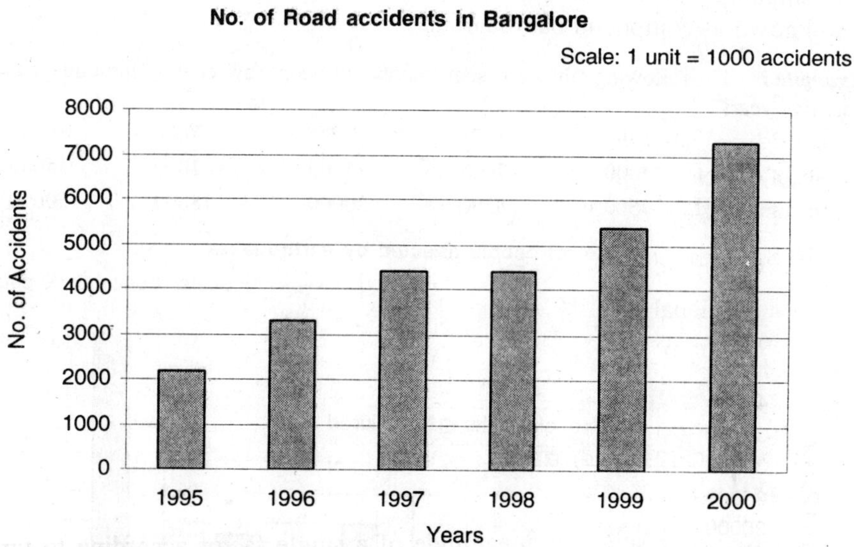

Example 3 Draw bar diagram for the following data.

Years	1996	1997	1998	1999	2000	2001
Scale ('000)	75	90	85	98	105	109

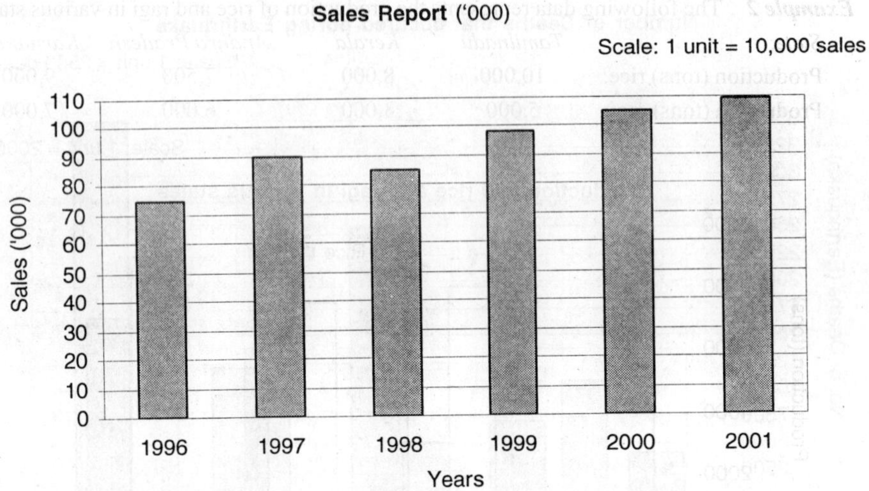

2. Multiple bar diagram

In a multiple bar diagram adjoining bars are drawn according to the number of factors and their heights in proportions to the values of the factors in the same order for each period or places. A constant distance is maintained between graphs of bars drawn for period or places.

Multiple bar diagram is preferred when to depict a number of related factors for comparison in various years or for number of places. Multiple bar diagram is also known as compound bar diagram.

Example 1 The following table represents number of people affected by earth quake for the past five years.

Years:	*1997*	*1998*	*1999*	*2000*	*2001*
No. of deaths	4000	7000	25000	12000	35000
No. of injured	7500	8500	20000	18000	50000

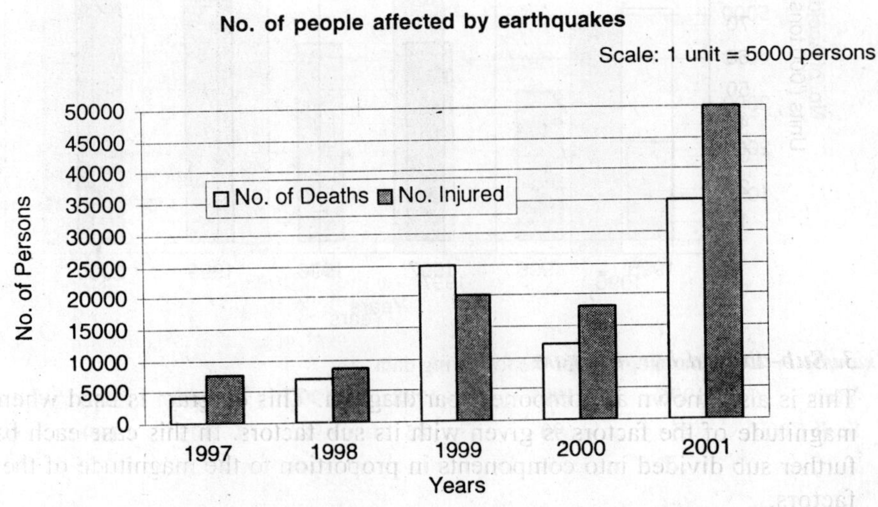

Example 2 The following data represents the production of rice and ragi in various states.

State:	*Tamilnadu*	*Kerala*	*Andhra Pradesh*	*Karnataka*
Production (tons) rice:	10,000	8,000	7,500	9,000
Production (tons) ragi:	5,000	3,000	6,000	7,000

Scale: 1 unit = 2000 tons

Production and rice and ragi in various states

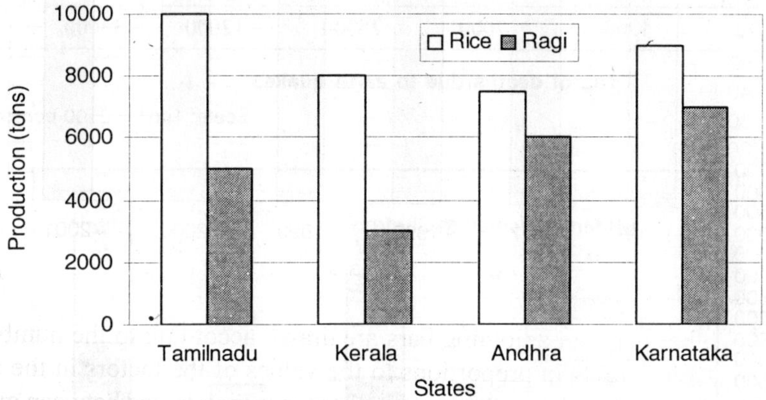

Example 3 Draw multiple bar diagram for the following data.

Years:	*1996*	*1997*	*1998*	*1999*
Production ('000 tons):	75	90	85	98
Sales ('000 tons):	70	93	80	93

Production and sales report ('000)

Scale: 1 unit = 10,000 tons

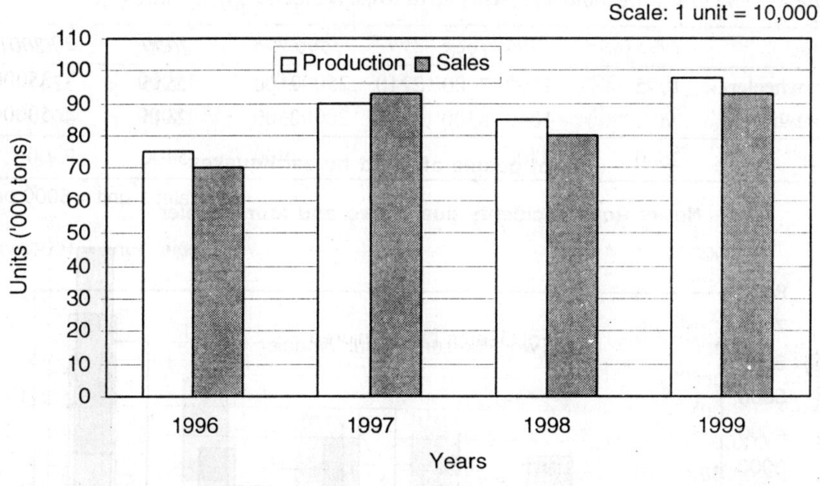

3. Sub-divided bar diagram

This is also known as component bar diagram. This diagram is used when the magnitude of the factors is given with its sub factors. In this case each bar is further sub divided into components in proportion to the magnitude of the sub factors.

Example 1 The following table represents No. of deaths due to earth quake for the past five years.

Years:	*1997*	*1998*	*1999*	*2000*	*2001*
No. of Men:	1700	3500	13000	4000	17000
No. of Women:	1500	3000	10000	6000	13000
No. of Children:	800	500	2000	2000	5000
Total	4000	7000	25000	12000	35000

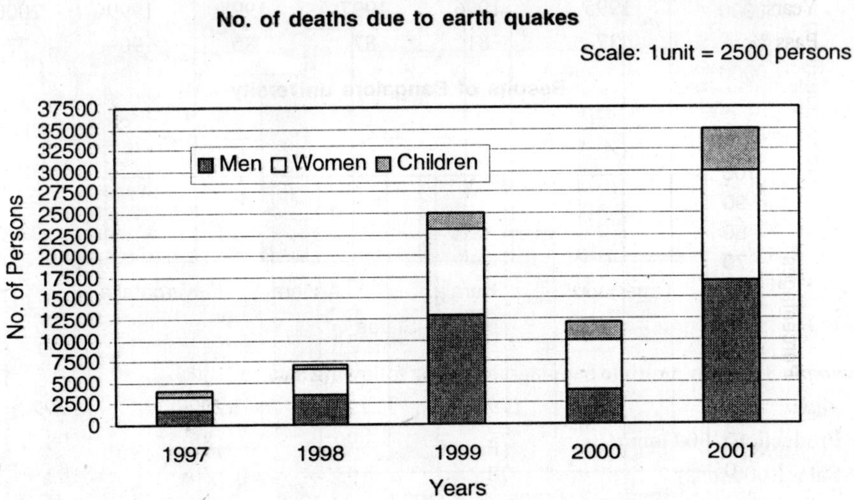

Example 2 The following table represents no of Road accident in Bangalore city.

Years:	*1995*	*1996*	*1997*	*1998*	*1999*	*2000*
Two wheeler:	1095	1460	2210	2100	2555	3285
Four wheeler:	905	1840	2190	2300	2845	4015
Total	2000	3300	4400	4400	5400	7300

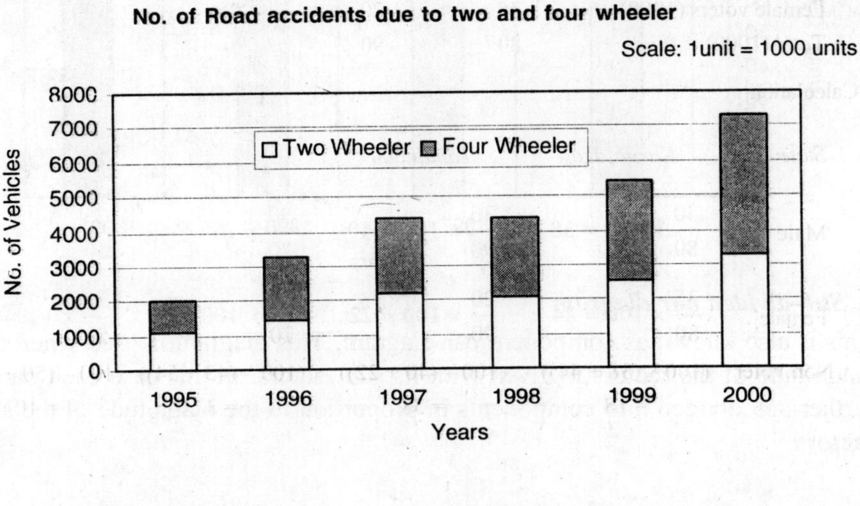

4. Percentage bar diagram

In case of subdivided bar diagram the actual value is represented, but here percentage is represented. In percentage bar diagram the magnitude of each bar is 100 and this bar is further subdivided into components in proportion to the percentage of the sub factors.

Example 1 The following table represents the pass percentage of students from Bangalore university.

Years:	1995	1996	1997	1998	1999	2000
Pass %:	37	81	87	85	90	77

Results of Bangalore university

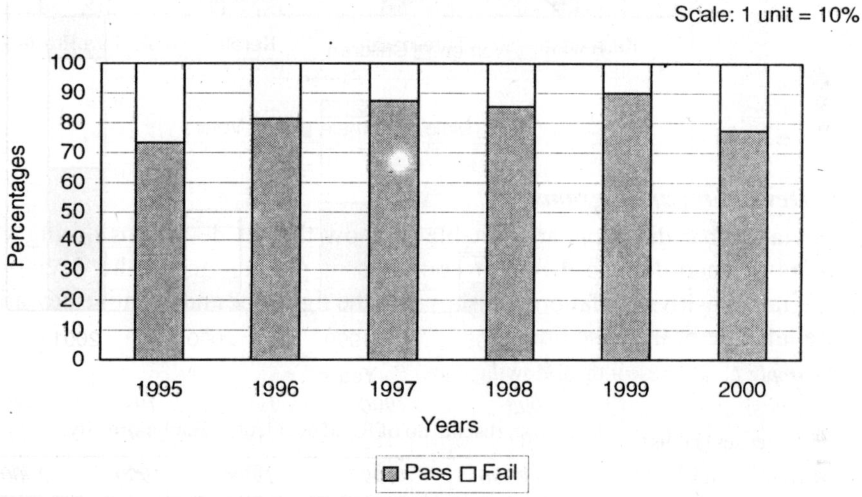

Example 2 Represented the following data using percentage bar diagram.

States:	*Karnataka*	*Tamilnadu*	*Kerala*	*Andhra Pradesh*
Male voters (1000)	30	36	30	25
Female voters (1000)	35	20	38	20
Total (1000)	80	90	70	50

Calculation

State	*Karnataka*	*Tamilnadu*	*Kerala*	*Andhra pradesh*
Male	$\frac{30}{80} \times 100 = 38$	$\frac{36}{90} \times 100 = 40$	$\frac{30}{70} \times 100 = 43$	$\frac{25}{50} \times 100 = 50$
Female	$\frac{35}{80} \times 100 = 44$	$\frac{20}{90} \times 100 = 22$	$\frac{38}{70} \times 100 = 54$	$\frac{20}{50} \times 100 = 40$
$\therefore$ Non voter	$(100-(38+44))$ $= 18$	$(100-(40+22))$ $= 38$	$(100-(43+54))$ $= 3$	$(100-(50+40))$ $= 10$

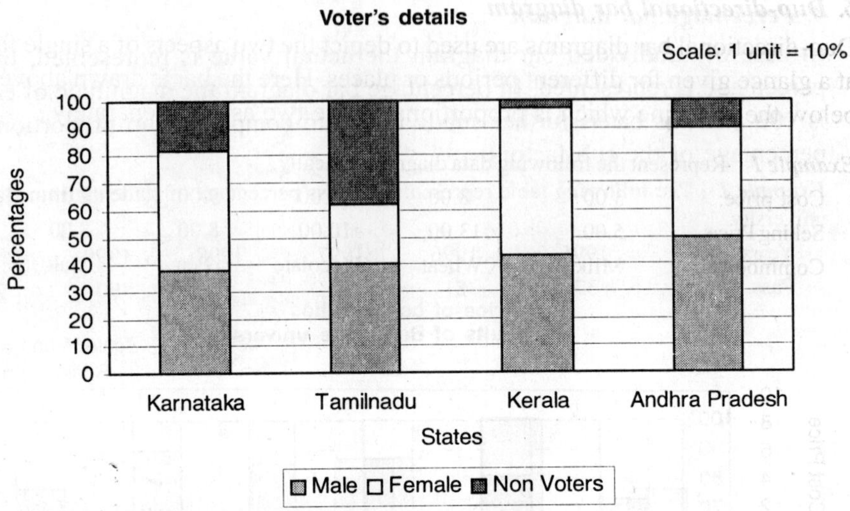

5. Deviation bar diagrams

Deviation bar diagrams are suitable to show the net deviations during various years or according to different countries or places, etc. In the deviation bar diagram, positive deviations are shown to the right side and negative deviations to the left side of the base line.

Example 1 Represent the following data.

Years:	*1995*	*1996*	*1997*	*1998*	*2000*
Expenses (lakhs):	10	15	18	20	30
Returns (lakhs):	11	13	17	25	40

Calculation

Years:	*1995*	*1996*	*1997*	*1998*	*1999*	*2000*
Expenses (lakhs):	10	15	18	20	27	30
Returns (lakhs):	11	13	17	25	25	40
Net results:	+1	–2	–1	5	–2	10

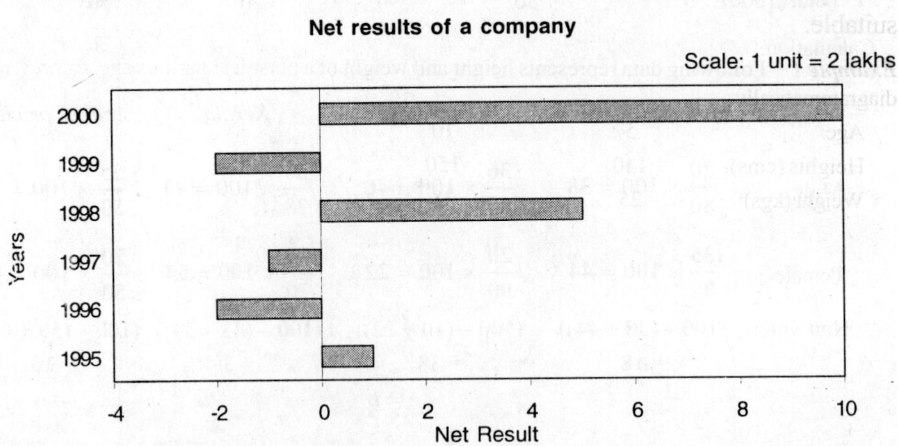

6. Duo-directional bar diagram

Duo-directional bar diagrams are used to depict the two aspects of a single factor at a glance given for different periods or places. Here the bar is drawn above and below the base line which is proportional to the two aspects separately.

Example 1 Represent the following data diagrammatically.

Cost price:	3.00	8.00	5.00	3.00	2.00
Selling Price:	5.00	13.00	10.00	8.00	8.00
Commodity:	Milk	Wheat	Chocolate	Pen	Salt

Price of commodities

Scale: 1 unit = Rs 2

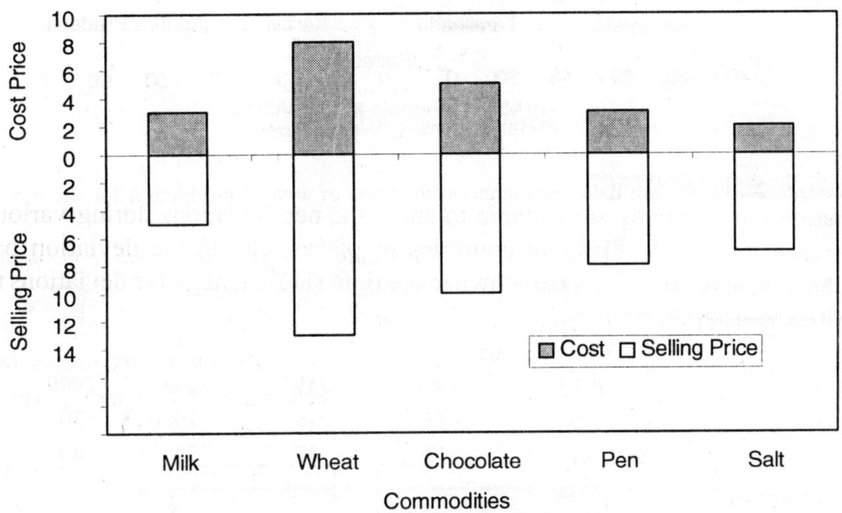

7. Paired bar diagram

When two related factors having different units of measurements are to be displayed for comparison in various periods or places paired bar diagram are suitable.

Example 1 Following data represents height and weight of a person at various age. Represent diagrammatically.

Age:	5	10	15	20
Heights (cms):	140	150	159	165
Weight (kgs):	25	38	50	57

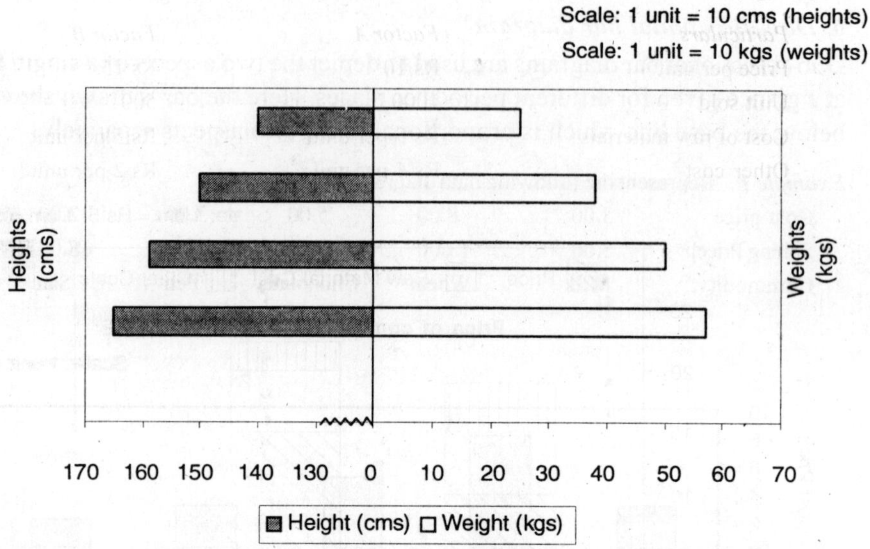

Example 2 Following data represents productions of area of production for the past five years.

States:	*Karnataka*	*Tamilnadu*	*Kerala*	*Andhra Pradesh*
Production (tons):	1500	1200	2000	2500
Area (hectare):	500	400	600	650

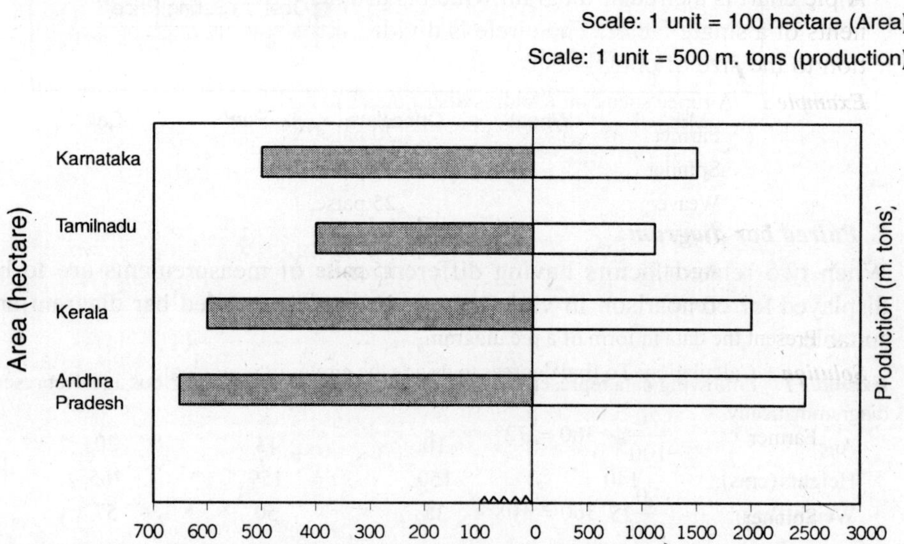

Two-dimensional Diagrams

1. Rectangle bar diagram

In rectangle bar diagram both the length and breadth are taken into consideration in drawing the diagram to represent the data.

Example Draw a rectangular diagram to represent the following information.

Particulars	*Factor A*	*Factor B*
Price per unit	Rs 10	Rs 15
Unit sold	1000	500
Cost of raw materials	Rs 5 per unit	Rs 5 per unit
Other cost	Rs 4 per unit	Rs 2 per unit

Scale: 1 cm = Rs 5; 1 cm = 500 units

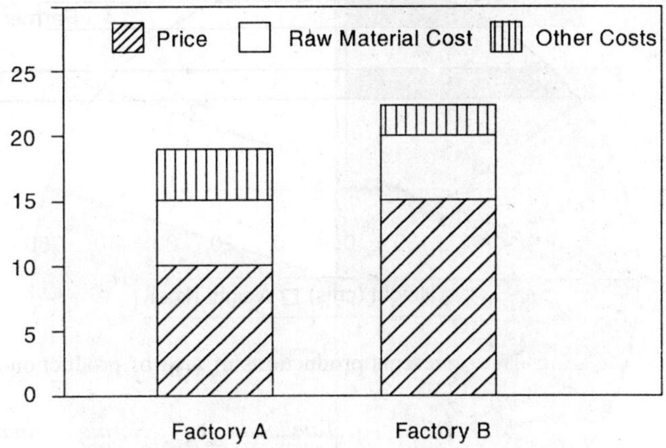

2. *Pie chart*

A pie chart is a circular diagram which is usually used for depicting the components of a single factor. The circle is divided into segment which are in proportion to the size of components.

Example 1 A rupee spend on Khadi is distributed as follows:

Farmer	20 paise
Spinner	30 paise
Weaver	25 paise
Dyer	10 paise
Agent	15 paise
Total	100 paise

Present the data in form of a pie diagram.

Solution Calculation: To find degrees to divide the circles into segment:

Farmer $\frac{20}{100} \times 360 = 72°$

Spinner $\frac{30}{100} \times 360 = 108°$

Weaver $\frac{25}{100} \times 360 = 90°$

Dyer $\frac{10}{100} \times 360 = 36°$

Agent $\frac{15}{100} \times 360 = 54°$

Total 360

A Rupee spent on Khadi

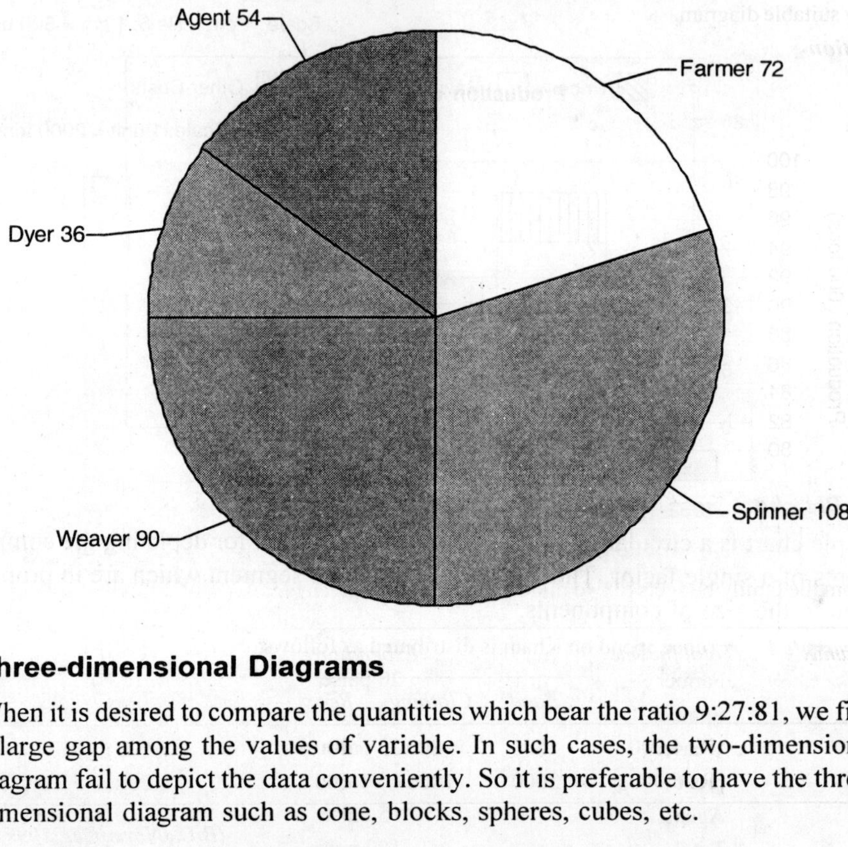

Three-dimensional Diagrams

When it is desired to compare the quantities which bear the ratio 9:27:81, we find a large gap among the values of variable. In such cases, the two-dimensional diagram fail to depict the data conveniently. So it is preferable to have the three-dimensional diagram such as cone, blocks, spheres, cubes, etc.

Non-dimensional Diagrams

*1. **Pictograms***: when the size of the data is very large, like population, we use pictograms. In pictograms, the data are displayed by the pictures of the items to which the data pertain.

*2. **Cartogram***: In cartogram instead of using pictures, different types of maps are used to present the data.

Solved Examples

1. Given below are the figures of production of a factory:

Year:	1992	1993	1994	1995	1996	1997
Production ('000 tones):	80	90	92	83	94	99

Draw suitable diagram.

Solution

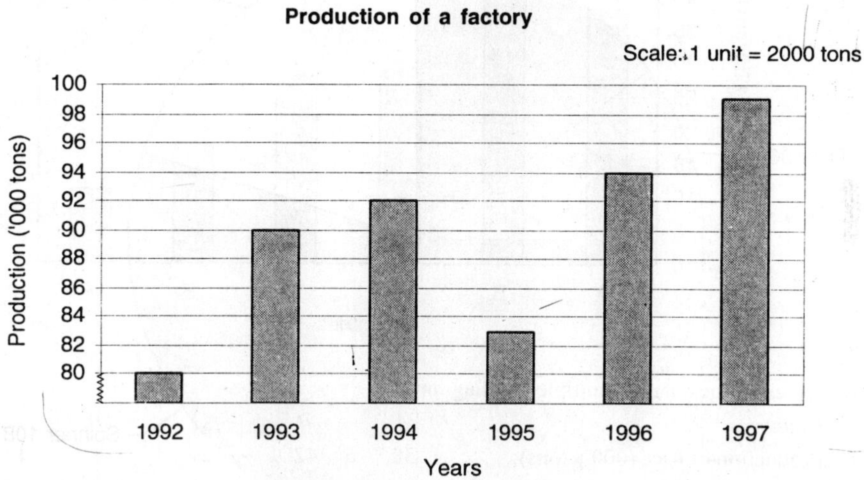

2. From the family budgets of two middle class families given below, draw a suitable diagram.

Family	*Total Income*	*Expenditure items*					
		Food	*Clothing*	*Rent*	*Fuel*	*Others*	*Total*
A	Rs 4,000	1500	800	400	600	1000	4300
B	Rs 6,000	2000	1000	600	800	1200	5600

(B.U., November, 1999)

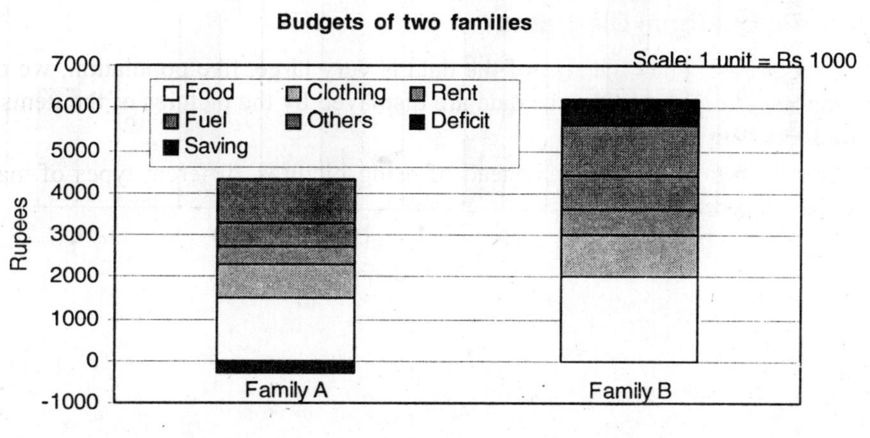

3. Represent the data by a simple bar diagram.

Countries:	A	B	C	D	E	F
Production of Rice (000's tons):	38	42	29	28	18	11

(B.U., September, 1998)

Production of various countries

Scale: 1 unit = 5000 tons

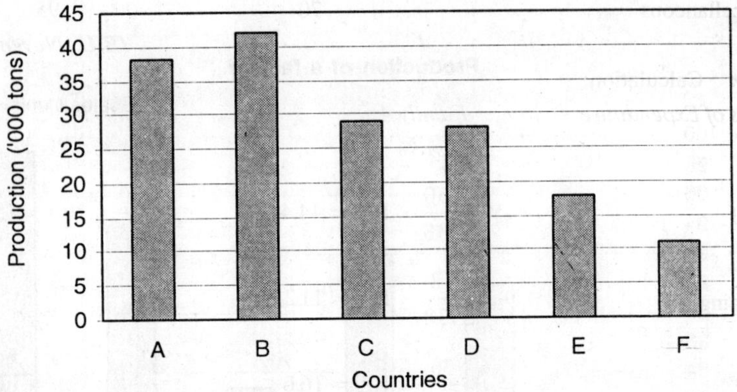

4. Represent the data by multiple bar diagram.

Countries:	*A*	*B*	*C*	*D*	*E*
Production of Rice (000's tons):	38	42	29	28	18
Production of Wheat (000's tons):	40	39	35	25	15

Comparative figure of production of rice and wheat

Scale: 1 unit = 5000 tons

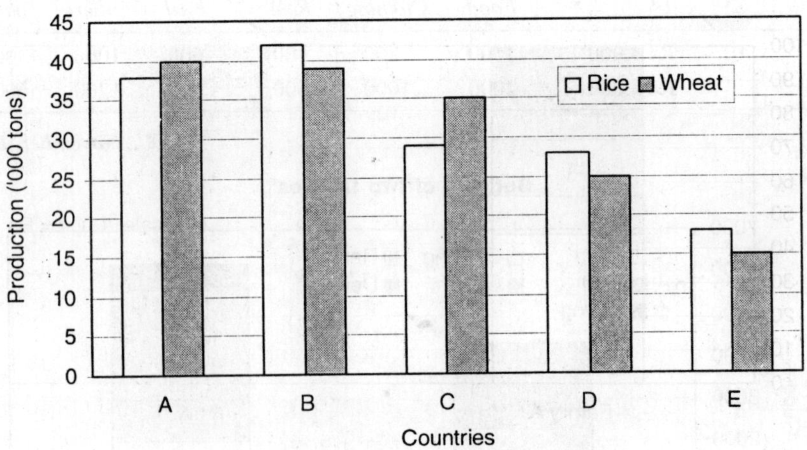

5. Represent by means of percentage bar diagram the following data.

Items of Expenditure	*Family A (Rs)*	*Family B (Rs)*
Food	150	350
Clothing	38	120
Rent	56	130
Education	24	68
Miscellaneous	70	95

(B.U., November, 1994)

Solution Calculation:

Items of Expenditure	*Family A* (Rs) %	*Family B* (Rs) %
Food	$\frac{150}{338} \times 100 = 44.4$	$\frac{350}{763} \times 100 = 45.9$
Clothing	$\frac{38}{338} \times 100 = 11.2$	$\frac{350}{763} \times 100 = 15.7$
Rent	$\frac{56}{338} \times 100 = 16.6$	$\frac{130}{763} \times 100 = 17$
Education	$\frac{24}{338} \times 100 = 7.1$	$\frac{68}{763} \times 100 = 8.9$
Miscellaneous	$\frac{70}{338} \times 100 = 20.7$	$\frac{95}{763} \times 100 = 12.5$

Diagram

Scale: 1 unit = 10%

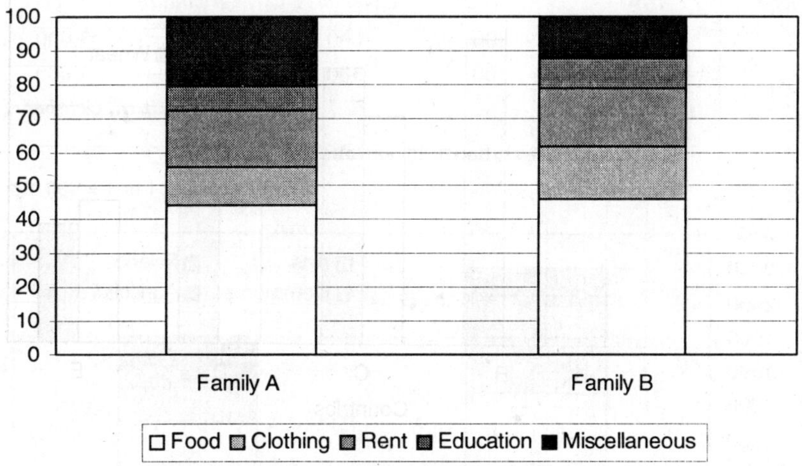

6. Represent the following data by a simple bar diagram.

Countries	*Production of sugar (quintals' 0000)*
China	38
Japan	40
Indonesia	25
India	20
Pakistan	08
Sri Lanka	08

(B.U., April, 1997)

Solution

Production of Sugar

Scale: 1 unit = 5 units

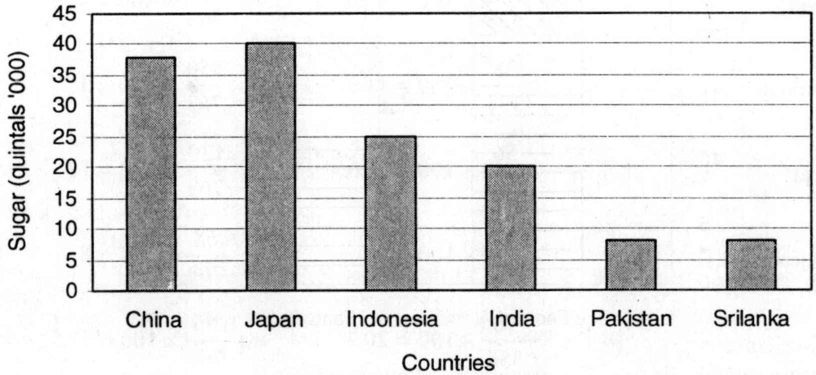

7. Construct a sub-divided bar diagram from the following data, regarding the number of students in 2 colleges.

College	*Arts*	*Science*	*Commerce*	*Agriculture*	*Total*
A	1,200	800	600	400	3,000
B	750	500	300	450	2,000

(B.U., October, 1995)

No. of students in various faculties

Scale: 1 unit = 500 students

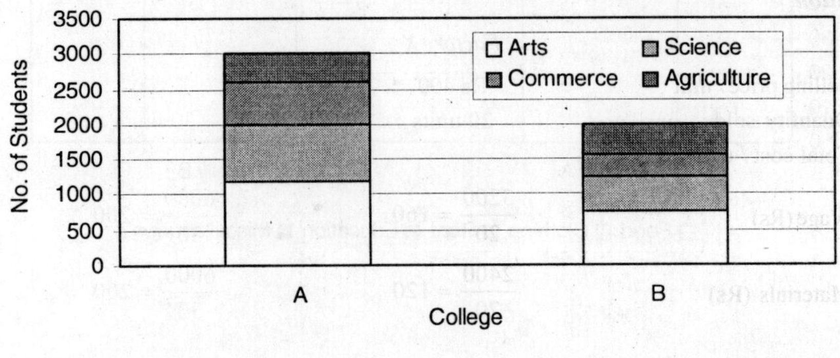

8. Draw a rectangular diagram to represent the following information:

Particulars	*Factory A*	*Factory B*
Price per unit	Rs 15	Rs 12
Units sold	1,000	1,200
Cost of raw materials	Rs 5 per unit	Rs 5 per unit
Other costs	Rs 4 per unit	Rs 3 per unit

(B.U., April, 1999)

Solution

Scale: 1 unit = Rs 5; 1 cm = 900 units

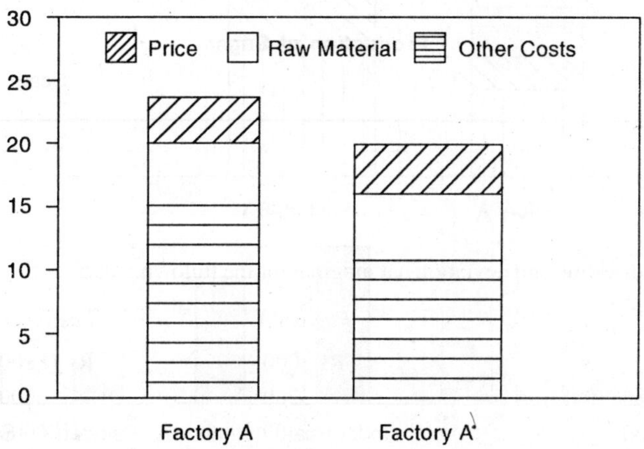

9. Draw a suitable diagram to represent the following information.

	Factory X	*Factory Y*
Selling price per unit	Rs 400	Rs 600
Quantity sold	20 units	30 units
Total cost		
Wages (Rs)	3200	6000
Materials (Rs)	2400	6000
Miscellaneous Expenses	1600	9000

(B.U., November, 1997)

Solution

	Factory X	*Factory Y*
Selling price / unit	Rs 400	Rs 600
Quantity sold	20 units	30 units
Total cost / unit		
Wage (Rs)	$\frac{3200}{20} = 160$	$\frac{6000}{30} = 200$
Materials (Rs)	$\frac{2400}{20} = 120$	$\frac{6000}{30} = 200$

Miscellaneous	$\frac{1600}{20} = 80$	$\frac{9000}{30} = 300$

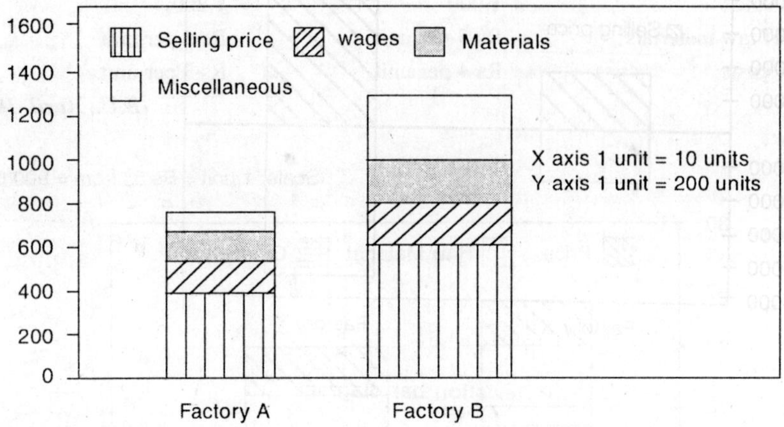

10. Draw a duo-direction and deviation bar diagram for the following data:

	Factory X	*Factory Y*
Selling price	Rs 8000	Rs 18000
Total cost wage (Rs)	3200	6000
Materials (Rs)	2400	6000
Miscellaneous expenses (Rs)	1600	9000

Solution

	Factory X	*Factory Y*
Selling price	Rs 8000	Rs 18000
Cost price	3200 + 2400 + 1600	6000 + 6000 + 9000
	= Rs 7200	= Rs 21000
Net result	8000 – 7200	18000 – 21000
	Rs 6000 (Profit)	Rs 3000 (Loss)

Duo-direction bar diagram

Scale: 1 unit = Rs 4000

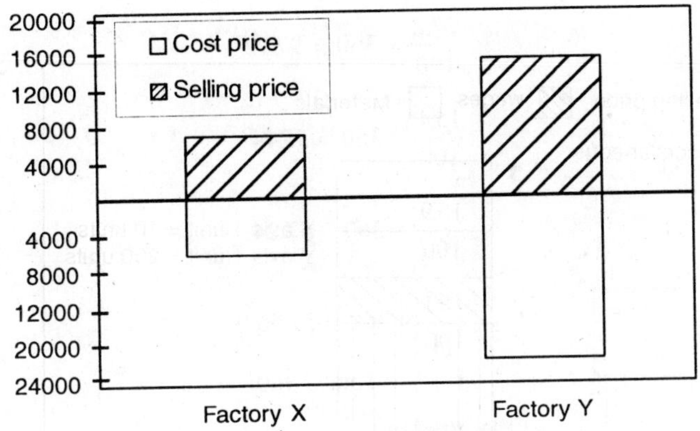

Deviation bar diagram

Scale: 1 unit = Rs 600

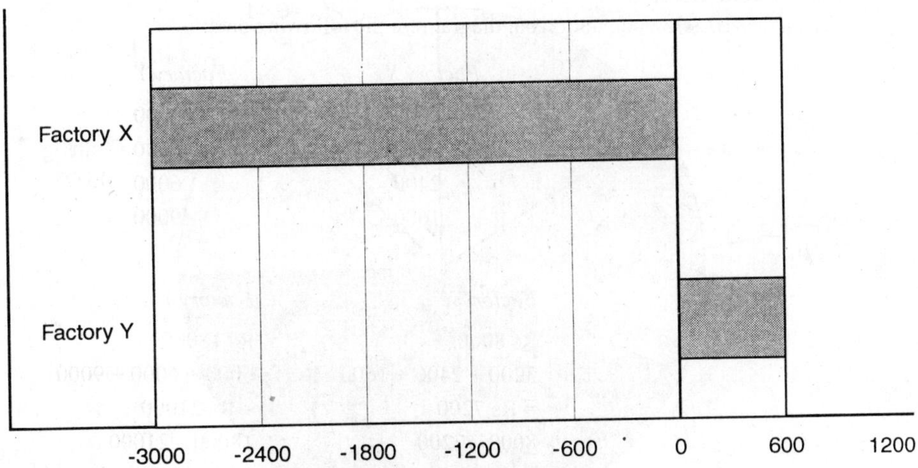

11. Draw the pie diagram for the following data.

Food	12.9%	Education	15.4%
Clothing	12.5%	Transport	15.9%
Oil	27.2%	Social service	16.1%

(B.U., October, 2001)

Solution

Calculation:

	Degree
Food	$\frac{12.9}{100} \times 360 = 46.44$

Clothing	$\frac{12.5}{100} \times 360 = 45.00$
Oil	$\frac{27.2}{100} \times 360 = 97.92$
Education	$\frac{15.4}{100} \times 360 = 55.44$
Transport	$\frac{15.9}{100} \times 360 = 57.24$
Social service	$\frac{16.1}{100} \times 360 = 57.96$
	Total = 360

Pie diagram

Scale: diameter 7.3 cm

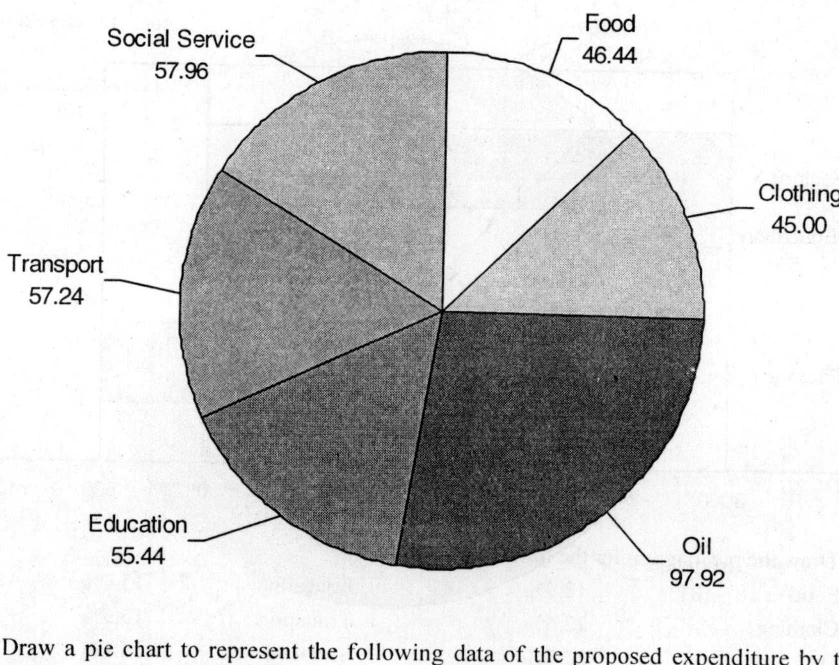

12. Draw a pie chart to represent the following data of the proposed expenditure by the Karnataka Governor for 1998–99

Items	*Expenditure (in lakhs)*
Agriculture	600
Industry	400
Irrigation	450
Education	300
Miscellaneous	250

(B.U., November, 1999)

Solution

Calculation

Agriculture	$\frac{600}{2000} \times 360 = 108$
Industry	$\frac{400}{2000} \times 360 = 72$
Irrigation	$\frac{450}{2000} \times 360 = 81$
Education	$\frac{300}{2000} \times 360 = 54$
Miscellaneous	$\frac{250}{2000} \times 360 = 45$
	Total = 360

Pie diagram

Scale: diameter 3"

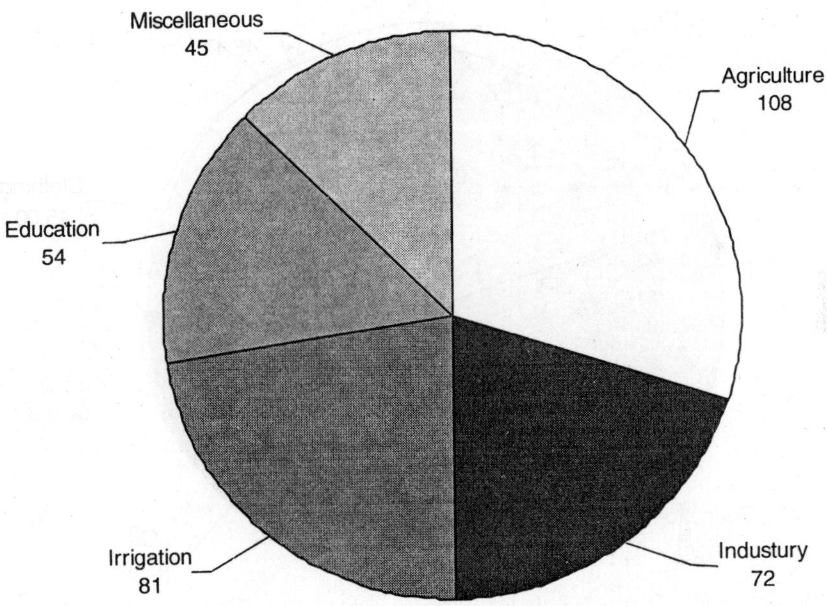

13. Draw a pie diagram to represent the following

Male	*Female*	*Girls*	*Boys*	*Total*
2000	1800	4200	2000	10,000

(B.U., September, 1998)

Solution

Calculation:

Male $\frac{2000}{10000} \times 360 = 72$

Female $\frac{1800}{10000} \times 360 = 64.8$

Girls $\frac{4200}{10000} \times 360 = 151.2$

Boys $\frac{2000}{10000} \times 360 = 72$

Total = 360

Pie diagram

Scale: Diameter = 3"

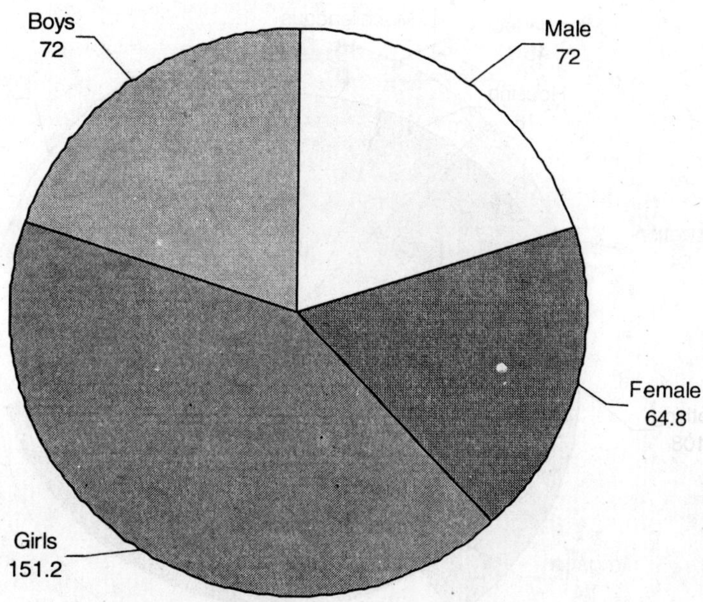

14. Show the following data by a pie diagram.

Items of expenditure	*% of total expenditure*
Food	60
Clothing	30
Housing	05
Miscellaneous	05

(B.U., October, 1996)

Solution

Calculation:

Food	$\frac{60}{100} \times 360 = 216$
Clothing	$\frac{30}{100} \times 360 = 108$
Housing	$\frac{5}{100} \times 360 = 18$
Miscellaneous	$\frac{5}{100} \times 360 = 18$
	Total = 360

Pie diagram

Scale: diameter 3"

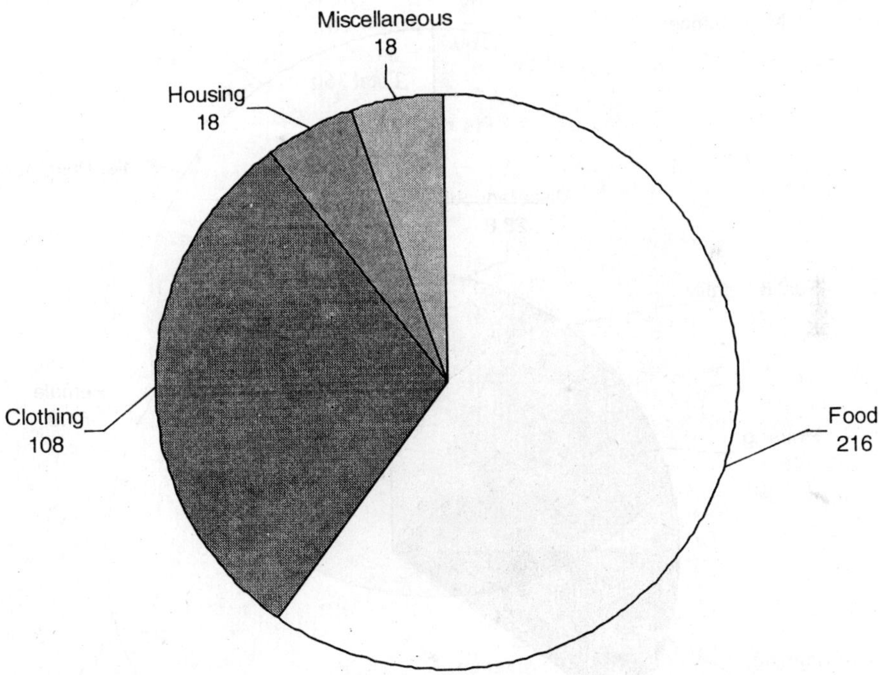

15. Show the following data of expenditure of an average working class family by a pie diagram.

Items of expenditure	*% of total expenditure*
Food	65
Clothing	10
Housing	12

Fuel and lighting	05
Miscellaneous	08

(B.U., April, 1995)

Solution

Calculation:

Food $\frac{65}{100} \times 360 = 234$

Clothing $\frac{10}{100} \times 360 = 36$

Housing $\frac{12}{100} \times 360 = 43.2$

Fuel and lighting $\frac{05}{100} \times 360 = 18$

Miscellaneous $\frac{08}{100} \times 360 = 28.8$

Total 360

Pie diagram

Scale: Diameter 3"

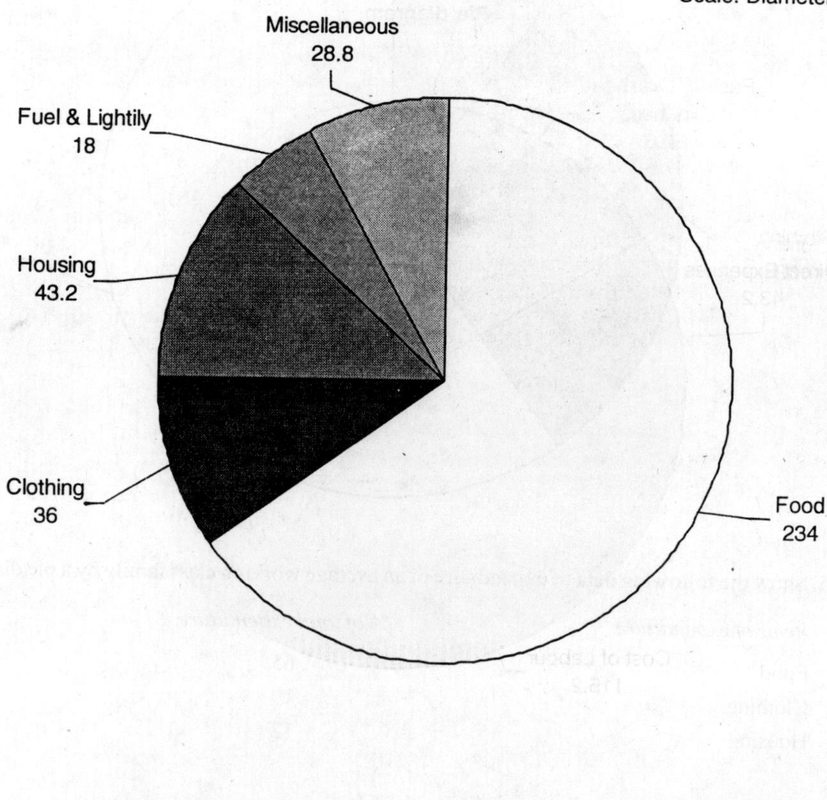

16. Draw a pie-chart to represent the following data relating to the production cost manufacture.

	Rs
Cost of materials	38,400
Cost of labour	30,720
Direct expenses of manufacture	11,520
Factory overhead expenses	15,360

(B.U., April, 1992)

Solution

Calculation:

Cost of materials $\frac{38400}{96000} \times 360 = 144$

Cost of labour $\frac{30720}{96000} \times 360 = 115.2$

Direct expenses of manufacture $\frac{11520}{96000} \times 360 = 43.2$

Factory overhead expenses $\frac{15360}{96000} \times 360 = 57.6$

Total 360

Pie diagram

Scale: diameter = 3"

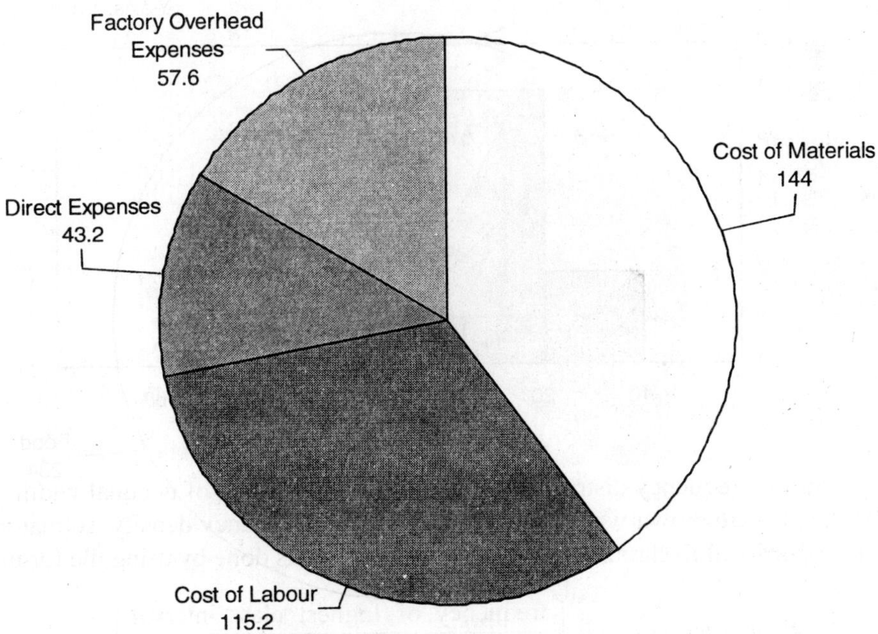

2.4 GRAPH

A graph is a display of points and lines. In a graph the paired values (x, y) co-ordinate of a point are plotted on a graph paper by suitably choosing the scales along abscissa (X-axis) and ordinates (Y-axis). The plotted points are joined by straight lines sequentially.

1. Types of graphs

Important types of graphs used for presentation of frequency distribution are

1. Histogram
2. Frequency polygon
3. Frequency curve
4. Ogives (Cumulative frequency curves).

A. Histogram

Histogram is a bar diagram which represents a frequency distribution with continuous classes. The width of all bars is equal to class interval and heights of the bars are proportional to the frequencies of the respective classes, i.e. the area of the rectangles are proportional to class frequency.

Example 1 Construct a histogram for the following frequency distribution.

C. I. :	10–20	20–30	30–40	40–50
Frequency:	2	6	4	2

Scale: X-axis: unit = 10 units;
Y-axis: 1 unit = 2 units

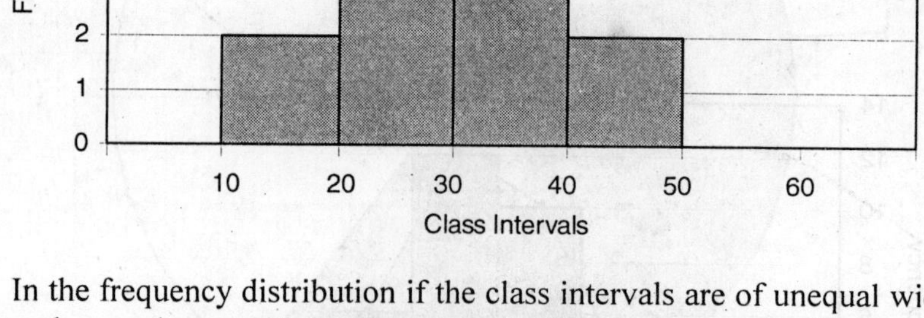

In the frequency distribution if the class intervals are of unequal width, the rectangles are drawn with height proportional to frequency density, so that area is proportional to class frequency. This adjustment is done by using the formula.

$$\text{Adjusted frequency} = \left(\frac{\text{frequency of highest class interval}}{\text{width of the highest class interval}}\right)$$

Multiplied by width of the lowest class interval.

Example 2 Draw a histogram for the following frequency distribution.

C. I. :	10–20	20–30	30–50	50–60
Frequency:	3	6	18	6

$$= \frac{18}{20} \times 10 = 9$$

∴ Frequency 30–50 is divided 30–40, 40–50 with frequency as 9.

Scale: X-axis: 1 unit = 10 units;
Y-axis: 1 unit = 3 units

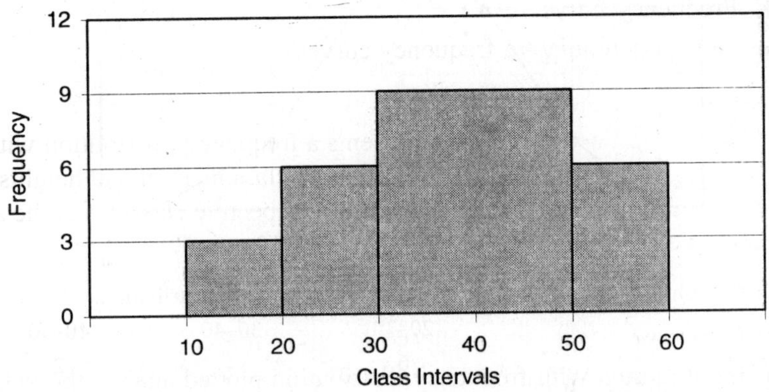

B. Frequency polygon

This is a graph with frequency density plotted against the values of the variable.

The frequency polygon can be obtained from the histogram by joining the mid points of the tops of the adjacent bars in order.

Example 1 Draw a frequency polygon for the given frequency distribution.

C. I. :	10–20	20–30	30–40	40–50	50–60	60–70
Frequency:	2	4	10	12	6	4

Scale: X-axis: 1 unit = 10 units;
Y-axis: 1 unit = 2 units

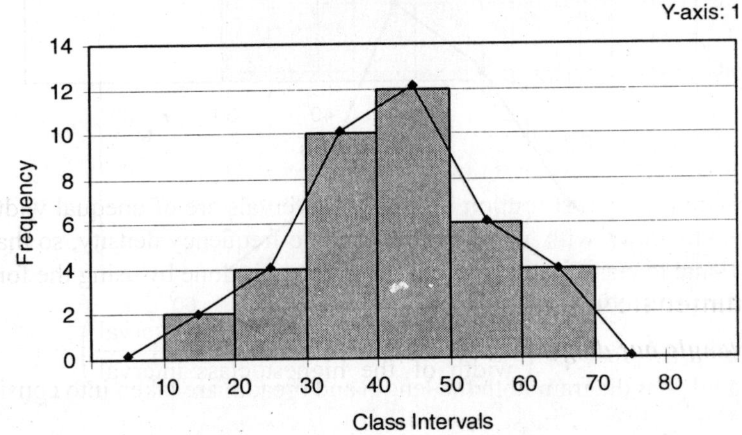

Example 2 Draw a frequency polygon for the following frequency distribution.

C. I. :	10–20	20–30	30–40	40–60
Frequency:	2	6	8	4

Scale: X-axis: 1 unit = 10 units;

Y-axis: 1 unit = 2 units

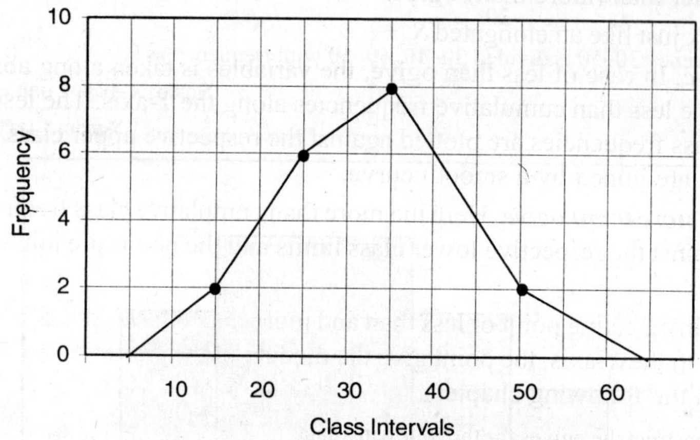

C. Frequency curve

This is a smooth graph with frequency distribution plotted against the value of the variable. A smoothened frequency polygon represents a frequency curve.

Example 1 Draw a frequency curve for the following data.

C. I. :	10–20	20–30	30–40	40–50
Frequency:	2	6	8	2

Scale: X-axis: 1 unit = 10 units;

Y-axis: 1 unit = 2 units

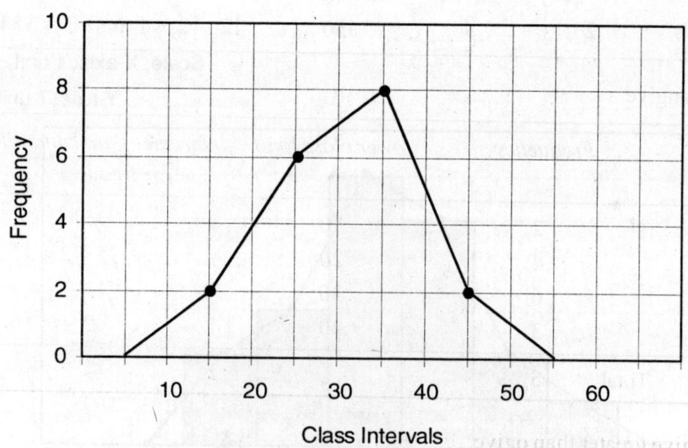

D. *Ogives*

For a given frequency distribution ogives are graph plotted for the variable values against their corresponding cumulative frequencies. There are two types of ogives.

1. Less than ogive.
2. Greater than (more than) ogive.

Ogive look just like an elongated *S*.

Less than ogive: In case of less than ogive, the variables is taken along abscissa (*X*-axis) and the less than cumulative frequencies along the *Y*-axis. The less than cumulative class frequencies are plotted against the respective upper class limits and the points are joined by a smooth curve.

Greater than (more than) ogive: Here the more than cumulative class frequencies are plotted against the respective lower class limits and the points are joined by a smooth curve.

From the intersecting point of less than and more than ogive if a line is drawn perpendicular to the *X*-axis, the point give the median of the distribution (we will discuss this in the following chapter).

Example 1 Construct the ogives for the following data.

C. I.:	10–20	20–30	30–40	40–50
Frequency:	2	4	6	3

Solution As shown in Graphs on the previous page

(a) Less than ogive

C. I	*Frequency*	*Upper class limit*	*Less than cumulative frequency*
10–20	2	20	2
20–30	4	30	2 + 4 = 6
30–40	6	40	2 + 4 + 6 = 12
40–50	3	50	2 + 4 + 6 + 3 = 15

(b) Greater than ogive

C. I	*Frequency*	*lower class limit*	*Greater than cumulative frequency*
10–20	2	10	15
20–30	4	20	15 – 2 = 13
30–40	6	30	13 – 4 = 9
40–50	3	40	9 – 6 = 3
Total	15		

(a) Less than ogive greater than ogive.

(a) Less than ogive

Scale: X-axis: 1 unit = 10 units;
Y-axis: 1 unit = 2 units

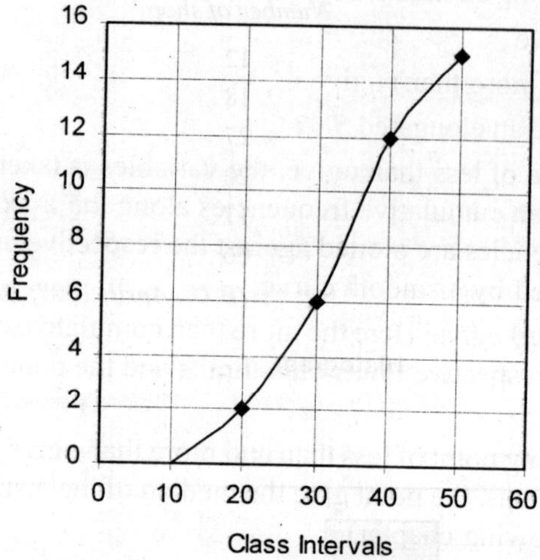

(b) Greater than ogive

Scale: X-axis: 1 unit = 10 units;
Y-axis: 1 unit = 2 units

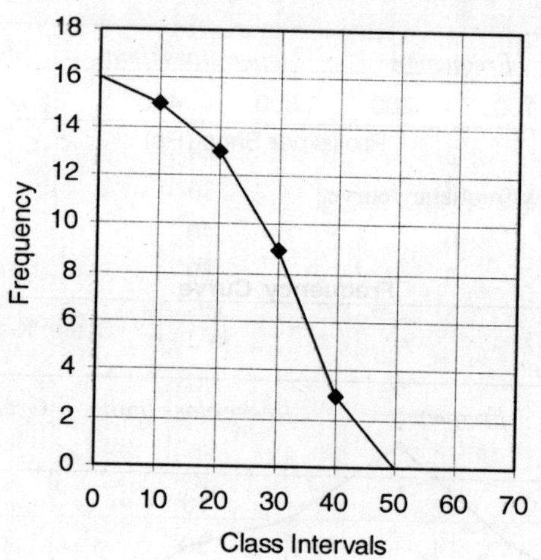

Utility of graphs: Graphs are used for further analysis: (i) Histograms are used to find the mode of the given frequency distribution (ii) Ogives are useful in finding out quartile, deciles.

Solved Examples

1. The monthly profits in rupees of 100 shops are distributed as follows: Draw a histogram.

Profits per shop	*Number of shop*
0–100	12
100–200	18
200–300	27
300–400	20
400–500	17
500–600	06

(B.U., April, 1997; B.U., April, 1998)

Solution

Histogram

Scale: X-axis: 1 unit = 100 units;
Y-axis: 1 unit = 5 units

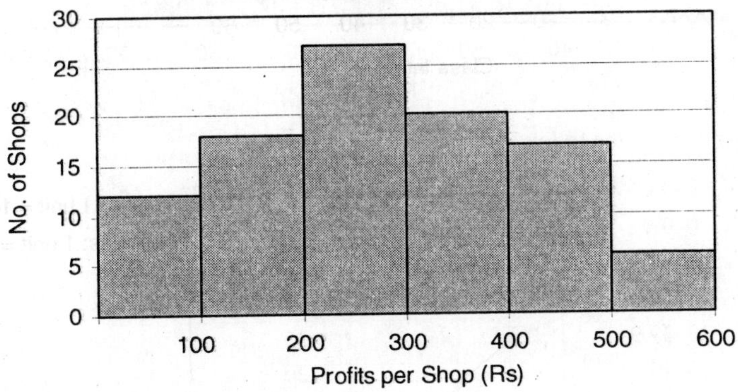

2. For the above data draw frequency curve.

Solution

Frequency Curve

Scale: X-axis: 1 unit = 100 units;
Y-axis: 1 unit = 5 units

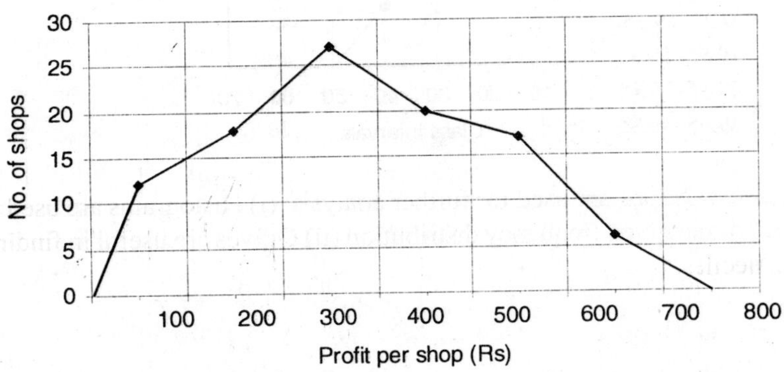

3. Given below are the marks obtained by 200 students of a university in statistics paper.

Marks:	40–50	50–60	60–70	70–80	80–90
No. of students:	32	42	55	40	31

Draw a histogram and frequency polygon.

Solution

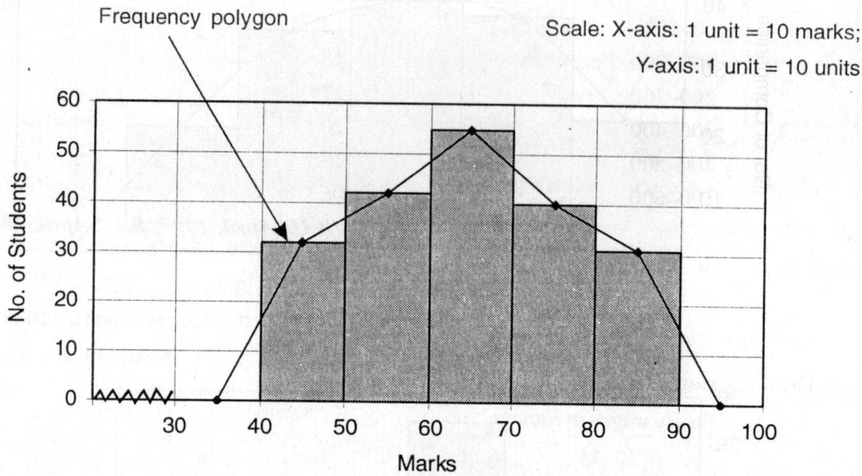

4. Draw a histogram from the following data

Profits (Rs)	*Number of companies*
0–9.9	22
10–19.9	38
20–29.9	40
30–39.9	35
40–49.9	20

Draw a histogram and frequency curve. ***(B.U., November, 1993)***

Solution

We have to convert the exclusive class intervals into inclusive type.

Profits (Rs)	*Number of Companies*
0.5–9.95	22
9.95–19.95	38
19.95–29.95	40
29.95–39.95	35
39.95–49.95	20

Scale: X-axis: 2 units = 10 units;
Y-axis: 1 unit = 10 units

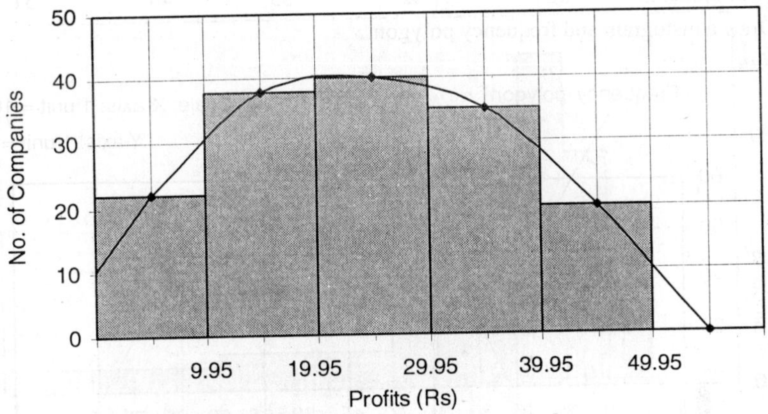

5. Draw a histogram for the following data.

Weekly wages in Rupees	*Number of workers*
10–15	7
15–20	19
20–25	27
25–30	15
30–40	12
40–60	12
60–80	8

Solution

Weekly wages in Rupees	*Number of workers*
10–15	7
15–20	19
20–25	27
25–30	15
30–35	6
35–40	6
40–45	3
45–50	3
50–55	3
55–60	3
60–65	2
65–70	2
70–75	2
75–80	2

Scale: X-axis: 1 unit = 5 units;
Y-axis: 1 unit = 5 units

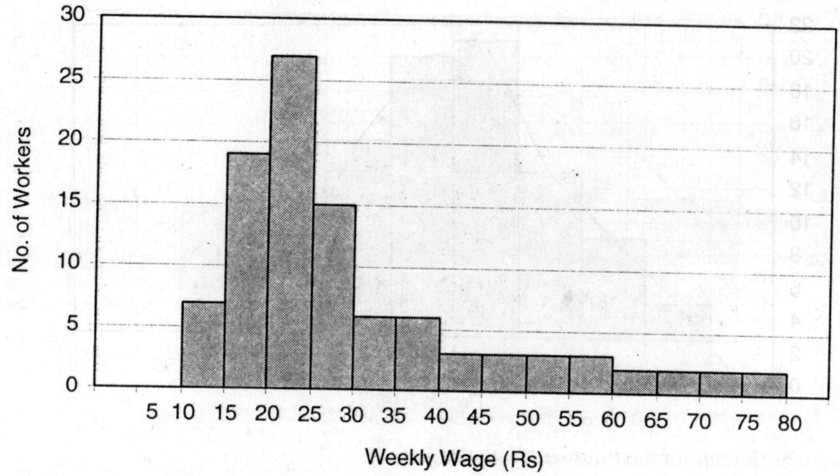

6. Given below is the distribution of profits (in 1000 rupees) cared by 94 book depots in Bangalore city.

Profit (Below):	20	30	40	50	60	70	80	90
No. of book depots:	5	14	27	48	68	83	91	94

Draw a histogram and frequency polygon.

Solution

Class intervals	*No. of book depots*
10–20	5
20–30	14 – 5 = 9
30–40	27 – 14 = 13
40–50	48 – 27 = 21
50–60	68 – 48 = 20
60–70	83 – 68 = 15
70–80	91 – 83 = 8
80–90	94 – 91 = 3

Scale: X-axis: 1 unit = 10 units;

Y-axis: 1 unit = 2 units

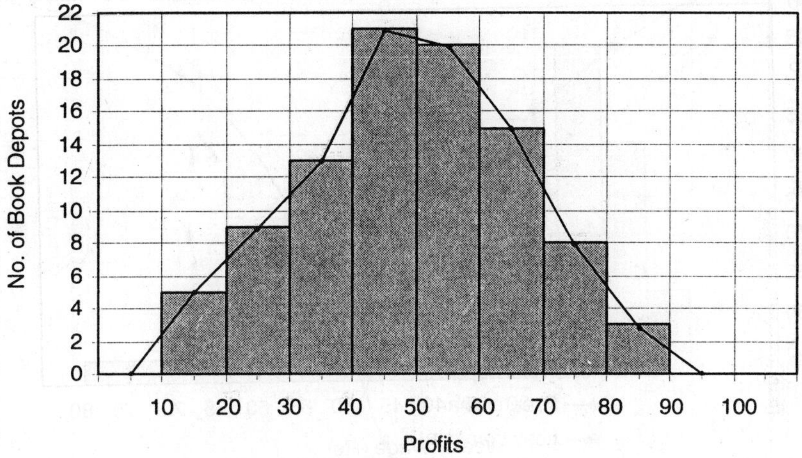

7. Draw an ogive cure from the data given below.

Profits (Rs in lakhs)	*No. of Shops*
10–20	6
20–30	8
30–40	12
40–50	18
50–60	25
60–70	16
70–80	8
80–90	5
90–100	2

Solution

Profits	*Number of Shops*	*Upper limit*	*Less than C. F*	*Lower limit*	*Greater than C. F*
10–20	6	20	6	10	100
20–30	8	30	6 + 8 = 14	20	94
30–40	12	40	14 + 12 = 26	30	86
40–50	18	50	26 + 18 = 44	40	74
50–60	25	60	44 + 25 = 69	50	56
60–70	16	70	69 + 16 = 85	60	31
70–80	8	80	85 + 8 = 93	70	15
80–90	5	90	93 + 5 = 98	80	7
90–100	2	100	98 + 2 = 100	90	2

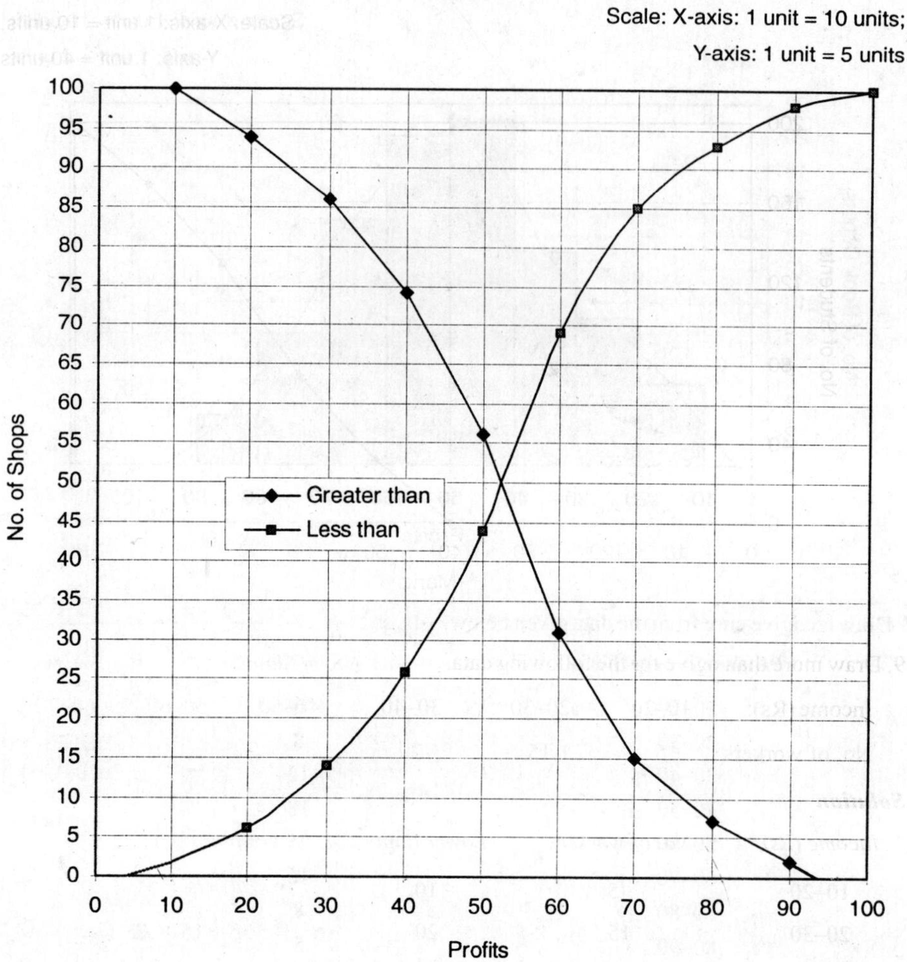

8. Draw less than ogives for the following data.

Marks:	40–50	50–60	60–70	70–80	80–90
No. of students	32	42	55	40	31

Solution

Marks	*No. of students*	*Upper limits*	*Less than C. F*
40–50	32	50	32
50–60	42	60	32 + 42 = 74
60–70	55	70	74 + 55 = 129
70–80	40	80	129 + 40 = 169
80–90	31	90	169 + 31 = 200

Scale: X-axis: 1 unit = 10 units;
Y-axis: 1 unit = 40 units

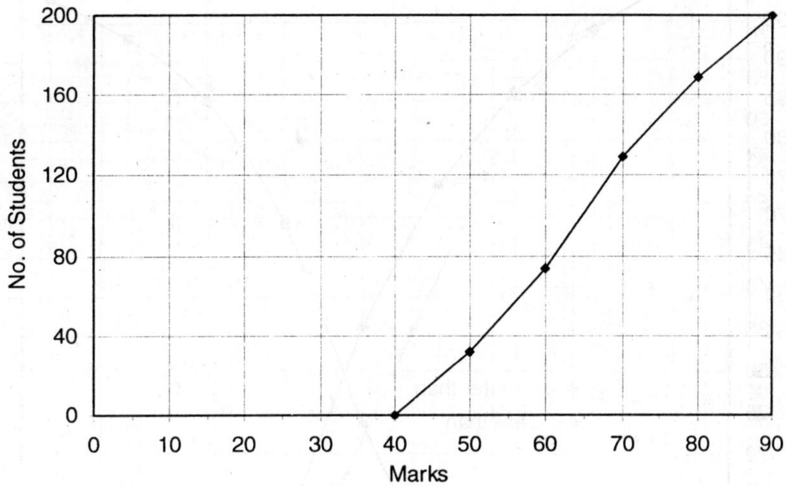

9. Draw more than ogive for the following data.

Income (Rs):	10–20	20–30	30–40	40–50	50–60	60–70
No. of workers	5	15	20	20	10	5

Solution

Income (Rs)	*No. of workers*	*Lower limits*	*Greater than C. F*
10–20	5	10	70 + 5 = 75
20–30	15	20	55 + 15 = 70
30–40	20	30	35 + 20 = 55
40–50	20	40	15 + 20 = 35
50–60	10	50	5 + 10 = 15
60–70	5	60	5

Scale: X-axis: 1 unit = 10 units;
Y-axis: 1 unit = 5 units

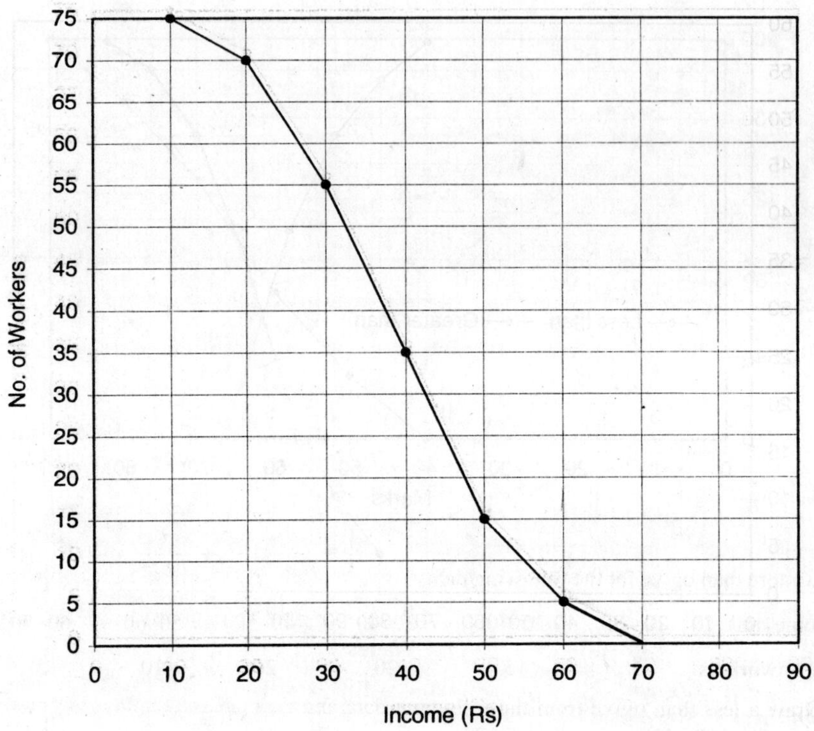

10. Draw an ogive of the following distribution.

Wage (in Rs):	70–80	80–90	90–100	100–110	110–120	120–130	130–140
No. of workers:	4	6	10	16	12	7	3

Solution

Wages	*Number of workers*	*Upper limits*	*Less than C.F.*	*Lower limits*	*Greater than C.F.*
70–80	4	80	4	70	58
80–90	6	90	10	80	54
Wages	*Number of workers*	*Upper limits*	*Less than C.F.*	*Lower limits*	*Greater than C.F.*
90–100	10	100	20	90	48
100–110	16	110	36	100	38
110–120	12	120	48	110	22
120–130	7	130	55	120	10
130–140	3	140	58	130	3

Scale: X-axis: 1 unit = 10 units;
Y-axis: 1 unit = 5 units

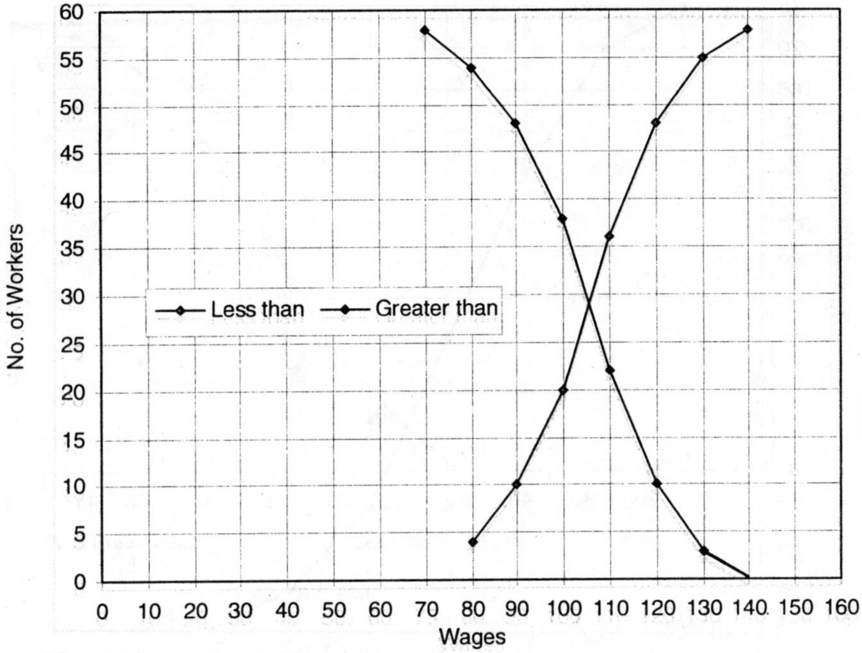

11. Draw a less than ogive from the following data and estimate the number of employees earning at least Rs 63 but less than Rs 75.

Wages:	50–60	60–70	70–80	80–90	90–100	100–110	110–120
No. of employees:	8	10	16	14	10	5	2

Solution

Wages	*No. of employees*	*Upper limits*	*Less than C. F.*
50–60	8	60	8
60–70	10	70	18
70–80	16	80	34
80–90	14	90	48
90–100	10	100	58
100–110	5	110	63
110–120	2	120	65

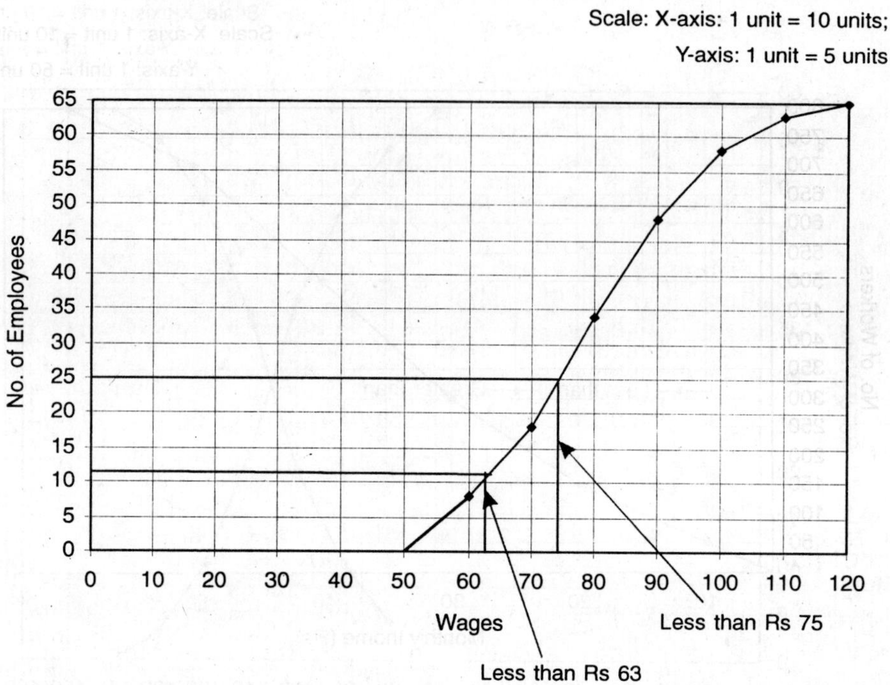

From the graph the number of employee earning at Rs 63 but less than Rs 75

$25 - 12 = 13$

12. Below are given the wages earned by workers per week in a certain factory.

Monthly income (Rs)	*Number of workers*
Upto Rs 10	50
Upto Rs 20	150
Upto Rs 30	300
Upto Rs 40	500
Upto Rs 50	700
Upto Rs 60	800

Draw an ogive.

Scale: X-axis: 1 unit = 10 units;

Y-axis: 1 unit = 50 units

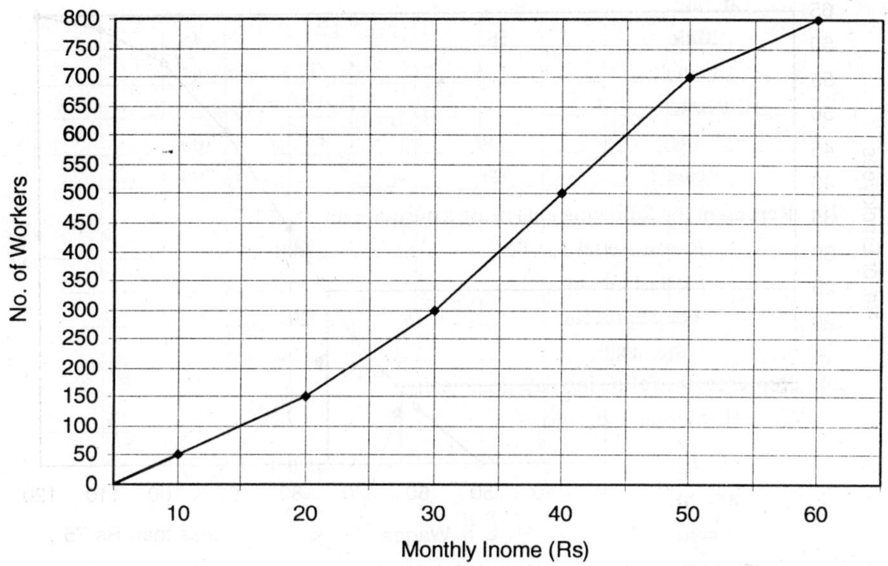

Exercises

1. What is a bar diagram?
2. What is a component bar diagram?
3. What is the significance of Duo-direction and paired bar diagram?
4. What is a pie diagram?
5. What is a histogram?
6. What is the purpose of ogives and histogram?
7. Represent the following data by a suitable diagram.

Year:	1997	1998	1999	2000	2001
Female:	40	42	45	60	59

8. Represent the following data by a suitable diagram which represent number of male and female workers in a factory.

Year:	*1997*	*1998*	*1999*	*2000*	*2001*
No. of male:	700	900	910	950	750
No. of Female:	200	300	350	450	500

9. Represent the following data diagrammatically.

	Town A	*Town B*
Literates		
Male	565	745
Female	435	255
Illiterates		
Male	250	300
Female	750	700

10. Represent the following data using a pie diagram:

Raw material	600
Cost of labour	350
Transportation	200
Other head	50

11. Represent the following data graphically.

a. Histogram b. Ogives

C. I.	*Frequency*
10–20	3
20–30	5
30–40	7
40–50	9
50–60	2
60–70	1

12. Draw a histogram for the following data.

Marks:	10–20	20–30	30–40	40–70	70–100
No. of students	10	12	18	12	6

13. For the above data draw a frequency polygon and frequency curve.

14. Draw a. Less than ogive.

b. Greater than ogive.

Wages:	2000–3000	3000–4000	4000–5000	5000–6000
No. of labours:	100	150	300	100

4

Measures of Central Tendency

Contents

- Mean, Median, Mode, Quartiles.

Formulae

Mean: $\dfrac{\text{Sum of the values}}{\text{Number of observation}}$

1. Discrete: $\overline{X} = \dfrac{\Sigma x}{N}$

2. Discrete frequency distribution and continuous distribution.

$$\overline{X} = \frac{\Sigma f x}{N}$$

3. Deviation method.

$$\overline{X} = a + \frac{\Sigma f d}{N} \quad d = \overline{X} - a$$

4. Step deviation method.

$$\overline{X} = a + \frac{h \Sigma f u}{N} \quad u = \frac{\overline{X} - a}{h}$$

Median:

1. Raw data:

   Median is $\left(\dfrac{N+1}{2}\right)$th item of the arrayed series.

2. Frequency distribution:

   Median is that value whose less than cumulative frequency is just greater than $N/2$.

3. Continuous frequency distribution:

$$M = l_0 + \left(\frac{N/2 - m}{f}\right) \times C$$

Mode:

1. Continuous frequency distribution

$$Z = l_0 + \frac{f_1 - f_0}{2 f_1 - f_0 - f_2}$$

2. Empirical formula: Mode = 3 median – 2 mean.

Combined mean: $\overline{X} = \dfrac{n_1 \overline{X}_1 + n_2 \overline{X}_2}{n_1 + n_2}$

Quartiles: $Q_r = l_0 + \dfrac{\left(r\, N/4 - m\right)}{f} \times c$

Points to Remember

1. In case of continuous frequency distribution the class interval should of exclusive type to calculate the median.
2. Mode may not be uniquely defined for some distribution.

INTRODUCTION

In the previous chapters we have seen how to present the given set of data for future analysis. Now let us focus on computing the average values of the data. Professor Bowley states that "Average are statistical constants which enable us to comprehend in a single effort the significance of the whole".

4.1 CHARACTERISTIC OF AN IDEAL AVERAGE

An ideal average should possess the following properties.

1. It should be rigidly defined.
2. It should be easy to calculate.
3. It should be least affected by extreme values.
4. It should be based on all observations.
5. It should be capable of further algebraic treatment so that it could be used in further analysis of the data.
6. It should be affected as little as possible by fluctuations of sampling.

4.2 AVERAGES

The most commonly used averages are:

- The Mean

- The Median
- The Mode

Now, let us discuss these in detail.

4.3 THE MEAN (ARITHMETIC MEAN)

Arithmetic mean of a set of data may be defined as the sum of the values divided by the number of values in the set. Mean is usually denoted by $\overline{X}$ for set of data $x_1, x_2 \ldots x_n$.

$$\overline{X} = \frac{\text{Sum of the values}}{\text{Number of observation}}$$

1. ***Individual Series*** (raw data)

Arithmetic mean of the value $x_1, x_2 \ldots x_n$ is

$$\overline{X} = \frac{x_1 + x_2 + \ldots x_n}{n} = \frac{\Sigma x}{n}$$

Example 1 Find the average of mark obtain by a student in five subjects.
40, 37, 43, 41, 35.

$$\text{Average} = \frac{40 + 37 + 43 + 41 + 35}{5} = \frac{196}{5}$$

$$= 39.2$$

2. ***Frequency distribution*** (discrete and continuous)

If the observations $x_1, x_2, x_3 \ldots x_n$ have frequencies $f_1, f_2, f_3 \ldots f_n$ the arithmetic mean is given by

$$\overline{X} = \frac{f_1 x_1 + f_2 x_2 + \ldots + f_n x_n}{f_1 + f_2 + \ldots + f_n} = \frac{\Sigma f X}{N}$$

Example 2 Discrete Series
Find the mean for the given frequency distribution:

X:	1	2	3	4	5
f:	3	5	12	9	1

Solution

X	*f*	*fX*
1	3	3
2	5	10
3	12	36
4	9	36
5	1	5
Total	30	90

$$\overline{X} = \frac{\Sigma f X}{N} = \frac{90}{30} = 3$$

The mean of the given frequency distribution is 3.

Example 3 Continuous Series

Calculate the mean of the given frequency distribution

Class intervals:	10–20	20–30	30–40	40–50
Frequency:	15	25	30	5

Solution

C. I	*f*	*Mid value x*	*fx*
10–20	15	15	225
20–30	25	25	625
30–40	30	35	1050
40–50	5	45	225
Total	75		2125

$$\text{Mean} = \frac{2125}{75}$$

$$= 28.333$$

The mean of the given frequency distribution is 28.333.

4.4 SHORT CUT METHOD

1. ***Deviation method*** Here we subtract a constant 'a' from the given set of observations. Now the original data is in the form

$$d = X - a$$

$$\therefore \text{Mean} = \frac{\Sigma f d}{N} + a \quad (\therefore x = d + a)$$

2. ***Step deviation method*** In addition to the subtraction the values are divided by a constant '*h*'

then $$u = \frac{X - a}{h}$$

Mean $$= \frac{h \Sigma f d}{N} + a \quad (\therefore X = h u + a)$$

i.e. $$\overline{X} = h\overline{u} + a$$

Example 4 Short cut method

1. ***Deviation method***

(a) Heights of six students are 160, 175, 169, 157, 162 and 165 cms. Find the arithmetic mean.

Solution

x	$d = X - a$
160	–2
175	13
169	7
157	–5
162	0
165	3
Total	16

Let $a = 162$

$$\text{Mean} = \frac{\Sigma d}{N} + a$$

$$= 164.66$$

(b) (Solution to data in Example 2)

Let $d = x - 3$

x	f	$d = x - 3$	fd
1	3	–2	–6
2	5	–1	–5
3	12	0	0
4	9	1	9
5	1	2	2

$$\text{Mean} = \frac{\Sigma f d}{N} + a$$

$$= 0 + 3$$ (***Ans***. 3)

2. ***Step deviation method***

(a) (Solution to data in Example 3)

Let $u = \dfrac{x - 25}{10}$

C. I	f	x	u	fu
10–20	15	15	–1	–15
20–30	25	25	0	0
30–40	30	35	1	30
40–50	5	45	2	10
Total	75			25

$$\text{Mean} = \frac{\Sigma f u}{N} \times h + a$$

$$= \frac{25}{75} \times 10 + 25$$

$$= 3.33 + 25$$

$$= 28.33.$$

Example 5 Missing frequencies

(a) The mean of the frequency distribution is given to 61. Find the missing frequency.

x:	40	50	60	70	80
f:	2	2	1	?	2

Solution Let *x* be the missing frequency.

x	*f*	*fx*
40	2	80
50	2	10
60	1	60
70	*x* (?)	70*x*

$$\text{Mean} = \frac{80 + 100 + 60 + 70x + 160}{2 + 2 + 1 + x + 2}$$

$$= \frac{400 + 70x}{7 + x}$$

The mean of the distribution is given to be 61.

$$\therefore\ 61 = \frac{400 + 70x}{7 + x}$$

$$61(7 + x) = 400 + 70x$$

$$427 + 61x = 400 + 70x$$

$$27 = 9x$$

$$\therefore\ x = \frac{27}{9} = 3$$

$\therefore$ The missing frequency is 3

(b) In the following frequency distribution of marks, one of the frequency is missing if the arithmetic mean of the distribution is 52 find the missing frequency.

Marks	0–20	20–40	40–60	60–80	80–100
No. of students	8	?	19	14	9

Solution Let *F* be the missing frequency:

Class	*Frequency f*	*Mid x*	$u = \dfrac{x - 30}{20}$	*fu*
0–20	8	10	–1	–8
20–40	*F*	20	0	0

Contd.

Class	*Frequency f*	*Mid x*	$u = \frac{x - 30}{20}$	*fu*
40–60	19	50	1	19
60–80	14	70	2	28
80–100	9	90	3	27
Total	50 + F			66

We have $\overline{X} = a + \frac{h\sum fu}{N}$

Give mean = 52

$$\therefore 52 = 30 + 20 \times \frac{66}{50 + F}$$

$$52 - 30 = \frac{1320}{50 + F}$$

$$22\,(50 + F) = 1320$$

$$22F = 1320 - 1100$$

$$F = \frac{220}{22} = 10$$

The missing frequency is $F = 10$.

Solved Problems

1. The arithmetic mean of 50 items was 100. At the time of calculation two items 180 and 90 were wrongly taken as 100 and 10 what is the correct mean.

(B. U., April/May, 1999)

Solution

Given: $n = 50$

$$\overline{X} = 100$$

we know $\overline{X} = \frac{\text{Sum of the value}}{\text{Number of value}}$

$$\therefore \text{Sum} = \overline{X} \times n$$

$$= 100 \times 50 = 5000$$

Now,

subtract the wrong entries and add the correct entries.

$$\therefore \text{Sum} = 5000 + (180 + 90) - (100 + 10)$$

$$= 5160$$

$$\therefore \text{Mean} = \frac{5160}{100}$$

$$= 51.6.$$

2. The arithmetic mean of 98 items is 50. Two items 60 and 70 were left out at the time of calculation. What is the correct mean of all the items?

(B. U., Nov./Dec., 1992)

Solution

Given $N = 98$

$\overline{X} = 50$

$\therefore$ Sum $= 98 \times 50$

$= 4900$

Two items are left out and the left out numbers are 60 and 70.

$\therefore n = 98 + 2 = 100$

Sum $= 4900 + 60 + 70$

$= 5130$

$$\therefore \text{Mean} = \frac{\text{Sum}}{n}$$

$$= \frac{5130}{100}$$

$= 51.3$

Correct mean is 51.3

3. Find the A.M of 12.4, 12.6, 12.9, 12.5, 12.3 and 12.7 cms.

$$\text{Mean} = \frac{\text{Sum of the value}}{\text{Number of value}}$$

$$= \frac{12.4 + 12.6 + 12.9 + 12.5 + 12.3 + 12.7}{6} = 75.4.$$

4. Find the mean of the following distribution number of students present in the class on different days.

No. of students	73	74	75	76	77	78	79	80	Total
No. of days	2	14	29	36	43	38	34	29	225

Solution

No. of students x	*No. of days* f	$d = x - 77$	fu
73	2	–4	–8
74	14	–3	–42
75	29	–2	–58
76	36	–1	–36
77	43	0	0
78	38	1	38
79	34	2	68
80	29	3	87
Total	225		49

Here $a = 77$

We know $\overline{X} = \frac{\Sigma f d}{N} + a$

$= \frac{49}{225} + 77$

$= 0.2178 + 77$

$= 77.2178$

The mean of the given distribution is 77.2178.

5. Find the mean for the following data.

Size	105	110	115	120	125	130	135	140
Frequency	8	23	34	35	27	20	9	4

Solution

Size x	*Frequency* f	$d = \frac{x - 120}{5}$	*fu*
105	8	–3	–24
110	23	–2	–46
115	34	–1	–34
120	35	0	0
125	27	1	27
130	20	2	40
135	9	3	27
140	4	4	16
Total	160		6

Here $a = 120$
$h = 5$

formula $\overline{X} = a + \frac{h\Sigma f u}{N}$

$= 120 + \frac{5 \times 6}{160}$

$= 120 + 0.1875$

Mean $= 120.1875$

6. Find the mean for the following discrete frequency distribution.

x	5	10	15	20	25	30	35	40	45	50	55	60	65	70
f	2	4	6	5	7	4	3	10	6	3	4	7	2	5

Solution

x	f	$u = \dfrac{x-40}{5}$	fu
5	2	–7	–14
10	4	–6	–24
15	6	–5	–30
20	5	–4	–20
25	7	–3	–21
30	4	–2	–8
35	3	–1	–3
40	10	0	0
45	6	1	6
50	3	2	6
55	4	3	12
60	7	4	28
65	2	5	10
70	5	6	30
Total	68		–28

Here $a = 40$, $h = 5$.

$$\text{Mean } \overline{X} = a + h\frac{\Sigma fu}{N}$$

$$= 40 + \frac{5 \times (-28)}{68}$$

$$= 40 - 2.0588$$

$$\text{Mean} = 37.9412$$

7. For the following distribution, find the mean.

Marks:	0–10	10–20	20–30	30–40	40–50
No. of students:	4	9	19	20	18

Solution

Marks C. I	No. of students f	x	$u = \dfrac{x-25}{10}$	fu
0–10	4	5	–2	–8
10–20	9	15	–1	–9
20–30	19	25	0	0
30–40	20	35	2	40
40–50	18	45	1	18
Total	70			41

Here $a = 25$, $h = 10$, $N = 70$

$$\text{Mean } \overline{X} = a + h\frac{\Sigma f u}{N}$$

$$= 25 + \frac{10 \times 41}{70}$$

$$= 25 + 5.857$$

$$= 30.857$$

8. The profit made by 130 firms in a year are as follows. Find the arithmetic mean.

Profits ('000) Rs:	1–2	2–3	3–4	4–5	5–6	6–7	7–8	Total
No. of firms:	5	27	36	28	20	11	3	130

Solution

Profits *C. I*	*No. of firms* *f*	*x* *('000)*	*d = x – 4.5*	*fd*
1–2	5	1.5	–3	–15
2–3	27	2.5	–2	–54
3–4	36	3.5	–1	–36
4–5	28	4.5	0	0
5–6	20	5.5	1	20
6–7	11	6.5	2	22
7–8	3	7.5	3	9
Total	130			–54

$d = x - 4.5$

$a = 4.5$

$$\text{Mean} = a + \frac{\Sigma f d}{N}$$

$$= 4.5 + \frac{(-54)}{130}$$

$$= 4.5 - .4154$$

$$= 4.0846$$

Mean = 4.0846

The mean profit made by firms is Rs 4084.6

9. From the following data, find the missing item.

x:	110	112	113	117	?	125	128	130
Frequency:	25	17	13	15	14	8	6	2

Mean wage = 115.86

(B. U., April, 1998)

Solution

x	Frequency (f)	fx
110	25	2750
112	17	1904
113	13	1465
117	15	1755
?(k)	14	$14k$
125	8	1000
128	6	768
130	2	260
Total	100	$9902 + 14k$

$$\text{Mean} = \frac{\text{Sum of the values}}{\text{No. of observation}}$$

Let the missing observation be k. Here mean wage is given to be 115.86.

$$\therefore\ 115.86 = \frac{9902 + 14k}{100}$$

$$11586 = 9902 + 14k$$

$$14k = 1684$$

$$k = 120.28$$

$$= 120$$

10. The arithmetic mean of the following series is 30.5 marks. Find the missing figure.

Marks:	10	20	?	40	50
Frequency:	8	10	20	15	7

(B. U., Oct./Nov, 2001)

Solution

Marks x	Frequency f	fx
10	8	80
20	10	200
?(k)	20	$20k$
40	15	600
50	7	350
Total	60	$1230 + 20k$

Let k be the missing observation.

Here mean is given to be 30.5 mark.

$$\text{Mean} = \frac{\text{Sum of the values}}{\text{No. of observations}}$$

$$30.5 = \frac{1230 + 20k}{60}$$

$$30.5 \times 60 = 1230 + 20k$$

$$1830 = 1230 + 20k$$

$$20k = 600$$

$$k = \frac{600}{20}$$

$$= 30$$

The missing value: 30 marks

4.5 PROPERTIES OF ARITHMETIC MEAN

The following are the properties of Arithmetic mean:

1. Algebraic sum of the deviation of a set of values from their arithmetic mean is zero.

   i.e., $\Sigma(X - \overline{X}) = 0$

2. Sum of the squared deviation of - set of values is minimum when deviation are taken around the arithmetic mean.
3. Combined mean: Let $\overline{X}_1$, be the mean of a set of data of size n_1 and let $\overline{X}_2$ be the mean of a set of data of size n_2. Then the combined mean is given by

$$\overline{X} = \frac{n_1 \overline{X}_1 + n_2 \overline{X}_2}{n_1 + n_2}$$

Solved Examples

Combined Mean

1. Mean weight of 150 students in a class is 60 kg. The mean weight of boys in the class is 70 kg's and that of girls is 55 kg's. Find the number of girls in the class.

(B. U., Oct./Nov, 1998)

Solution

Let the number of boys be k (n_1)

Total number of students: 150 $(n_1 + n_2)$

Number of girls: $150 - k$ (n_2)

$\overline{X}_1$ (mean weight of boys) = 70 kgs

$\overline{X}_2$ (mean weight of girls) = 55 kgs

$\overline{X}$ (combined weight) = 60 kgs

Combined mean formula

$$\overline{X} = \frac{n_1 \overline{X}_1 + n_2 \overline{X}_2}{n_1 + n_2}$$

$$60 = \frac{k.70 + (150 - k)\,55}{150}$$

$150 \times 60 = 70k + 8250 - 55k$

$9000 = 8250 - 15k$

$750 = 15k$

$\therefore\ k = 50$

The number boys in class is 50, the number of girls = (150 – 50)

∴ The number of girls is 100.

2. The mean wage of a group of workers which consists of men and women is Rs 66. In the group, mean wage of men is Rs 70 and the mean wage of women is Rs 50. Find the percentage of women in the group.

Solution

Let the total no. of employee be 100 i.e. $n_1 + n_2 = 100$

No. of female be k (n_1)

∴ No. of male is $(100 - k)$ (n_2)

$\overline{X} = 66,\ \overline{X}_1 = 50,\ \overline{X}_2 = 70$

we know that

$$\overline{X} = \frac{n_1 \overline{X}_1 + n_2 \overline{X}_2}{n_1 + n_2}$$

$$66 = \frac{k \times 50 + (100 - k)\,70}{100}$$

$6600 = 50k + 7000 - 70k$

$= 7000 - 20k$

$20k = 400$

$k = 20$

The percentage of women in the group is 20.

4.6 MERITS AND DEMERITS OF MEAN

Merits:

1. It is rigidly defined.
2. It is easily computed.
3. It is based on all observations.
4. It is amendable for further mathematical treatments.
5. Of all averages, arithmetic mean is affected least by fluctuations of sampling.

Demerits:

(i) It is affected by extreme values

(ii) Mean cannot be found in the case of open end classes.

(iii) It cannot be determined by inspection nor can be located graphically.

(iv) It cannot be obtained if one data is missing.

4.7 MEDIAN

Median is the value of the variable that divides the ordered set of values into two equal halves. It is a most preferable measure of location for asymmetric distribution.

4.8 COMPUTATION OF MEDIAN

1. Individual (raw) series and discrete frequency distribution:

Median for an arrayed series is given by $M = \left(\frac{n+1}{2}\right)$th value.

Example 1 The following are the marks scored by 11 students in an examination. Find the median mark.

Marks: 49, 87, 73, 62, 15, 29, 75, 19, 23, 41, 39.

Solution

Arrange the given data in ascending order:

Array: 15, 19, 23, 29, 39, 41, 49, 62, 73, 75, 87.

Here $n = 11$

Median is

$$M = \left(\frac{n+1}{2}\right)\text{th item.}$$

$$= \left(\frac{11+1}{2}\right)\text{th item}$$

$$= 6\text{th item}$$

$$= 41$$

∴ the median of the given series is 41 marks.

Example 2 Find the median height from the following heights of 8 students.

Height (cms): 59, 60, 72, 53, 42, 45, 49, 71.

Solution

Arrange the given data in ascending order:

Array: 42, 45, 49, 53, 59, 60, 71, 72.

Here $n = 8$

Median is $M = \left(\frac{n+1}{2}\right)$th item

$= \left(\frac{8+1}{2}\right)$th item

$= (4.5)$th item

$=$ 4th item + .5 (5th item – 4th item)

$= 53 + .5\,(59 - 53)$

$= 53 + .5\,(6)$

$= 56$

The median height is 56 cms.

Example 3 For the following distribution find the median:

x:	120	121	122	123	124	125
f:	2	4	7	14	21	22

Solution

Group	x	f	Cumulative frequency
1	120	2	2
2	121	4	$2 + 4 = 6$
3	122	7	$6 + 4 + 7 = 13$
4	123	14	$2 + 4 + 7 + 14 = 27$
5	124	21	$2 + 4 + 7 + 14 + 21 = 48$
6	125	22	$2 + 4 + 7 + 14 + 21 + 22 = 70$

Median = value of $\left(\frac{N+1}{2}\right)$th item

$= \left(\frac{70+1}{2}\right)$th item

$= 35.5$

Now from the cumulative frequency column we notice that the 35.5th item belongs to 5th group. The median is the value (x) that corresponds to 5th group. Hence median = 124 or median value is given by that value of x whose $c.f$ is just greater than $N/2$.

2. Continuous frequency

Find $\left(\frac{N}{2}\right)$th value of the given frequency distribution. The class interval whose cumulative frequency is just greater than or equal to $\left(\frac{N}{2}\right)$th value is know as the median class, the exact value of the median is calculated using the formula.

$$M = l_0 + \left(\frac{\left(N/2 - m\right) \times c}{f}\right)$$

Where l_0 = lower than of the median class

m = cumulative frequency corresponding to the class preceding the median class

c = class width of the median class

f = frequency of the median class.

N = total frequency.

Example 4 Find the median wage of the following distribution:

Wages (in Rs):	20–30	30–40	40–50	50–60	60–70
No. of laboures:	3	5	20	10	5

Solution

Wage (in Rs)	No. of labourers	c.f.
20–30	3	3
30–40	5	8 (m)
40–50	20 (f)	28
50–60	10	38
60–70	5	43

Here $\frac{N}{2} = \frac{43}{2} = 21.5$

∴ Median class is that C. I. Whose *c.f.* is just greater than 21.5, here 28

∴ Median class is 40–50.

Using formula

$$M = l_0 + \left(\frac{\left(N/2 - m\right) \times c}{f}\right)$$

$l_0 = 40,\ m = 8,\ f = 20,\ c = 10$

$$M = 40 + \frac{10}{20} \times (21.5 - 20)$$

$$= 40 + 6.75$$

$$= 46.75$$

The median wage is Rs 46.75.

Example 5 Missing observation

An incomplete frequency distribution is given as follows:

Variable:	10–20	20–30	30–40	40–50	50–60	60–70	70–80	Total
Frequency:	12	30	?	65	?	25	18	229

Given that the median value is 46. Determine the missing frequencies using the median formula.

Solution

Let the missing frequency of class 30–40 be f_1 and that of class 50–60 be f_2.

Given that median = 46

∴ The median class is 40–50.

Using the formula

$$M = l_0 + \left(\frac{N}{2} - m\right) \times \frac{c}{f}$$

we get

$$46 = 40 + \left(\frac{114.5 - (42 + f_1)}{65}\right) \times 10$$

$$46 - 40 = \frac{72.5 - f_1}{65} \times 10$$

$$6 = \frac{725 - 10f_1}{65}$$

$$10f_1 = 725 - 390$$

$$10f_1 = 335$$

$$10f_1 = 33.5 = 34 \; [\because \text{ frequency is integer number}]$$

given that total is 229

$$229 = 12 + 30 + f_1 + 65 + f_2 + 25 + 18$$

$$= 150 + f_1 + f_2$$

$$f_1 + f_2 = 229 - 150$$

$$= 79$$

we know that $f_1 = 34$

$$\therefore f_2 = 79 - 34$$

$$= 45$$

The missing frequencies are $f_1 = 34$ and $f_2 = 45$.

4.9 GRAPHICAL METHOD

We have discussed about obtaining the median graphically in Chapter 2. Now let us illustrate the same by means of an example.

Example 6 Find the median wage of the following distribution graphically:

Wages (in Rs):	20–30	30–40	40–50	50–60	60–70
No. of laboures:	3	5	20	10	5

Solution

Wage (in Rs)	No. of labourers	Less than *c.f.*	More than *c.f.*
20–30	3	3	43
30–40	5	8	40
40–50	20	28	35
50–60	10	38	15
60–70	5	43	5

Step 1: Plot less than and greater then ogives.

Step 2: From the intersecting point, draw a line perpendicular to X-axis.

Step 3: The value corresponding to the point that touches the X-axis is the median.

Scale: X-axis: 1 unit = 10 units;
Y-axis: 1 unit = 5 units

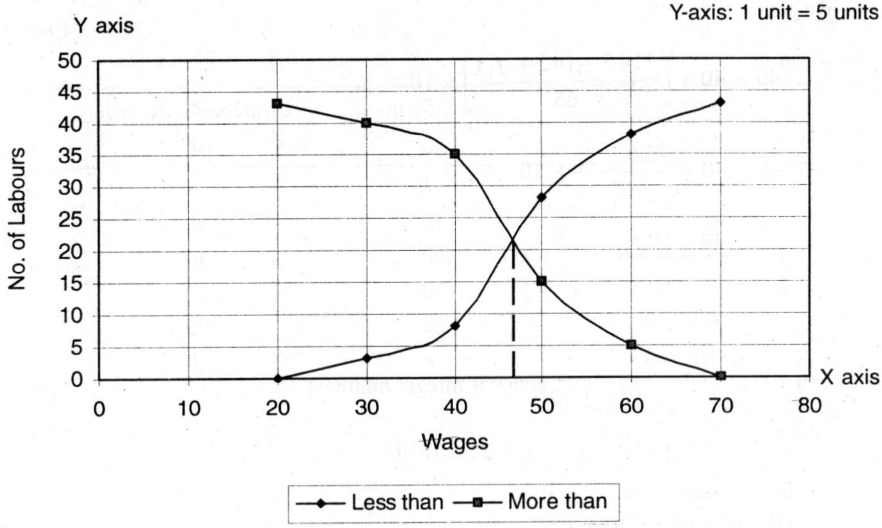

From the graph we get the median value = 47.

4.10 MERITS AND DEMERITS OF MEDIAN

Merits:

1. It is rigidly defined.
2. It is easy to understand and compute.
3. It is not affected by extreme values.
4. It can be computed while dealing with a distribution with open end classes.
5. It can be computed graphically and also can be located by simple inspection.
6. Median is the only average to be used while dealing with qualitative characteristics.

Demerits:

1. It is not based on all observations.
2. It is not suitable for fuller mathematical treatments
3. It is relatively less stable than mean.

Solved Examples

1. Find the median of the following:

Weekly wages:	60–70	70–80	80–90	90–100	100–110	110–120
No. of workers:	5	15	20	30	20	8

(B. U., April, 1996)

Solution

Weekly wages C. I.	No. of workers f	Cumulative frequency *c.f.*
60–70	5	5
70–80	15	20
80–90	20	40
90–100	30	70
100–110	20	90
110–120	8	8

Here $N = 98,\ \dfrac{N}{2} = 49$

$\therefore$ The median class is 90–100

$l_0 = 90,\ m = 40,\ f = 30,\ c = 10$

using the formula

$$M = l_0 + \frac{\left(\frac{N}{2} - m\right)}{f} \times c$$

$$= 90 + \frac{(49 - 40)}{30} \times 10$$

$$= 90 + 3$$

$$= 93$$

Median $= 93$

2. Draw an ogive curve from the data given below and locate the median.

Profits : (Rs in lakhs)	10–20	20–30	30–40	40–50	50–60	60–70	70–80	80–90	90–100
No. of shops:	6	8	12	18	25	16	8	5	2

(B. U., Oct./Nov., 2001)

Solution

Profits C. I.	No. of shops f	Less than c.f.	Greater than c.f.
10–20	6	6	100
20–30	8	14	94
30–40	12	26	86
40–50	18	44	74
50–60	25	69	56
60–70	16	85	31
70–80	8	93	15
80–90	5	98	7
90–100	2	100	2

From the graph we get the median value as 53.

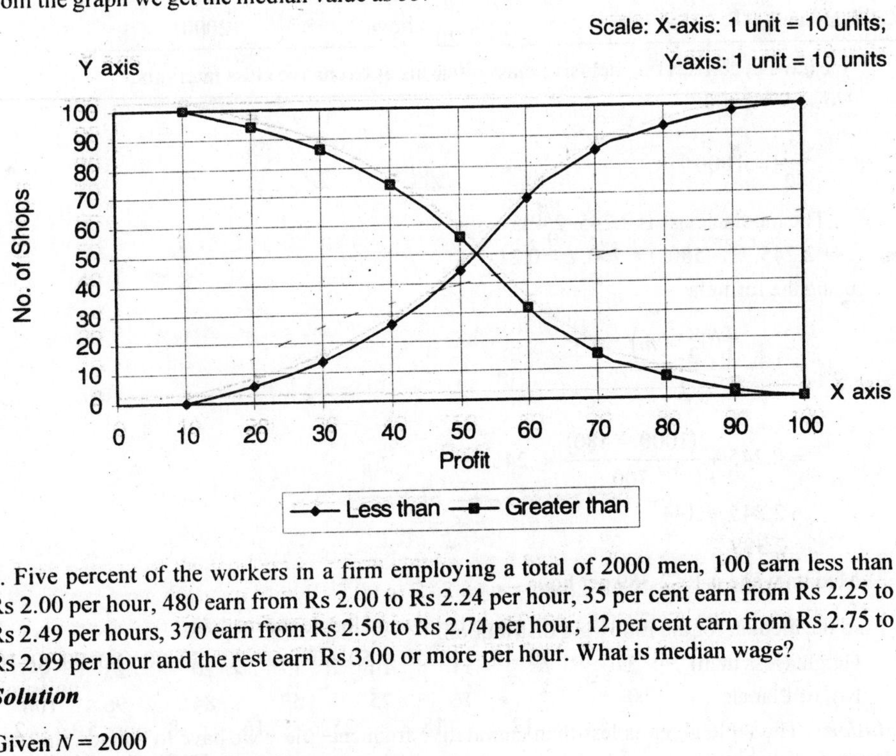

3. Five percent of the workers in a firm employing a total of 2000 men, 100 earn less than Rs 2.00 per hour, 480 earn from Rs 2.00 to Rs 2.24 per hour, 35 per cent earn from Rs 2.25 to Rs 2.49 per hours, 370 earn from Rs 2.50 to Rs 2.74 per hour, 12 per cent earn from Rs 2.75 to Rs 2.99 per hour and the rest earn Rs 3.00 or more per hour. What is median wage?

Solution

Given $N = 2000$

100 earn less than Rs 2.00 per hour

480 earn from Rs 2.00 to Rs 2.24 per hour

$35\% = \frac{35}{100} \times 2000 = 700$ earn from Rs 2.25 to Rs 2.49 per hour

370 earn from Rs 2.50 to Rs 2.74 per hour

$12\% = \frac{12}{100} \times 2000 = 240$ earn from Rs 2.75 to Rs 2.99 per hour

$2000 - (100 + 480 + 700 + 370 + 240) = 110$ earn above Rs 3.8.

Let us first tabulate the given data:

Wages (C. I.)	No. of workers f	C. I.	*c.f.*
<– 2.00	100	<– 1.995	100
2.00–2.24	480	1.995–2.245	580 (m)
2.25–2.49	700 (f)	2.245–2.495	1280
2.50–2.74	370	2.495–2.745	1650
2.75–2.99	240	2.745–2.995	1890
above 3.00	110	above 2.995	2000

We have to convert the inclusive class intervals as exclusive class intervals.

Now, $N = 2000$

$$\frac{N}{2} = 1000$$

∴ The median class is 2.245–2.495

$l_0 = 2.245$, $m = 580$, $f = 700$, $c = 0.24$

using the formula

$$M = l_0 + \frac{\left(N/_2 - m\right)}{f} \times c$$

$$= 2.245 + \frac{(1000 - 580)}{700} \times .24$$

$$= 2.245 + .144$$

$$= 2.389$$

Median wage is Rs 2.389 per hour.

4. Find the median for the following distribution.

Height (less than):	5	8	11	14	17	20	23	26
No. of Plants:	0	7	16	25	57	84	96	100

Solution The table given is less than cumulative frequency now we have to convert this to frequency distribution.

Height (x)	No. of Plant (f)	cumulative frequency *c.f.*
< 5	0	0
5–8	7	7
8–11	16 – 7 = 9	16
11–14	25 – 16 = 9	25
14–17	57 – 25 = 32	57
17–20	84 – 57 = 27	84
20–23	96 – 84 = 12	96
23–26	100 – 96 = 4	100

Here $N = 100$, $\frac{N}{2} = 50$

The median class is 14–17.

$l_0 = 14$, $m = 25$, $f = 32$, $c = 3$

Using the formula

$$M = l_0 + \frac{\left(N/2 - m\right)}{f} \times c$$

$$= 14 + \frac{(50 - 25)}{32} \times 3$$

$$= 16.3438$$

median height is 16.3438.

5. Find the median for the following distribution.

Income above	Rs 1000	1500	2000	2500	3000	3500	4000
No. of person	80	62	33	16	6	2	0

Solution The table represents greater then cumulative frequency, now we have to convert it to frequency distribution.

Income C. I.	No. of persons (f)	Less then cumulative frequency (*c.f.*)
1000–1500	80 – 62 = 18	18
1500–2000	62 – 33 = 29	47
2000–2500	33 – 16 = 17	64
2500–3000	16 – 6 = 10	74
3000–3500	6 – 2 = 4	78
3500–4000	2 – 0 = 2	80

Here $N = 80$, $\frac{N}{2} = 40$

$\therefore$ Median class is 1500–2000.

$l_0 = 1500$, $f = 29$, $m = 18$, $c = 500$

using the formula

$$\text{Median } (M) = l_0 + \frac{(N/2 - m)}{f} \times c$$

$$= 1500 + \frac{(40 - 18)}{29} \times 500$$

$$= 1500 + 379.31$$

$$= 1879.31$$

The median value is 1879.31.

4.11 MODE

Mode is the value of the variable that most frequently occurs in a given series. In other word mode is the value which has the highest frequency.

Mode is not necessarily unique for some distribution. We will discuss about unimodal and bimodal case.

4.12 UNIMODAL

(A) Computation of mode

1. Individual series and discrete frequency distribution:

Mode is the value which corresponds to the highest frequency. This value can be obtain be mere inspection.

Example 1 Calculate the mode for the give data.

No. of children / People: 8, 3, 0, 1, 0, 2, 3, 3, 2, 0, 2, 5, 2, 1, 3, 2, 4, 2, 3, 9, 3, 6, 2, 0, 0, 5, 1, 2, 7, 2, 2.

Solution

x	Tally mark	frequency	x	Tally mark	frequency
0	~~1111~~ 1	6	5	11	2
1	11	2	6	1	1
2	~~1111~~ 1111	9	7	1	1
3	~~1111~~ 1	6	8	1	1
4	1	1	9	1	1

Here we see that the highest frequency is 9, which corresponds to 2.

$\therefore$ mode = 2 children / couple.

Example 2 Locate the mode from the following frequency distribution.

Marks:	40	50	60	70	80	90
Frequency:	10	8	20	7	3	10

Solution Here 20 is the highest frequency which corresponds to value 60. Hence mode is 60 marks.

2. Continuous frequency distribution

The class corresponding to the highest frequency is known as the modal class now to find the exact value, we use the formula.

$$Z = l_0 + \frac{(f_1 - f_0) \times c}{2 f_1 - f_0 - f_2}$$

Where l_0 = Lower limit of modal class.

f_1 = Frequency of the modal class.

f_0 = Frequency of the class proceeding to the modal class.

f_2 = Frequency of the class succeeding to the modal class.

c = Frequency of the modal class

Example 3 Locate the mode from the following frequency distribution:

C.I.:	10–20	20–30	30–40	40–50	50–60
x:	12	35	40	25	10

Solution

C. I.	*x*
10–20	12
20–30	35
30–40	40
40–50	25
50–60	10

Here the highest frequency is 40.

30–40 is the modal class.

$l_0 = 30$

$f_1 = 40, f_0 = 35, f_2 = 25, c = 10$

using the formula

$$\text{Mode} = l_0 + \frac{(f_1 - f_0) \times c}{2 f_1 - f_0 - f_2}$$

$$= 30 + \frac{(40 - 35) \times 10}{80 - 35 - 25}$$

$$= 30 + \frac{5}{20} \times 10$$

$$= 32.5$$

Mode is 32.5

3. Graphical Method

Prepare a histogram (refer chapter 2: Histogram). Consider the bar having maximum height and the bars adjoining to its left and right.

Now, join the top left corner of the highest bar to the top left corner of the right side bar and top right corner of the highest bar to top right corner of the bar to its left. From the point of intersection of the highest bar, draw a line perpendicular to X-axis, the value corresponding to the point touching X-axis is the mode.

Example 4 Locate the mode graphically for the given frequency distribution.

C. I. :	10–20	20–30	30–40	40–50	50–60	60–70
f:	12	35	40	25	10	6

Solution

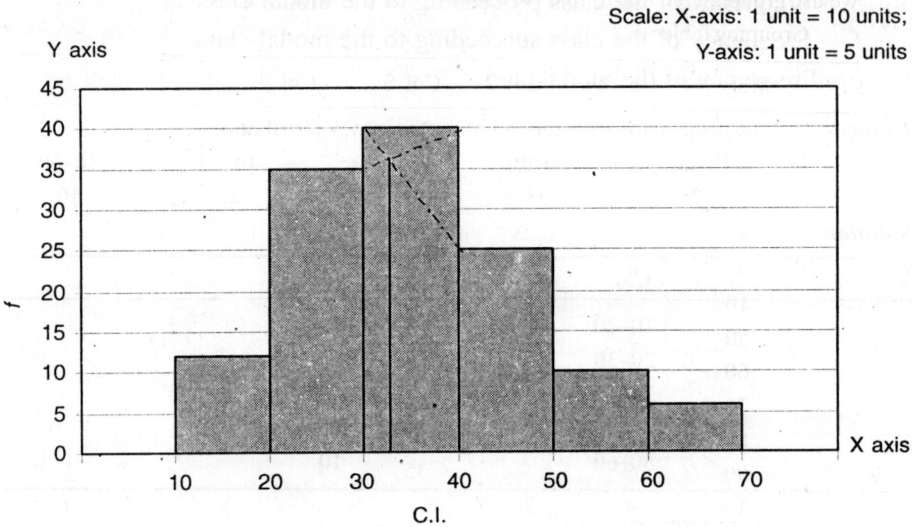

Mode value from the graph is 32.

4.13 BI-MODAL SERIES

In certain series, we see more than one peak points, in those case we can't find mode directly, we use a method known as grouping method.

Grouping method consist of a grouping table and analysis table.

Grouping table consists of six columns.

Column 1: Given frequencies.

Column 2: Add the frequencies two by two.

Column 3: Leave the first frequency, add the frequencies two by two.

Column 4: Add the frequencies three by three.

Column 5: Leave the first frequency and add three by three.

Column 6: Leave the first two frequencies and add the rest three by three.

In each column note the maximum frequencies.

Analysis table: Now in the analysis table against each class interval draw a tally mark for each column maximum frequency combination. The class corresponding to the maximum tally marks is the mode (in case of discrete series) or modal class (in case of continuous series).

Example 5 Obtain mode for the following frequency distribution.

Wage:	5	6	7	8	9	10	50	60	70	80	90	100
f:	10	2	20	3	8	4	3	7	17	19	16	4

Solution In the above problem we see that the highest value is 20 which corresponds to 7, but we also see that 80 is also having highest peak and its adjacent values are more concentrated. So we use grouping method to find the mode.

Grouping table

Wage	Col. 1	Col. 2	Col. 3	Col. 4	Col. 5	Col. 6
5	10					
6	2	12				
7	20		22	32		
8	3	23			25	
9	8		11			31
10	4	12		15		
50	3		7		15	
60	7	10				14
70	17		24	27		
80	19	36			43	
90	16		35			52
100	4	20		39		

Analysis table

Wage	Col. 1	Col. 2	Col. 3	Col. 4	Col. 5	Col. 6	Total
5							
6							
7	1						1
8							
9							
10							
50							
60					1		1
70		1			1	1	3
80		1	1	1	1	1	5
90			1	1		1	3
100				1			1

In column (1) highest values corresponds to 7. So it is mark 1, in column 2 highest values is 36 which is combined value frequency of 70 and 80. So in the table it is marked 1 against 70 and 80, and it goes on so. We see that 80 has the highest value ∴ 80is the mode.

Mode is 80.

Example 6 Calculate the mode by grouping method.

C. I.:	5–10	10–15	15–20	20–25	25–30
x:	32	35	25	34	12

Solution Grouping table

C. I.	Col. 1	Col. 2	Col. 3	Col. 4	Col. 5	Col. 6
5–10	32					
10–15	35	67				
15–20	25		60	92		
20–25	34	59			94	
25–30	12		46			71

Analysis table

C. I.	Col. 1	Col. 2	Col. 3	Col. 4	Col. 5	Col. 6	Total
5–10		1		1			2
10–15	1	1	1	1	1		5
15–20			1	1	1	1	4
20–25					1	1	2
25–30						1	1

Explanation as in Example 5: Here the modal class is 10–15. The mode is given by

$$Z = l_0 + \frac{(f_1 - f_0) \times c}{2\,f_1 - f_0 - f_2}$$

$l_0 = 10, f_1 = 35, f_0 = 32, f_2 = 25, c = 5$

$$Z = 10 + \frac{(35 - 32) \times 5}{70 - 32 - 25}$$

$$= 10 + \frac{3}{13} \times 5$$

$$= 10 + \frac{15}{13}$$

$$= 10 + 1.15 = 11.15$$

∴ Mode is 11.15.

4.14 EMPIRICAL FORMULA

Mode can be found out approximately based upon the empirical formula. Empirical formula is used when mode is ill-defined.

Mode = 3 median – 2 mean.

Example 7 Calculate mode using empirical formula for the given frequency distribution.

C. I.:	10–20	20–30	30–40	40–50	50–60	60–70
f:	12	35	40	25	10	6

Solution

C. I.	f	x	$u = \frac{x-35}{10}$	*c.f.* (less than)	*fu*
10–20	12	15	–2	12	–24
20–30	35	25	–1	47	–35
30–40	40	35	0	87	0
40–50	25	45	1	112	25
50–60	10	55	2	122	20
60–70	6	65	3	128	18
Total	128				4

$$\text{Mean} = A + \frac{h\Sigma u}{N}$$

$$= 35 + \frac{10 \times 4}{128}$$

$$= 35 + 0.3125$$

$$= 35.3125$$

$$\text{Now } \frac{N}{2} = \frac{128}{2} = 64$$

$\therefore$ The median class is (30–40)

$$\text{Median} = l_0 + \frac{\left(N/2 - m\right)}{f} \times c$$

$$= 30 + \frac{(64 - 47)}{40} \times 10$$

$$= 30 + 4.25$$

$$= 34.25$$

Mode = 3 median – 2 mean

$$= 3 \times 34.25 - 2 \times 35.3125$$

$$= 102.75 - 70.625$$

$$= 32.125$$

Mode: 32.125.

4.15 MERITS AND DEMERITS OF MODE

Mode also has the same merits and demerits as that of median. In addition to that mode has one more demerit, that is, mode is ill-defined in certain frequency distribution.

Merits:

1. Mode is easy to calculate and understand.
2. It can be located by mere iinspection and can be obtained graphically.
3. It is not affected by extreme values.
4. It is convenient to obtain in case of open end classes.

Demerits:

1. It is not rigidly defind. It is ill defined.
2. It is not based on all observations.
3. It is not suitable for further mathematical treatments.
4. As compared with mean, mode is affected to a greater extent by the fluctuation of sampling.

Solved Examples

1. Determine the mode graphically from the following series:

Week wages in Rupees	Number of workers
10–15	7
15–20	19
20–25	27
25–30	15
30–40	12
40–60	12
60–80	8

(B. U., April, 1992)

Solution Construct a histogram from gives data.

Mode = 24.0 graphically.

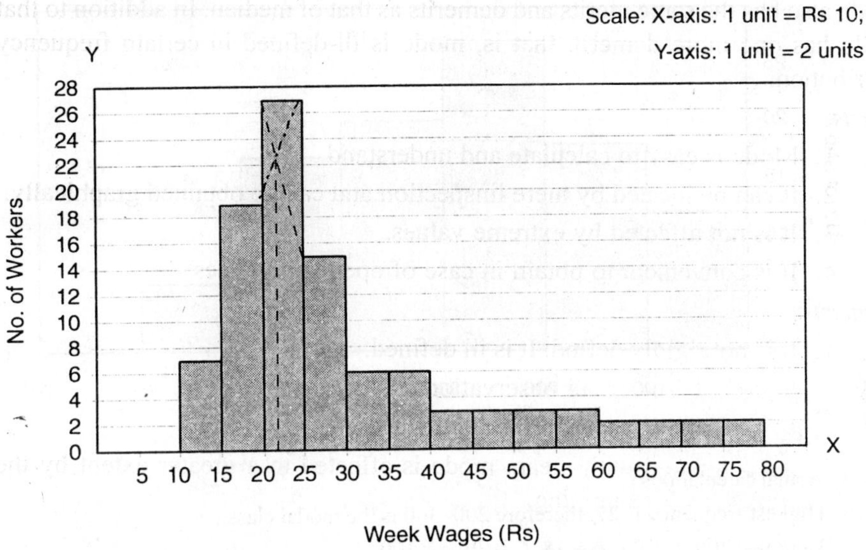

2. Determine mode graphically and verify this value with actual calculations.

Profits (Rs)	No. of shops
0–100	12
100–200	18
200–300	27
300–400	20
400–500	17
500–600	6

(B. U., April, 1997, 1998)

Solution

Graphical method

From the histogram we find that mode is = 256.

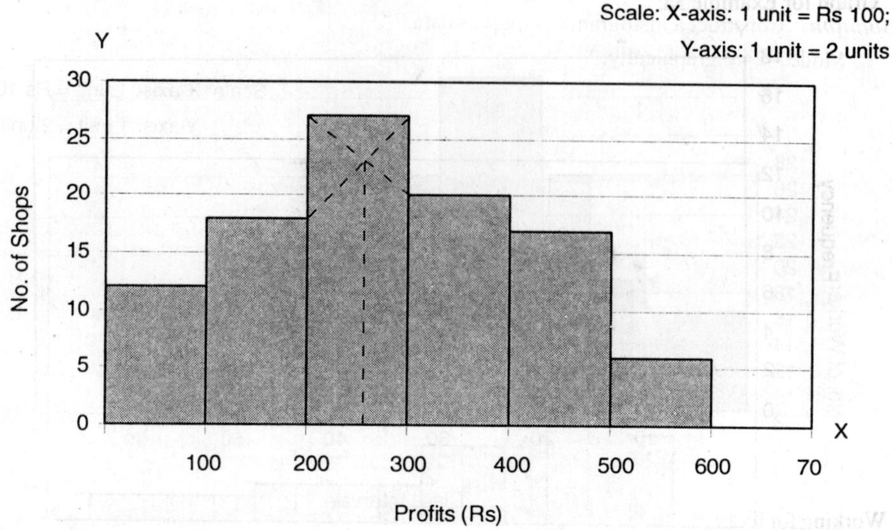

Actual calculations

Highest frequency is 27, therefore 200–300 is the modal class.

Here $l_0 = 200, f_1 = 27, f_0 = 18, f_2 = 20, c = 100$

$$Z = l_0 + \frac{(f_1 - f_0) \times c}{2\,f_1 - f_0 - f_2}$$

$$= 200 + \frac{(27 - 18) \times 100}{2 \times 27 - 18 - 20}$$

$$= 200 + \frac{9 \times 100}{16}$$

$$= 200 + 0.5625 \times 100$$

$$= 200 + 56.25 = 256.25$$

3. Draw a histogram and locate mode from the following and verify with actual calculation.

Class interval:	0–10	10–20	20–30	30–40	40–50
Frequency:	5	8	17	12	8

Solution

(a) Graphical method

From the histogram we find the mode = 26.4 (Refer graph on the next page).

(b) Actual calculation

C. I.	Frequency
0–10	5
10–20	8
20–30	17
30–40	12
40–50	8

Graph for Example 3a:

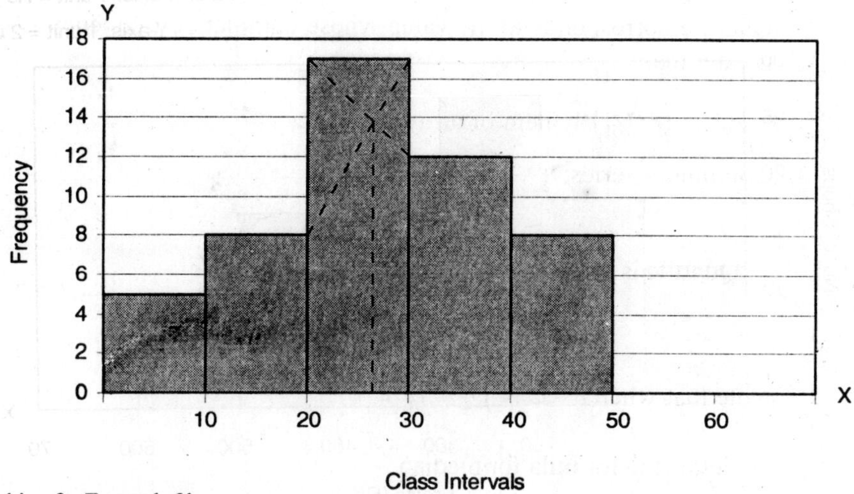

Working for Example 3b:

Here the highest frequency is 17, therefore the modal class is 20–30.

$\therefore\ l_0 = 20, f_1 = 17, f_0 = 8, f_2 = 12, c = 10$

using the formula

$$Z = l_0 + \frac{(f_1 - f_0) \times c}{2 f_1 - f_0 - f_2}$$

$$= 20 + \frac{(17 - 8) \times 10}{34 - 8 - 12}$$

$$= 20 + \frac{9}{14} \times 10$$

$$= 20 + 0.642 \times 10$$

26.42

Mode is 26.42.

4.16 QUARTILES

Quartiles divides a given frequency distribution in to four equal parts. There are three quartiles namely Q_1, Q_2, and Q_3. Quartiles are useful in determining the shape of a distribution.

3.16 CALCULATION OF QUARTILES

1. Individual series

$$Q_r = \left(\frac{r(n+1)}{4}\right)\text{th item of arrayed series.}$$

2. Discrete frequency distribution

   Quartile corresponds to the value whose cumulative frequency is just greater than:

   $$Q_r = \left(r\,{}^{N}\!/_{4}\right)\text{th item of arrayed series.}$$

3. Continuous series:

   $$\text{Quartile is given by } Q_r = l_o + \frac{\left(r\left({}^{N}\!/_{4}\right) - m\right)}{f} \times c$$

   $$\text{Note that when } r = 2 \text{ is } Q_2 = l_o + \frac{\left({}^{N}\!/_{2} - m\right)}{f} \times c$$

   Which is the formula for median

   $\therefore$ Q_2 is median.

Example 1 Calculate the quartiles of the given data:
50, 45, 33, 60, 43, 37, 39.

Solution Arrange the given data 33, 37, 39, 43, 45, 50, 60.

$$Q_1 = \left(\frac{n+1}{4}\right)\text{th}$$

$$= \left(\frac{7+1}{4}\right)\text{th}$$

= 2th item

$Q_1 = 37$

$$Q_2 = \left(\frac{2(n+1)}{4}\right)\text{th}$$

= 4th item

$Q_2 = 43$

$$Q_3 = \left(\frac{3(n+1)}{4}\right)\text{th}$$

= 6th item

$Q_3 = 50$

Example 2 For the following distribution find the quartiles.

x:	120	121	122	123	124	125
f:	2	4	7	14	21	22

Solution

x	f	$c.f.$
120	2	2
121	4	6
122	7	13
123	14	27
124	21	48
125	22	70
		Total $N = 70$

$$Q_1 = \left(\frac{N}{4}\right)\text{th item}$$

$$= \left(\frac{70}{4}\right)\text{th item}$$

$$= 17.5\text{th item}$$

$$Q_1 = 123$$

$$Q_2 = \left(2 \times \frac{N}{4}\right)\text{th item}$$

$$Q_2 = \left(\frac{70}{2}\right)\text{th item}$$

$$= 35\text{th item}$$

$$Q_2 = 124$$

$$Q_3 = \left(3 \times \frac{N}{4}\right)\text{th item}$$

$$= \left(70 \times \frac{3}{4}\right)\text{th item}$$

$$= 52.5\text{th item}$$

$$Q_3 = 125$$

Example 3 For the following distribution find the quartiles.

C. I.:	0–10	10–20	20–30	30–40	40–50
f:	5	8	17	12	8

Solution

C. I.	f	$c.f.$
0–10	5	5
10–20	8	13
20–30	17	30
30–40	12	42
40–50	8	50

Here $Q_1: \frac{N}{4} = \frac{50}{4} = 12.5$

∴ First quartiles class is 10–20.

$$Q_1 = 10 + \frac{(12.5 - 5)}{8} \times 10$$

$$= 19.375$$

here $Q_2: \frac{N}{2} = \frac{50}{2} = 25$

∴ Second quartile class is 20–30.

$$Q_2 = 20 + \frac{(25 - 13)}{17} \times 10$$

$$= 20 + 7.0588$$

$$= 27.0588$$

here $Q_3: 3\frac{N}{4} = \frac{50 \times 3}{4} = 37.5$

∴ Third quartile class is 30–40.

$$Q_3 = 30 + \frac{(37.5 - 30)}{12} \times 10$$

$$= 30 + 6.25$$

$$= 36.25$$

Hence $Q_1 = 19.375$

$Q_2 = 27.0588$

$Q_3 = 36.25$

Extra Worked Problems

1. The following is the distribution of percentage alcohol content in 60 bottles of a particular medicine. Calculate the mean, median and the mode.

Percentage:	7.0–7.1	7.1–7.2	7.2–7.3	7.3–7.4	7.4–7.5
No. of bottles:	6	14	19	14	7

Solution

Percentage C. I.	No. of bottles f	x	$d = x - 7.25$	fd	Less than $c.f.$
7.0–7.1	6	7.05	–.2	–1.2	6
7.1–7.2	14	7.15	–.1	–1.4	20
7.2–7.3	19	7.25	0	0	39
7.3–7.4	14	7.35	.1	1.4	53
7.4–7.5	7	7.45	.2	1.4	60
Total	60			0.2	

Mean:

$$\overline{X} = \frac{\Sigma f d}{N} + A \text{ Here } A = 7.25$$

$$= \frac{0.2}{60} + 7.25$$

$$= 7.2533$$

Median:

$$\frac{N}{2} = \frac{60}{2} = 3$$

∴ 7.2–7.3 is the median class.

Here $l_0 = 7.2$, $m = 20$, $f = 19$, $c = 0.10$

$$M = l_0 + \left(\frac{N/2 - m}{f}\right) \times c$$

$$= 7.2 + \frac{(30 - 20)}{19} \times .1$$

$$= 7.2 + .0526$$

Median = 7.2526.

Mode:

Here the highest frequency is 19.

∴ The modal class is 7.2–7.3.

$$Z = l_0 + \frac{(f_1 - f_0) \times c}{2 f_1 - f_0 - f_2}$$

Where $l_0 = 7.2, f_1 = 19, f_0 = 14 f_2 = 14, c = 0.1$

$$= 7.2 + \frac{(19 - 14) \times 0.1}{2 \times 19 - 14 - 14}$$

$$= 7.2 + \frac{5}{10} \times 0.1 = 7.2 + 0.05$$

Mode = 7.25

2. Calculate the mean and median from the following data:

Marks	No of Students
0–10	5
10–20	10
20–30	25
30–40	30
40–50	20
50–60	10

(B. U., April, 1995)

Solution

Marks C. I.	No. of students f	x	$u = \frac{x - 35}{10}$	fu	Less than c.f.
0–10	5	5	–3	–15	5
10–20	10	15	–2	–20	15
20–30	25	25	–1	–25	40
30–40	30	35	0	0	70
40–50	20	45	1	20	90
50–60	10	55	2	20	100
Total	100			–20	

Mean:

$A = 35, h = 10$

$$\overline{X} = \frac{h \sum f u}{N} + A$$

$$= 10 \times \frac{-20}{100} + 35$$

$$= -2 + 35$$

$$= 33$$

Median:

$$N = 100, \quad \frac{N}{2} = 50$$

Median class is 30–40

$l_0 = 30, m = 40, f = 30, c = 10$

$$M = l_0 + \left(\frac{N/2 - m}{f}\right) \times c$$

$$= 30 + \left(\frac{50 - 40}{30}\right) \times 10$$

$$= 30 + 3.333$$

$$= 33.333.$$

3. Calculate mean and median of the following.

Income (Rs)	10–20	20–30	30–40	40–50	50–60	60–70
No. of workers:	5	15	20	20	10	5

(B. U., October, 1996)

Solution

Income (Rs) C. I.	No. of workers f	Mid value x	$u = \dfrac{x-35}{10}$	fu	Less than $c.f.$
10–20	5	15	–2	–10	5
20–30	15	25	–1	–15	20
30–40	20	35	0	0	40
40–50	20	45	1	20	60
50–60	10	55	2	20	70
60–70	5	65	3	15	75
Total	75			30	

Mean:

$A = 35, h = 10$

$$\overline{X} = A + \frac{(\Sigma f u) \times h}{N}$$

$$= 35 + \frac{10 \times 30}{75}$$

$$= 35 + 4$$

Mean = 39

Median:

$$N = 75, \frac{N}{2} = 37.5$$

The median class: 30–40.

$l_0 = 30$, $m = 20$, $f = 20$, $c = 10$

$$M = l_0 + \left(\frac{N/2 - m}{f}\right) \times c$$

$$= 30 + \frac{(37.5 - 20)}{20} \times 10$$

$$= 30 + 8.75$$

Median = 38.75.

4. Draw an ogive curve from the data given below.

Profits Rs in lakhs	No. of shop
10–20	6
20–30	8
30–40	12
40–50	18
50–60	25
60–70	16

Contd

Profits Rs in lakhs	No. of shop
70–80	8
80–90	5
90–100	2

(B. U., Oct./Nov., 2001)

Locate first quartiles, third quartiles and second quartiles graphically.

C. I.	x	Less than
10–20	6	6
20–30	8	14
30–40	12	26
40–50	18	44
50–60	25	69
60–70	16	85
70–80	8	93
80–90	5	98
90–100	2	100

Scale: X-axis: 1 unit = 10 units;

Y-axis: 1 unit = 10 units

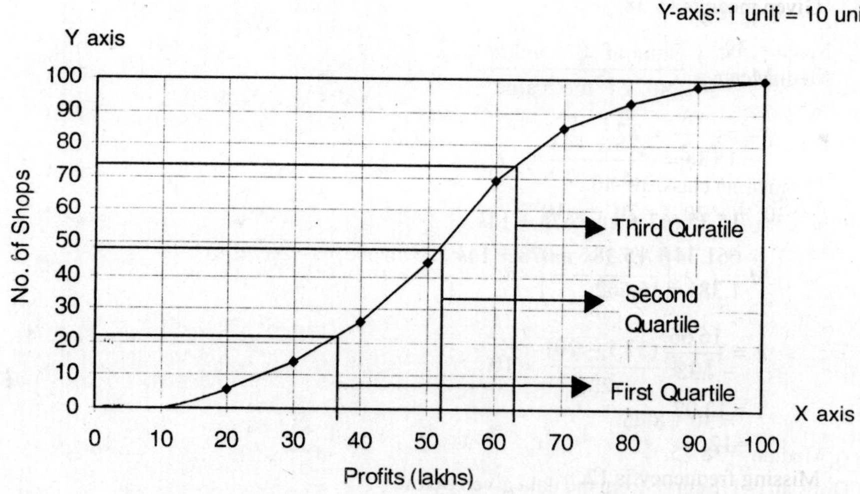

Draw less then ogive to find first and third quartile.

$$\text{First quartile} = \left(\frac{100}{4}\right)\text{th} = 25\text{th item} = 39$$

$$\text{Third quartile} = \left(\frac{3}{4}\,100\right)\text{th} = 75\text{th item} = 64$$

Median:

$$\text{Second quartile} = \left(\frac{100}{2}\right)\text{th} = 50\text{th item} = 53$$

5. From the following data find the missing frequency when mean is 15.38.

Central value:	10	12	14	16	18	20
Frequency:	3	7	?	20	8	5

(B. U., April/May, 2001)

Solution

Central value x	Frequency f	fx
10	3	30
12	7	84
14	?(k)	$14k$
16	20	320
18	8	144
20	5	100
Total	$43 + k$	$678 + 14k$

Given mean is 15.38

$$\text{Mean} = \frac{\text{Sum of the value}}{\text{No. of the value}}$$

$$15.38 = \frac{678 + 14k}{43 + k}$$

$$15.38\,(43 + k) = 678 + 14k$$

$$661.34 + 15.38k = 678 + 14k$$

$$1.38k = 16.662$$

$$k = \frac{16.66}{1.38}$$

$$= 12.07$$

$$= 12.$$

Missing frequency is 12.

6. The following data gives the distribution of marks scored by a set of students. Find the mean and quartiles of the distribution.

Percentage of Marks	20 & more	30 & more	40 & more	50 & more	60 & more	70 & more
No. of students	80	72	60	27	10	0

Solution

Convert the given greater than table to frequency distribution:

Percentage of marks (C. I)	No. of students (*f*)	*x*	$u = \frac{x - 35}{10}$	*fu*	Less than *c.f.*
20–30	8	25	–2	–16	8
30–40	12	35	–1	–12	20
40–50	33	45	0	0	53
50–60	17	55	1	17	70
60–70	10	65	2	20	80
Total				9	

Mean:

$A = 45, h = 10$

$$\overline{X} = A + \frac{\Sigma f u}{N} \times h$$

$$= 45 + \frac{9 \times 10}{80}$$

$$= 45 + 1.125$$

$$= 46.125$$

Quartiles

$$Q_1 : \frac{N}{4} = \frac{80}{4} = 20$$

First quartiles class 30–40

$l_0 = 30, m = 8, f = 12, c = 10$

$$Q_1 = l_0 + \frac{(N/4 - m)}{f} \times c$$

$$= 30 + \frac{(20 - 8)}{12} \times 10$$

$$= 40$$

$$Q_2 : \frac{N}{2} = \frac{80}{2} = 40$$

Second quartiles class 40–50

$l_0 = 40, m = 20, f = 33, c = 10$

$$Q_2 = l_0 + \frac{(N/2 - m)}{f} \times c$$

$$= 40 + \frac{(40 - 20)}{33} \times 10$$

$$= 46.06$$

Q_3 Third quartiles class 50–60

$$Q_3 = 50 + \frac{(60-53)}{17} \times 10$$

$$= 54.117$$

7. Find median and mode.

Income (above Rs):	1000	1500	2000	2500	3000	3500	4000
No. of person:	80	62	33	16	6	2	0

Solution Let us convert the greater then table to frequency distribution.

Income (C.I.)	No. of persons (f)	Cumulative frequency Less than
1000–1500	18	18
1500–2000	29	47
2000–2500	17	64
2500–3000	10	74
3000–3500	4	78
3500–4000	2	80
Total	80	

Median:

$$N = 80, \quad \frac{N}{2} = 40$$

The median class is 1500–2000.

$l_0 = 1500$, $m = 18$, $f = 29$, $c = 500$

$$M = l_0 + \frac{(N/2 - m)}{f} \times c$$

$$= 1500 + \frac{(40-18)}{29} \times 500$$

$$= 1500 + 379.310$$

Median = 1879.310

Mode:

The highest frequency is 29, $\therefore$ the modal class is 1500–2000.

$l_0 = 1500, f_1 = 29, f_0 = 18, f_2 = 17$

$$Z = l_0 + \frac{(f_1 - f_0) \times c}{2 f_1 - f_0 - f_2}$$

$$= 1500 + \frac{(29-18) \times 500}{2 \times 29 - 18 - 17}$$

$$= 1500 + 0.478 \times 500 = 1500 + 239 = 1739.$$

Mode is 1739.

Exercises

1. Explain briefly the measures of central tendency.
2. Bring out the characteristic of an ideal average.
3. Define mean.
4. Mention any three properties of Arithmetic mean.
5. State any three merits of mean.
6. Define median.
7. State any three merits of median.
8. Define mode.
9. How do you calculate mode when it is ill-defined.
10. Find A.M of 12.4, 12.6, 12.9, 12.1, 12.3 cms. **[Ans. 12.46 cms]**
11. Find A.M of 3, 5, 12, 0, 4, 1, 3 and 7. **[Ans. 4.375]**
12. In an office there are 84 employees. Their salaries are as given.

Salary (Rs)	2430	2590	2870	3390	4720	5160
Employees	4	28	31	16	3	2

Find the mean salary of the employees. **[Ans. Rs 2975.36]**

13. Find the arithmetic mean of the given data.

Mark (x)	75	50	47	43	25
Frequency (f)	3	1	1	3	2

[Ans. 50.1]

14. Find the mean and the quartiles for the following data.

Size:	105	110	115	120	125	130	135	140
Frequency:	8	23	34	35	27	18	9	4

[Ans. 120.07, 113.25, 119.50, 125.93]

15. A survey of 128 smokers revealed the following frequency distribution of daily expenditure on smoking. Find the mean daily expenditure.

Expenditure (Rs)	10–20	20–30	30–40	40–50	50–60	60–70	70–80
No. of Smokers	23	44	35	12	9	3	2

[Ans. Rs 31.64]

16. Calculate the mean for the following frequency distribution.

Class interval:	0–8	8–16	16–24	24–32	32–40	40–48
Frequency	8	7	16	24	15	7

[Ans. 25.404]

17. The mean of marks scored by 30 girls of a class is 44%. The mean marks of 50 boys is 42%. Find the mean for the whole class.

[Ans. 42.75%]

18. A set of 20 values have mean 54. Another set of values have mean 60. If the combined mean is 56, how many values are there in the latter set?

[Ans. 10]

19. Obtain the missing frequency if the mean of the series is 10.5 years.

Age (years):	8	9	10	11	12	13	14	Total
Children:	13	23	25	?	20	12	2	120

[*Ans.* 25]

20. Obtain the median for the following frequency distribution.

x:	1	2	3	4	5	6	7	8	9
f:	8	10	11	16	20	25	15	9	6

[*Ans.* 5]

21. Locate the median for the following frequency distribution.

Marks:	45	50	52	57	60	65	70
Frequency:	4	7	8	5	3	2	5

[*Ans.* 52]

22. Find the median and the quartiles.

C. I.:	18–20	20–22	22–23	23–24	24–26	26–30	30–35
f:	4	10	9	17	20	15	5

Verify the answer graphically.

[*Ans.* 24, 22.67, 25.6]

23. Find the mode and the quartiles of the following frequency distribution.

Wage (Rs):	below 30	30–50	50–70	70–90	90–110	110–130	above 130
No. of workers:	3	25	34	34	12	2	2

[*Ans.* 50, 66.50, 82.90]

24. Graphically find the mode of the following frequency distribution.

Marks:	0–10	10–20	20–40	40–60	60–70
No. of students:	8	10	15	12	4

25. Calculate the value of the mean, median and mode from the following.

Monthly Rent (in Rs)	Number of families
20–40	6
40–60	9
60–80	11
80–100	14
100–120	20
120–140	5

26. Given below is the distribution of profits (in 000 Rs) earned by 94 book depots in Bangalore city. Calculate Mean and Median.

Profit (Below):	20	30	40	50	60	70	80	90
No. of book depots:	5	14	27	48	68	83	91	94

(B. U., November, 1997)

(***Hint***: Convert the less than frequency to frequency distribution.)

5

Measures of Dispersion

Contents

Formulae

1. Range $= L - S$

2. Coefficient of Range $= \dfrac{L - S}{L + S}$

3. Quartile Deviation (Q. D) $= \dfrac{Q_3 - Q_1}{2}$

4. Co-efficient of Quartile Deviation $= \dfrac{Q_3 - Q_1}{Q_3 + Q_1}$

5. Mean Deviation (M. D): $= \dfrac{\Sigma f|X - A|}{N}$ continuous and discrete case.

   $\dfrac{\Sigma |X - A|}{n}$ raw data.

6. Co-efficient of Mean Deviation $= \dfrac{M.D}{A}$

7. Standard Deviation (S. D):

   (a) For raw data $\sqrt{\dfrac{\Sigma X^2}{n} - \left(\dfrac{\Sigma X}{n}\right)^2}$

(b) For discrete and continuous data:

$$\sigma = \sqrt{\frac{\Sigma f X^2}{N} - \left(\frac{\Sigma f X}{N}\right)^2}$$

8. Co-efficient of Variance: $C.V = \frac{S.D}{\overline{X}} \times 100$

Key Points

1. Range, Q.D, M.D, S.D are absolute measure and their coefficients are relative measures.
2. S.D satisfies almost all the characteristic of an ideal measure of dispersion.
3. Coefficient of variance is an ideal measure to compare the consistency of two series.

INTRODUCTION

Averages gives us an idea about the central value of the distribution, but it does not provides an information about the scattering of values in the distribution. For e.g.: Consider the series (i) 7, 8, 9, 10, 11 (ii) 3, 6, 9, 12, 15 (iii) 1, 5, 9, 13, 17 for all the series $n = 5$ and mean is 9, but we can observe that the values are not scattered in the same manner, hence it can be observed that averages does not reflects in the scattering of the values. So we need certain measures which tells about the scattering of the values, those measures are termed as measures of dispersion.

5.1 MEASURES OF DISPERSION

The following are the measures of dispersion.

1. Range
2. Quartile deviation
3. Mean deviation
4. Standard deviation
5. Coefficient of variation

5.2 CHARACTERISTICS FOR AN IDEAL MEASURE OF DISPERSION

The following are main requisites of an ideal measure of dispersion.

1. It should be rigidly defined.
2. It should be based on all observation.
3. It should be easily computed
4. It should be amenable for further mathematical analysis.
5. It should be least affected by fluctuation of sampling.

5.3 RANGE (R)

The difference between the largest and smallest values of a set of data is known as its range.

$$\text{Range} = L - S$$

5.4 CO-EFFICIENT OF RANGE

It is the ratio of difference between the largest and smallest values to the sum of the largest and smallest value.

$$\text{Coefficient of range: } \frac{L - S}{L + S}$$

Example 1 Calculate the range and co-efficient of range for the following data.

Marks: 50, 75, 60, 23, 57, 90, 95, 29, 30, 37.

Solution In the given set of data, the highest value = 95

Lowest value = 23

$$\text{Range} : R = 95 - 23$$

$$= 72$$

$$\text{Co-efficient of } R = \frac{L - S}{L + S}$$

$$= \frac{95 - 23}{95 + 23} = \frac{72}{118}$$

$$= 0.610$$

5.5 QUARTILE DEVIATION (Q.D)

Quartile deviation is half of the difference between third quartile and first quartile.

$$\text{Q.D} = \frac{1}{2}(Q_3 - Q_1)$$

It is an absolute measure of dispersion.

5.6 CO-EFFICIENT OF QUARTILE DEVIATION

It is the ratio of difference between the third and first quartiles to the sum of the third and first quartiles.

$$\text{Co-efficient of Q. D} = \frac{Q_3 - Q_1}{Q_3 + Q_1}$$

Example 2 In a frequency distribution, the three quartiles are 128, 147, and 157. Calculate the quartiles deviation and its co-efficient.

Solution Given $Q_1 = 128$, $Q_3 = 157$

$$Q.D = \frac{1}{2}(Q_3 - Q_1)$$

$$= \frac{1}{2}(157 - 128)$$

$$= 14.5$$

$$\text{Co-efficient of Q. D} = \frac{Q_3 - Q_1}{Q_3 + Q_1}$$

$$= \frac{157 - 128}{157 + 128} = \frac{29}{285}$$

$$= 0.1018$$

5.7 MEAN DEVIATION (M.D)

Mean deviation is the average of the absolute deviations taken from an average A (usually mean, median or mode).

1. ***Individual Series***

$$\text{M.D} = \frac{1}{N} \Sigma |X - A|$$

Where A is mean, median or mode.

2. ***For Grouped Data*** (***Discrete*** and ***Continuous***)

$$\text{M.D} = \frac{1}{N} \Sigma f\,|X - A|$$

5.8 CO-EFFICIENT OF MEAN DEVIATION

It is the ratio of mean deviation to the average A.

$$\text{Co-efficient of M.D} = \frac{\text{Mean Deviation}}{A} \times 100$$

This gives the percentage of the coefficient.

Example 3 Calculate mean deviation from mode and also its co-efficient.

Size of shoes: 3, 5, 3, 4, 2, 1, 5, 3, 3, 4.

Solution Mode of the give set of observation is 3 as 3 is having the highest frequency.

X	$\|X-A\|$
3	0
5	2
3	0
4	1
2	1
1	2

Contd.

X	$\|X-A\|$
5	2
3	0
3	0
4	1
Total	9

Here A is mode = 3

Mean deviation from mode.

$$\text{M.D (Z)} = \frac{\Sigma|X - A|}{N}$$

$$= \frac{9}{10}$$

$$= 0.9$$

$$\text{Co-efficient of M.D (Z)} = \frac{M.D\,(2)}{Z}$$

$$= \frac{9}{3}$$

$$= 0.3$$

Example 4 Calculate mean deviation from median and also its co-efficient for the following frequency distribution.

x:	1	2	3	4	5	6	7
f:	3	6	5	17	9	7	3

Solution

x	f	$c.f.$	$\|X-A\|$	$f\|X-A\|$
1	3	3	3	9
2	6	9	2	12
3	5	14	1	5
4	17	31	0	0
5	9	40	1	9
6	7	47	2	14
7	3	50	3	9
Total	50			58

Here median deviation about median is taken $\therefore A$ = median

$$N = 50, \therefore \frac{N}{2} = 25$$

$\therefore$ Median = 4

$$\text{Mean deviation about median} = \frac{\Sigma f|X - A|}{N}$$

$$= \frac{58}{50}$$

$$= 1.16$$

$$\text{Co-efficient of M.D (M)} = \frac{M.D\,(M)}{M}$$

$$= \frac{1016}{4}$$

$$= 0.29$$

Example 3 Calculate the mean deviation about the mean and its co-efficient for the following data:

C. I.:	10–20	20–30	30–40	40–50	50–60
f:	5	7	12	3	3

Solution

C. I.	f	x	$u = \frac{X - 35}{10}$	fu	$\|X-A\|$	$f\|X-A\|$
10–20	5	12	–2	–10	17.333	86.665
20–30	7	25	–1	–7	7.333	51.331
30–40	12	35	0	0	2.667	32.004
40–50	3	45	1	3	12.667	38.001
50–60	3	55	2	6	22.667	68.001
Total	30			–8		276.002

Here mean deviation is taken about mean

A = Mean

$$\text{Mean} = 35 + \frac{(-8)}{30} \times 10$$

(using the formula $\overline{X} = A + h\,\dfrac{\Sigma f u}{N}$)

$$= 35 - 2.667$$

$$= 32.333$$

$$\text{Mean deviation about mean} = \frac{\Sigma f\left(X - \overline{X}\right)}{N}$$

$$= \frac{276.002}{30}$$

$$= 9.2$$

$$\text{Co-efficient of M.D} = \frac{M.D(\overline{X})}{\overline{X}}$$

$$= \frac{9.2}{32.333}$$

$$= 2.845$$

5.9 STANDARD DEVIATION (σ)

The average of the square of the deviations taken from mean is called variance, and the positive square root of the variance is called standard deviation.

A. Direct Method

1. ***Individual Series (Raw Data)***

$$\sigma = \sqrt{\frac{1}{n}\Sigma\left(X - \overline{X}\right)^2}$$

For computation the following formula may be used

$$\sigma = \sqrt{\frac{1}{n}\left(\Sigma X^2\right) - \left(\overline{X}\right)^2}$$

Example 1 Calculate the standard deviation for the following set of data.

5, 7, 9, 10, 11, 13, 15.

Solution

	X	X^2
	5	25
	7	49
	9	81
	10	100
	11	121
	13	169
	15	225
Total	57	770

$$\text{Mean} = \frac{\Sigma X}{n}$$

$$= \frac{57}{7}$$

$$= 8.1429$$

$$\sigma = \sqrt{\frac{1}{n}\left(\Sigma X^2\right) - \left(\overline{X}\right)^2}$$

$$= \sqrt{\frac{770}{7} - (8.1429)^2}$$

$$= \sqrt{110 - 66.3068}$$

$= 6.61$

2. ***Grouped data*** (***Discrete*** and ***Continuous***)

$$\sigma = \sqrt{\frac{\Sigma f(X - \overline{X})^2}{N}}$$

for the computation the following formula may be used

$$\sigma = \sqrt{\frac{\Sigma f X^2}{N} - \left(\frac{\Sigma f X}{N}\right)^2}$$

$$= \sqrt{\frac{\Sigma f X^2}{N} - (\overline{X})^2}$$

B. Short Cut Method

We will discuss the problem using short cut method.

Step deviation method

Let $u = \dfrac{X - A}{h}$

Then $S.D\ (X) = h \times S.D\ (u)$

Note Standard deviation is not affected by change in origin.

i.e. if $d = X - A$

then $S.D\ (X) = S.D\ (d)$

Example 2 Find the standard deviation of the following data.

x:	1	2	3	4	5
f:	3	5	12	9	1

Solution

x	f	$d = X - 3$	d^2	fd	fd^2
1	3	–2	4	–6	12
2	5	–1	1	–5	5
3	12	0	0	2	0
4	9	1	1	9	9
5	1	2	4	2	4
Total	30			0	30

$$\sigma = \sqrt{\frac{\Sigma f d^2}{N} - \left(\frac{\Sigma f d}{N}\right)^2}$$

$$= \sqrt{\frac{30}{30} - (0)^2}$$

$$= \sqrt{1}$$

$$= 1$$

Example 3 Calculate the mean of the given frequency distribution.

Class intervals:	10–20	20–30	30–40	40–50
Frequency:	15	25	30	5

Solution

C. I.	f	Mid value	$u = \frac{X-35}{10}$	u^2	fu	fu^2
10–20	15	15	–2	4	–30	60
20–30	25	25	–1	1	–25	25
30–40	30	35	0	0	0	0
40–50	5	45	1	1	5	5
Total	75				–50	90

$$\sigma = \sqrt{\frac{\Sigma f u^2}{N} - \left(\frac{\Sigma f u}{N}\right)^2}$$

$$= \sqrt{\frac{90}{75} - \left(\frac{-50}{75}\right)^2}$$

$$= \sqrt{1.2 - 0.444}$$

$$= \sqrt{0.756}$$

$$= 0.8695$$

$S.D\,(u) = .8695$

$S.D\,(x) = h \times S.\,D\,(u)$

$= 10 \times .8695$ ($\because$ here $h = 10$)

$= 8.695$

5.10 CO-EFFICIENT OF VARIATION (C.V)

Co-efficient of variation is the ratio of the standard deviation to the mean. It is usually expressed in percentage.

Co-efficient of variation is given by:

$$C.V = \frac{S.D}{\text{Mean}} \times 100$$

$$= \frac{\sigma}{\overline{X}} \times 100$$

This is a suitable measure to compare any two series. A series is more consistent if its C.V is less compared to the other. It is a relative measure.

Example 4 The following are the scores secured by two batsmen *I* and *E* is a series of ten innings between India and England. Find the more consistent batsman.

I:	15	110	6	75	7	19	119	36	84	70
E:	49	10	19	40	4	51	37	48	13	60

Solution

Batsman *I*		Batsman *E*	
X	X^2	*Y*	Y^2
15	225	49	2401
110	12100	10	100
6	36	19	361
75	5625	40	1600
7	49	4	16
19	361	51	2601
119	14161	37	1396
36	1296	48	2304
84	7056	13	169
70	4900	60	3600
Total 541	45448	331	14521

$$\text{Mean } I = \frac{541}{10} = 54.1$$

$$\sigma_I = \sqrt{\frac{\sum X^2}{n} - \left(\frac{\sum X}{n}\right)^2}$$

$$= \sqrt{\frac{45448}{10} - \left(\frac{54}{10}\right)^2}$$

$$\sigma_I = 40.224$$

$$C.V_I = \frac{S.D}{\text{Mean}} \times 100 = \frac{40.224}{54.1} \times 100$$

$$= 74.35\%$$

$$\text{Mean } E = \frac{331}{10} = 33.1$$

$$\sigma_E = \sqrt{\frac{\sum Y^2}{n} - \left(\frac{\sum Y}{n}\right)^2}$$

$$= \sqrt{\frac{14521}{10} - \left(\frac{331}{10}\right)^2}$$

$$\sigma_E = 18.880$$

$$C.V_E = \frac{18.880}{33.1} \times 100$$

$$= 57.039\%$$

Batsman E is more consistent as the C.V of E is less that C.V of I.

Example 5 Two workers on the same job show the following results over a period of time

	Worker A	Worker B
Mean time of completion of work	30	25
S.D of time of completion	6	4

1. Which worker appears to be more consistent?
2. Which worker appears to be faster is completing the job?

Solution

$$C.V_A = \frac{S.D}{\text{Mean}} \times 100 = \frac{6}{30} \times 100 = 20\%$$

$$C.V_B = \frac{S.D}{\text{Mean}} \times 100 = \frac{4}{25} \times 100 = 16\%$$

1. $C.V_B$ is less than $C.V_A$ therefore worker B is more consistent.
2. $C.V_A$ is greater than $C.V_B$ therefore worker A appears to be faster is completing the job.

Example 6 An analysis of monthly wages paid to the workers in two firms A and B belonging to the same industry, gave the following results.

	Firm A	Firm B
No. of wage-earners	986	548
Average monthly wages	Rs 52.5	Rs 47.5
Variance of distribution of wage	100	121

1. Which firm A or B has a larger wage bill?
2. In which firm A or B is there greater variability in individual wages.
3. Calculate the average monthly wages of all the workers in the firm A and B taken together.

Solution

1. Total wages in firm A is

Sum = Mean wage No. of people

= 52.5 × 986

= 51765

Total wages is Rs 51765

Total wages in firm B is

Sum = mean wage × No. of people

= 47.5 × 548

= 26030

Total wages is Rs 26030.

2. $C.V_A = \frac{S.D}{\text{Mean}} \times 100$

given variance = 100

$\therefore S.D = \sqrt{100} = 10$

$C.V_B = \frac{S.D}{\text{Mean}} \times 100$

given variance = 121

$\therefore S.D = \sqrt{121} = 11$

$C.V_A = \frac{10}{52.5} \times 100$	$C.V_B = \frac{11}{47.5} \times 100$
$= 19.05\%$	$= 23.16\%$

As $C.V_B$ is greater than $C.V_A$, firm B has greater variability in individual wages.

3. Combined mean $= \frac{n_1 \overline{X}_A + n_2 \overline{X}_B}{n_1 + n_2}$

$n_1 = 986, n_2 = 548$

$\overline{X}_A = 52.5, \ \overline{X}_B = 47.5$

$$\overline{X} = \frac{986 \times 52.5 + 548 \times 47.5}{986 + 548}$$

$$= \frac{51765 + 26030}{1534}$$

$= 50.714$

Averages monthly wages of all the workers in the firm A and B taken together is Rs 50.714.

Solved Examples

1. Calculate the quartile deviation and its co-efficient.

Time in minutes	Frequency
5–9.9	10
10–14.9	36
15–19.9	62
20–24.9	72
25–29.9	20

Solution We have to convert the exclusive class interval to inclusive type

	Time in minutes	Frequency	Less than cumulative frequency
	4.95–9.9	10	10
	9.95–14.9	36	46
Q_1 Class	14.95–19.9	62	108
Q_3 Class	19.95–24.9	72	180
	24.95–29.9	20	200

$$Q_1 : \frac{N}{4} = 50$$

∴ 14.95–19.95 is the first quartile class

$$Q_1 = l_0 + \frac{(N/4 - m)}{f} \times c$$

Where $l_0 = 14.95$, $c = 5$, $m = 46$, $f = 62$

$$= 14.95 + \frac{(50 - 46)}{62} \times 5$$

$$= 14.95 + .3225$$

$$= 15.2725$$

$$Q_3 : 3 \times \frac{N}{4} = 150 = 150$$

$\therefore$ **19.95–24.95 is the third quartile class.**

$$Q_3 = l_0 + \frac{\left(3\,{}^{N}\!/_{4} - m\right)}{f} \times c$$

Where $l_0 = 19.95$, $m = 108$, $f = 72$, $c = 5$

$$= 19.95 + \frac{(150 - 108)}{72} \times 5$$

$$= 19.95 + 2.9167$$

$$= 22.8667$$

Quartile deviation $= \dfrac{Q_3 - Q_1}{2}$

$$= \frac{22.8667 - 15.2725}{2}$$

$$= \frac{7.5942}{2}$$

$$= 3.7971$$

Co-efficient of Q.D $= \dfrac{Q_3 - Q_1}{Q_3 + Q_1}$

$$= \frac{22.8667 - 15.2725}{22.8667 + 15.2725}$$

$$= \frac{7.5942}{38.1392}$$

$$= 0.1991$$

2. Calculate the quartile deviation and its co-efficient from the following data.

C. I.:	10–15	15–20	20–25	25–30	30–35	35–40	40–45	45–50
f:	4	12	16	22	10	8	6	4

(B. U., November, 1994)

Solution

C. I.	f	Less than $c.f.$
10–15	4	4
15–20	12	16
20–25	16	32
25–30	22	54
30–35	10	64
35–40	8	72
40–45	6	78
45–50	4	82

Given $N = 82$

$$\frac{N}{4} = 20.5$$

$\therefore$ 20–25 is first quartile class

$$Q_1 = l_0 + \frac{\left(N/4 - m\right)}{f} \times c$$

$l_0 = 20$, $m = 16$, $f = 16$, $c = 5$

$$= 20 + \left(\frac{20.5 - 16}{16}\right) \times 5$$

$$= 20 + 1.4063$$

$$= 21.4063$$

Now $3\frac{N}{4} = 61.5$

$\therefore$ 30–35 is third quartile class

$$Q_3 = l_0 + \frac{\left(3\,N/4 - m\right)}{f} \times c$$

$l_0 = 30$, $m = 54$, $f = 10$, $c = 5$

$$= 30 + \frac{(61.5 - 54)}{10} \times 5$$

$$= 30 + 3.75$$

$$= 33.75$$

$$\text{Quartile deviation (Q.D)} = \frac{Q_3 - Q_1}{2}$$

$$= \frac{33.75 - 21.4063}{2}$$

$$= \frac{12.3437}{2}$$

$$= 6.1719$$

$$\text{Co-efficient of Q.D} = \frac{Q_3 - Q_1}{Q_3 + Q_1}$$

$$= \frac{33.75 - 21.4063}{33.75 + 21.4063}$$

$$= \frac{12.3437}{55.1563}$$

$$= 0.2238$$

3. Find quartile deviation and its co-efficient from the following data.

Marks (X):	23	29	34	38	47	53
No. of students:	4	9	11	18	23	31

(B. U. October, 1995)

Solution

Marks (X)	No. of students	Cumulative frequency ($c.f.$)
23	4	4
29	9	13
34	11	24
38	18	42
47	23	65
53	31	96

Here $N = 96$

$$\therefore \frac{N}{4} = \frac{96}{4} = 24$$

$$\therefore Q_1 = 34$$

and $$3\frac{N}{4} = \frac{3 \times 96}{4} = 42$$

$$\therefore Q_3 = 53$$

$$\text{Quartile deviation (Q.D)} = \frac{Q_3 - Q_1}{2}$$

$$= \frac{53 - 34}{2}$$

$$= \frac{19}{2}$$

$$= 9.5$$

$$\text{Co-efficient of Q.D} = \frac{Q_3 - Q_1}{Q_3 + Q_1}$$

$$= \frac{53 - 34}{53 + 34}$$

$$= \frac{19}{87}$$

$$= 0.2184$$

4. Calculate the mean deviation (from median) from the following data and also calculate its co-efficient.

Wages per week:	10–20	20–30	30–40	40–50	50–60	60–70	70–80
No. of workers:	4	6	10	20	10	6	4

(B. U., April, 1995)

Solution

Wages/week C. I.	No. of workers (f)	$c.f.$	X	$[X-A_1]$	$f[X-A]$
10–20	4	4	15	27.5	110
20–30	6	10	25	17.5	105
30–40	10	20	35	7.5	75
40–50	20	30	45	2.5	50
50–60	10	40	55	12.5	125
60–70	6	46	65	22.5	135
70–80	4	50	75	32.5	130
Total	50				730

Here $N = 50$, $\frac{N}{4} = 25$

$\therefore$ (40–50) is the median class.

$$\text{Median} = l_0 + \left(\frac{N/2 - m}{f}\right) \times c$$

$l_0 = 40$, $m = 20$, $f = 20$, $c = 10$

$$= 40 + \left(\frac{25 - 20}{20}\right) \times 10$$

$$= 40 + 2.5$$

$$= 42.5$$

Here mean deviation is about median $\therefore A =$ Median.

$$\text{M.D }(M) = \frac{\Sigma f(X - A)}{N}$$

$$= \frac{730}{50}$$

$$= 14.6$$

$$\text{Co-efficient M.D } (M) = \frac{M.D}{M}$$

$$= \frac{14.6}{42.5}$$

$$= .3435$$

5. (a) Calculate the value of co-efficient of mean deviation (from median) of the following data.

Marks:	10–20	20–30	30–40	40–50	50–60	60–70	70–80	80–90
No. of students	2	6	12	18	25	20	10	7

(B. U., September, 2001)

Solution

(a)

Marks C.I.	No. of student f	X	$c.f.$	$[X-A]$	$f[X-A]$
10–20	2	15	2	39.8	79.6
20–30	6	25	8	29.8	178.8
30–40	12	35	20	19.8	237.6
40–50	18	45	38	9.8	176.4
50–60	25	55	63	.2	5
60–70	20	65	83	10.2	204
70–80	10	75	93	20.2	202
80–90	7	85	100	30.2	211.4
Total	100				1294.8

Here A = Median

Median $N = 100, \frac{N}{2} = 50$

∴ 50–60 is the median class.

$$M = l_0 + \left(\frac{N/2 - m}{f}\right) \times c$$

$l_0 = 50, m = 38, f = 25, c = 10$

$$= 50 + \left(\frac{50 - 38}{25}\right) \times 10$$

$$= 50 + 4.8$$

$$= 54.8$$

$$A = 54.8$$

$$\text{M.D } (M) = \frac{\Sigma f (X - A)}{N}$$

$$= \frac{1294.8}{100}$$

$$= 12.948$$

$$\text{Co-efficient of M.D } (M) = \frac{M.D}{M}$$

$$= \frac{12.948}{54.8}$$

$$= 0.2363$$

(b)

Marks C. I.	No. of student f	X	$u = \frac{X - 55}{10}$	$[X-A]$	$f[X-A]$	fu
10–20	2	15	–4	39.3	78.6	–8
20–30	6	25	–3	29.3	175.8	–18
30–40	12	35	–2	19.3	231.6	–24
40–50	18	45	–1	9.3	167.4	–18
50–60	25	55	0	0.7	17.5	0
60–70	20	65	1	10.7	214.0	20
70–80	10	75	2	20.7	207	20
80–90	7	85	3	30.7	214.9	21
Total	100				1306.8	–7

Here A = Mean

Mean:-

$$\overline{X} = \left(A + \frac{h \Sigma f du}{N} \right)$$

$$= 55 - \frac{7 \times 10}{100}$$

$$= 55 - 0.7$$

$$= 54.3$$

$$A = 54.3$$

$$\text{M.D } (\overline{X}) = \frac{\Sigma f (X - A)}{N}$$

$$= \frac{1306.8}{100}$$

$= 13.068$

$$\text{Co-efficient of M.D } (\overline{X}) = \frac{M.D}{\overline{X}}$$

$$= \frac{13.068}{54.3}$$

$= 0.24066$

6. The following table gives the frequency distribution of marks obtained by 150 students in a certain examination.

Marks:	0–10	10–20	20–30	30–40	40–50	50–60	60–70	70–80
No. of students:	7	10	20	40	30	28	10	5

From the above distribution, calculate the standard deviation and the co-efficient of variation.

(B. U., April, 1992)

C. I.	f	X	$u = \frac{X-35}{10}$	u^2	fu	fu^2
0–10	7	5	–3	9	–21	63
10–20	10	15	–2	4	–20	40
20–30	20	25	–1	1	–20	20
30–40	40	35	0	0	0	0
40–50	30	45	1	1	30	30
50–60	28	55	2	4	56	112
60–70	10	65	3	9	30	90
70–80	5	75	4	16	20	80
Total	150				75	435

$$S.D(u)\sigma_u = \sqrt{\frac{\Sigma f u^2}{N} - \left(\frac{\Sigma f u}{N}\right)^2}$$

$$= \sqrt{\frac{435}{150} - \left(\frac{75}{150}\right)^2}$$

$$= \sqrt{2.9 - 0.25}$$

$$= \sqrt{2.65}$$

$= 1.6279$

$S.D\,(X) = h \times \text{S.D}\,(u)$ here $h = 10$

$= 10 \times 1.6279$

$= 16.279$

$$\text{Mean} = \frac{\Sigma f u}{N} + A$$

$$= \frac{75}{150} + 35$$

$$= .25 + 35$$

$$= 35.25$$

$$\text{Co-efficient of Variance} = \frac{S.D}{\text{Mean}} \times 100$$

$$= \frac{16.279}{35.25} \times 100$$

$$= 46.18\%$$

7. Following table gives data on time in minutes for an operation taken by 200 workers.

Time in minutes	Frequency
5–9.9	10
10–14.9	36
15–19.9	62
20–24.9	72
25–29.9	20

Compute arithmetic mean and standard deviation.

(B. U., November, 1993)

Solution

Time in min. C. I.	Frequency f	Mid value X	$u = \frac{X - 17.45}{5}$	u^2	fu	fu^2
4.95–9.95	10	7.45	–2	4	–20	40
9.95–14.95	36	12.45	–1	1	–36	36
14.95–19.95	62	17.45	0	0	0	0
19.95–24.95	72	22.45	1	1	72	72
24.95–29.95	20	27.45	2	4	40	80
Total	200				56	228

$$\text{Mean:-} \quad = A + \frac{h\sum f u}{N}$$

$$= 17.45 + \frac{5 \times 56}{200}$$

$$= 17.45 + 1.4$$

$$\text{Mean} \quad = 18.85$$

$$\text{Standard deviation} = \sqrt{\frac{\sum f u^2}{N} - \left(\frac{\sum f u}{N}\right)^2}$$

$$= \sqrt{\frac{228}{200} - \left(\frac{56}{200}\right)^2}$$

$$= \sqrt{1.14 - (0.2)^2}$$

$$= \sqrt{1.14 - 0.0784}$$

$$= \sqrt{1.0616}$$

$$= 1.0303$$

$$\text{S.D}(X) = h \times \text{S.D}(u)$$

$$= 5 \times 1.0303$$

$$= 5.1515$$

8. The number runs scored by cricketers A and B during a test series consisting of 5 match show below for each of the 10 innings.

Cricketer A:	5	26	97	76	112	89	6	108	24	16
Cricketer B:	51	47	36	60	58	39	44	42	71	50

Find which cricketer may be considered more consistent.

(B. U., April 1995)

Solution

	Cricketer A		Cricketer B	
	A	A^2	B	B^2
	5	25	51	2601
	26	676	47	2209
	97	9406	36	1296
	76	5776	60	3600
	112	12544	58	3364
	89	7921	39	1521
	6	36	44	1936
	108	11664	42	1764
	24	576	71	5041
	16	256	50	2500
Total	559	48883	498	25832

Cricketer A

$$\sigma_A = \sqrt{\frac{\sum A^2}{n} - \left(\frac{\sum A}{n}\right)^2}$$

$$= \sqrt{\frac{48883}{10} - \left(\frac{559}{10}\right)^2}$$

Cricketer B

$$\sigma_B = \sqrt{\frac{\sum B^2}{n} - \left(\frac{\sum B}{n}\right)^2}$$

$$= \sqrt{\frac{25832}{10} - \left(\frac{498}{10}\right)^2}$$

$= \sqrt{4888.3 - 3124.81}$

$= \sqrt{1763.49}$

$= 41.994$

Mean $= 55.9$

$$C.V_A = \frac{S.D}{\overline{X}} \times 100$$

$$= \frac{41.994}{55.9} \times 100$$

$= 75.123\%$

$= \sqrt{2583.2 - 2480.04}$

$= \sqrt{103.16}$

$= 10.1568$

Mean $= 49.8$

$$C.V_B = \frac{S.D}{\overline{X}} \times 100$$

$$= \frac{10.1568}{49.8} \times 100$$

$= 20.395\%$

$C.V_B$ is less than $C.V_A$ therefore cricketer B is more consistent.

9. Following are the marks obtained by two students A and B.

A:	40	45	50	30	20	60
B:	20	60	80	10	20	60

If the consistency of performance is the criterion for awarding a prize, who should get the prize?

(B. U., October, 1996)

Solution

	Students *A*		Students *B*	
	A	A^2	*B*	B^2
	40	1600	20	400
	45	2025	60	3600
	50	2500	80	6400
	30	900	10	100
	20	400	20	400
	60	3600	60	3600
Total	245	11025	250	14500

$$\text{Mean } \left(\overline{A}\right) = \frac{\Sigma A}{n}$$

$$= \frac{245}{6}$$

$= 40.833$

$$\sigma(A) = \sqrt{\frac{\Sigma A^2}{n} - \left(\frac{\Sigma A}{n}\right)^2}$$

$$\text{Mean } \left(\overline{B}\right) = \frac{\Sigma B}{n}$$

$$= \frac{250}{6}$$

$= 41.667$

$$\sigma(B) = \sqrt{\frac{\Sigma B^2}{n} - \left(\frac{\Sigma B}{n}\right)^2}$$

$$= \sqrt{\frac{11025}{6} - \left(\frac{245}{6}\right)^2}$$

$$= \sqrt{1837.5 - 1667.33}$$

$$= \sqrt{170.17}$$

$$= 13.0449$$

$$C.V_A = \frac{S.D}{\overline{A}} \times 100$$

$$= \frac{13.0449}{40.833} \times 100$$

$$= 31.95\%$$

$$= \sqrt{\frac{14500}{6} - \left(\frac{250}{6}\right)^2}$$

$$= \sqrt{2416.667 - 1736.138}$$

$$= \sqrt{680.529}$$

$$= 26.087$$

$$C.V_B = \frac{S.D}{\overline{B}} \times 100$$

$$= \frac{26.087}{41.667} \times 100$$

$$= 62.60\%$$

$C.V_A$ is less than $C.V_B$ therefore students A is more consistent. The prize should be awarded to student A.

10.

Particulars	Factory A	Factory B
No. of employees	50	100
Average wages per head Per month	Rs 1,200	Rs 850
Variance of wages per Head per month	81	64

In which factory is there great variation in the distribution of wage per employee? Which factory pays more wages?

(B. U., April, 1999)

Solution

$$C.V_A = \frac{S.D}{\text{Mean}} \times 100$$

Variance = 81

$\therefore$ S.D = 9

$$C.V_A = \frac{9}{50} \times 100$$

$$= 18\%$$

$$C.V_B = \frac{S.D}{\text{Mean}} \times 100$$

Variance = 64

$\therefore$ S.D = 8

$$C.V_B = \frac{8}{100} \times 100$$

$$= 8\%$$

$C.V_A$ is greater than $C.V_B$, therefore for factory A has greater variation.

Total pay of A = Mean × No. of Employees

= 1,200 × 50

= Rs 60000

Total pay of B = Mean × No. of Employees

= 850 × 100

= Rs 85000

Factory B pays more wages.

11. An analysis of the monthly wages paid to the workers in firm A and B belonging to the same industry gives the following results?

	Firm A	Firm B
No. of workers	500	600
Average monthly wages	Rs 186	Rs 175
Variance of distribution of wages	Rs 81	Rs 100

Which firm has larger wage bill? In which firm is there greater variability in individual wage?

(B. U., November, 1999)

Solution

Total wage (A) = Mean × No. of workers

= 186 × 500

= Rs 93000

Total wage (B) = Mean × No. of workers

= 175 × 600

= Rs 105000

Firm B has larger wage bill.

$$C.V_A = \frac{S.D}{\text{Mean}} \times 100$$

Variance = 81

S.D = 9

$$C.V_A = \frac{9}{186} \times 100$$

= 4.838%

$$C.V_B = \frac{S.D}{\text{Mean}} \times 100$$

Variance = 100

S.D = 10

$$C.V_B = \frac{10}{175} \times 100$$

= 5.714%

$C.V_B$ is greater than $C.V_A$, hence firm B has greater variability.

12. The following are some of the particulars of the distribution of weights of boys and girls in a class.

	Boys	Girls
Number	100	50
Mean weight	60 kgs	45 kgs
Variance	9	4

a. Find the standard deviation.

b. Which of the two distribution is more variable?

(B. U., April, 2001)

Solution

S.D of boy = $\sqrt{\text{Variance}}$

= $\sqrt{9}$

= 3

S.D of girls = $\sqrt{\text{Variance}}$

= $\sqrt{4}$

= 2

$$C.V_{Boys} = \frac{S.D}{\text{Mean}} \times 100 \qquad C.V_{Girls} = \frac{S.D}{\text{Mean}} \times 100$$

$$= \frac{3}{60} \times 100 \qquad = \frac{2}{45} \times 100$$

$$= 5\% \qquad = 4.44\%$$

$C.V_{boys}$ is greater than $C.V_{girls}$, the distribution of weights of boys have greater variability.

13. If C.V = 17.75% $\overline{X}$ = 58.35. Find S.D.

we know that $C.V = \frac{S.D}{\overline{X}} \times 100$

$$17.75\,\% = \frac{S.D}{\overline{X}} \times 100$$

$$S.D = 58.35 \times 0.1775$$

$$= 10.357$$

Exercises

1. State the need for measures of dispersion.
2. Mention various measures of dispersion.
3. What are the desired qualities of a good measure of dispersion.
4. What is range? What is co-efficient of range.
5. Define quartile deviation and its co-efficient.
6. Define S.D.
7. Mention any three merits of S.D.
8. What is the corresponding relative measure of standard deviation. Give the formula. (***Hint***: Co-efficient of variance)
9. For a distribution C.V is 22.5% and the value of arithmetic mean is 7.5, find the value of S.D.

   (***Hint***: $C.V = \frac{S.D}{\overline{X}} \times 100$, S.D 1.6875)

10. If $C.V = 3.99\%$, σ = 2,68. Find mean.

    (***Ans. Mean: 67.168***)

11. Obtain the range and its co-efficient.

    Marks: 10 45 65 74 88 98.

    (***Ans. 88, 0.815***)

12. Find Q.D and its co-efficient for the following data.

    X: 45 25 35 60 70 50 65 47 80 90 95.

    (***Ans. Q.D = 20, Co-efficient Q.D, 0.333***)

13. For the following data, find the quartile deviation and the co-efficient of quartile deviation.

No. of apples/ basket:	36	43	37	30	35	38	38	29	35	36	35	32.

(Ans. 2.5, 0.0709)

14. For the following discrete grouped data calculate Q.D and its co-efficient.

C.I.	10	30	50	70	90
f:	5	7	15	20	20

(Ans. 10, 0.25)

15. For the following data calculate Q.D and its co-efficient.

Students present:	75	76	77	78	79	80
No. of days:	13	24	27	21	11	2

(Ans. 1, 0.01299)

16. Calculate the Q.D and its co-efficient for the following data.

Income (Rs):	2500	3000	3250	3500	3750	4000	6000
No. of persons:	20	14	6	26	9	13	4

(Ans. Rs 375, 0.1111)

17. For the following frequency distribution, calculate Q.D and its co-efficient.

C.I.:	0–20	20–40	40–60	60–80	80–100
f:	5	7	15	20	20

(Ans. 14.285, 0.312)

18. Find the Q.D and its co-efficient for the following distribution of marks scored by students.

Marks:	0–19	20–39	40–49	50–59	60–69	70–79	80–99
Student:	15	31	19	15	8	7	7

(Ans. 15.4465, 0.3702)

19. The following is the distribution of I.Q of 100 children. Find Q.D and its co-efficient.

I.Q (less than):	80	90	95	100	105	110	130
No. of children:	2	16	33	55	77	95	100

(Ans. 5.95, 0.0603)

20. Calculate the quartile deviation and its co-efficient for the following data regarding percentage marks.

Marks (more than):	0	40	50	60	70	80	85	90	95
No. of students:	100	80	60	35	18	6	3	2	1

(Ans. 11.69, 0.22)

21. In a frequency distribution, the three quartiles are 25, 50 and 75. Calculate the Q.D and its co-efficient.

(Ans. 25, 0.5)

22. Calculate the mean deviation about the mean for the following data and find its co-efficient.

X: 10, 12, 14, 15, 15, 13, 15, 10, 12, 14.

(Ans. 1.6, 0.1231)

23. Calculate mean deviation (median).

T.V sets sold: 18, 16, 16, 19, 12, 14, 20.

(Ans. 16, 2.143)

24. Obtain the mean deviation (mode) for the following frequency distribution and also calculate its co-efficient.

Marks:	40	50	60	70	80	90
Frequency:	10	8	20	10	8	9

(Ans. 12.4615, 0.2077)

25. The following is the distribution of age at the time of first delivery of 65 women.

Age (years):	18–22	22–26	26–30	30–34	34–38
No. women:	20	30	11	3	1

Find the mean deviation from the mean and its co-efficient.

(Ans. 2.462, 0.1026)

26. Obtain the mean deviation (mode) for the following continuous frequency distribution and calculate its co-efficient.

Marks:	40–50	50–60	60–70	70–80	80–90
Frequency:	10	8	20	10	8

(Ans. 9.37, 0.1487)

27. Calculate the standard deviation for the following data and obtain C.V.

Marks: 74 47 90 37 84 51 40 66 54 59

(Ans. 17.10, 28.4%)

28. Find the S.D for the following frequency distribution and find its C.V.

x:	10	11	12	13	14
f:	3	12	18	12	3

(Ans. 1, 8.333)

29. For the following distribution find S.D and C.V.

Pressure (mm):	78–80	80–82	82–84	84–86	88–90
No. of min.:	3	15	26	23	4

(Ans. 2.336, 2.7875%)

30. The following data relates to the salary of workers of two factories.

	Factory A	Factory B
Number of workers	80	110
Mean	Rs 2320	Rs 2160
S.D	Rs 120	Rs 122

1. In which factory is the total expenditure on salary more?
2. In which factory is average salary of workers more?
3. In which factory is variation in salary more?

(Ans. Factory B, Factory A, Factory B)

6

Correlation Analysis

Content

- Types of correlation
- Scatter diagram
- The covariance
- The correlation co-efficient
- Probable error
- Spearman's rank correlation

Formulae

1. Covariance

$$\sigma_{xy} = \frac{\Sigma\,(X - \overline{X})\,(Y - \overline{Y})}{N}$$

2. Correlation co-efficient

(a) $$r\,(x, y) = \frac{\Sigma\,(X - \overline{X})\,(Y - \overline{Y})}{\sqrt{\Sigma\,(X - \overline{X})^2\;\Sigma\,(Y - \overline{Y})^2}}$$

(b) $$r\,(x, y) = \frac{n\Sigma XY - \Sigma X \Sigma Y}{\sqrt{n\Sigma X^2 - (\Sigma X)^2}\,\sqrt{n\Sigma Y^2 - (\Sigma Y)^2}}$$

(c) $$r\,(x, y) = \frac{Cov\,(X, Y)}{S.D(X)\;S.D(Y)}$$

(d) When deviations are taken from actual mean.

$$r\,(x, y) = \frac{\Sigma uv}{\sqrt{\Sigma u^2}\,\sqrt{\Sigma v^2}}$$

3. Probable error

$$P.E(r) = 0.6745 \frac{1 - r^2}{\sqrt{n}}$$

4. Spearman's rank correlation

(a) $\rho = 1 - \frac{6\sum d^2}{n(n^2 - 1)}$

(b) $\rho = 1 - \frac{6\sum\left(d^2 + \frac{\sum m(m^2 - 1)}{12}\right)}{n(n^2 - 1)}$

Point to Remember

1. If the value of r is '+'. There is a positive correlation between the variables if it is negative then negative correlation.
2. If $r = 0$ there is no correlation.
3. Value of r is not affected by change in scale and origin.
4. The range of r is -1 to $+1$.

INTRODUCTION

In the field of Business, we very often come across the existence of relationship between two or more variables. For example, the demand and price of commodities, we note that as the demand increase, the price increase or as the demand decrease the price decrease. Now to study about the relationship (Correlation) between variables, we need a parameter, that parameter is known as correlation co-efficient.

6.1 TYPES OF CORRELATION

Let X, Y, be the two variable under study. We note that there exist, three types of relationship between these two variables (i) they are directly proportional i.e. when one increases (decreases) the other increases (decreases); (ii) they are inversely proportional, i.e. when one increases (decreases) the other decreases (increases); (ii) no relation. These three situations can be stated as follows Statistically (i) positively correlated (ii) negatively correlated (iii) no-correlation.

6.2 SCATTER DIAGRAM

It is the simplest way of the diagrammatic representation of bivariate data. Let X and Y be the two variable under study with n observation each. If we plot (X_1, Y_1)

$(X_2, Y_2) \ldots (X_n, Y_n)$ on a graph sheet, the resulting diagram is known as a scatter diagram. The scatter diagram gives us a vague idea about the correlation between these two variables X and Y.

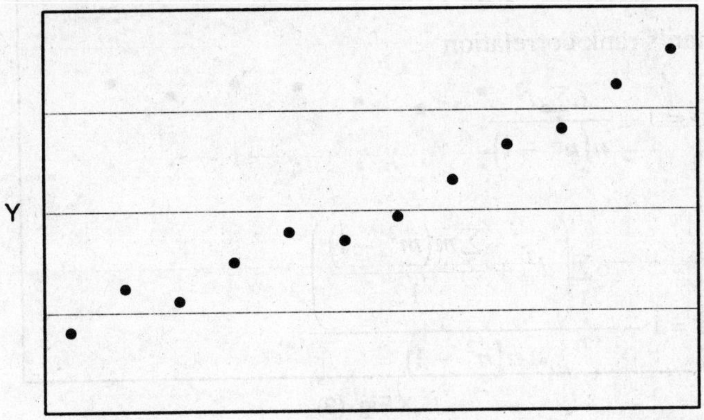

X Fig. (1)

Fig. (1) represents that there is a positive correlation between the variables X and Y.

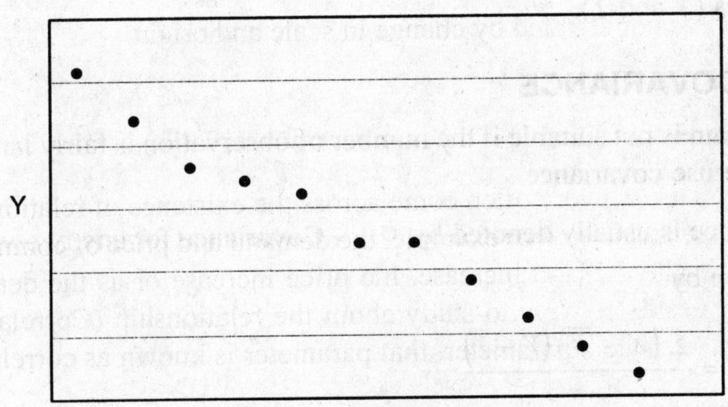

X Fig (2)

Fig. (2) represents that there is a negative correlation between the variables X and Y.

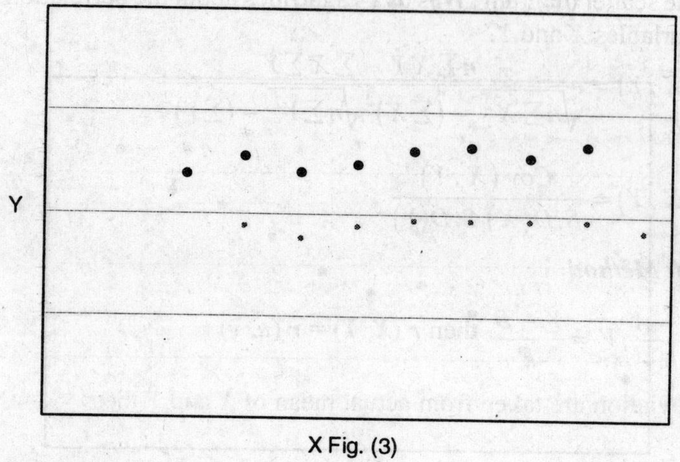

Fig. (3)

In Figure (3) we are not able to recognize a pattern, i.e., there is no correlation between X and Y.

Thus a scatter diagram help us to get an idea of the correlation between the two variables (X and Y).

6.3 THE COVARIANCE

Scatter diagram is not suitable if the number of observation is fairly large in such situation, we use covariance.

Covariance is usually denoted by σ_{xy}. Covariance for any two variables X and Y is given by

$$\sigma_{XY} = \frac{\Sigma\left(X - \overline{X}\right)\left(Y - \overline{Y}\right)}{n}$$

If σ_{XY} = '+' then there is a positive correlation between X and Y variate. If σ_{XY} = '–' (minus) then X and Y are negatively correlated, if $\sigma_{XY} = 0$ then there is no linear relation between X and Y.

6.4 THE CORRELATION CO-EFFICIENT

Covariance states only the nature of the relationship between the variables, but we also want to know how far these variables are associated with each other, we use correlation co-efficient to find the same. Correlation co-efficient is a measure of degree or extent of linear relationship between two variables X and Y. Correlation co-efficient was defined by Karl Person in 1890.

The correlation co-efficient for (X, Y) is given by:

$$r(X, Y) = \frac{\Sigma(X - \overline{X})(Y - \overline{Y})}{\sqrt{\Sigma(X - \overline{X})^2 \Sigma(Y - \overline{Y})^2}}$$

$$r(X, Y) = \frac{n\Sigma XY - \Sigma X \Sigma Y}{\sqrt{n\Sigma X^2 - (\Sigma X)^2}\sqrt{n\Sigma Y^2 - (\Sigma Y)^2}}$$

$$r(X, Y) = \frac{Cov(X, Y)}{S.D(X)\, S.D(Y)}$$

A. Short Cut Method

Let $U = \dfrac{X - A}{h}, V = \dfrac{Y - B}{k}$ then $r(X, Y) = r(u, v)$

If the deviation are taken from actual mean of X and Y then $\sum u = \sum v = 0$.

$$\therefore r(X, Y) = r(u, v) = \frac{n\sum uv}{\sqrt{n\sum u^2}\sqrt{n\sum v^2}}$$

$$= \frac{\sum uv}{\sqrt{\sum u^2}\sqrt{\sum v^2}}$$

(Range of the correlation co-efficient is (–1, 1) i.e. $-1 \le r \le 1$).

Example 1 You are given the following data relating to aptitude scores and productivity index..

Aptitude scores (X):	9	18	18	20	20	23
Productivity index (Y):	33	23	33	42	29	32

Find the co-efficient of correlation between aptitude scores and productivity index.

(B. U., April, 2001)

Solution

X	Y	$u = \dfrac{X-18}{1}$	$v = \dfrac{Y-32}{1}$	uv	u^2	v^2
9	33	–9	1	–9	81	1
18	23	0	–9	0	0	81
18	33	0	1	0	0	1
20	42	2	10	20	4	100
20	29	2	–3	–6	4	9
23	32	5	0	0	25	0
Total 108	192	0	0	5	114	192

Here $\overline{X} = \frac{\sum X}{N} = 18$

$\overline{Y} = \frac{\sum Y}{N} = 32$

$n = 6$

$$r(x, y) = r(u, v) = \frac{\sum uv}{\sqrt{\sum u^2}\sqrt{\sum v^2}}$$

$$= \frac{5}{\sqrt{114}\sqrt{192}}$$

$$= \frac{5}{10.677 \times 13.856}$$

$$= \frac{5}{147.94}$$

$$= 0.0337$$

Example 2 Find the correlation co-efficient for the following data.

x:	4	6	8	10	12
y:	3	6	9	12	15

Solution

x	y	$u = \frac{X-8}{2}$	$v = \frac{Y-9}{3}$	uv	u^2	v^2
4	3	–2	–2	4	4	4
6	6	–1	–1	1	1	1
8	9	0	0	0	0	0
10	12	1	1	1	1	1
12	15	2	2	4	4	4
Total 40	45	0	0	10	10	10

$r(x, y) = r(u, v)$

$$r(u, v) = \frac{\sum uv}{\sqrt{\sum u^2}\sqrt{\sum v^2}}$$

$$= \frac{10}{\sqrt{10} \times \sqrt{10}}$$

$$= \frac{10}{10}$$

$$= 1$$

$r(x, y) = 1$

Which shows that there is a perfect positive linear correlation between x and y.

6.5 PROBABLE ERROR AND ITS SIGNIFICANCE.

Probable error of correlation co-efficient is a measure of reliability of r, which is given by the formula.

$$P.E\,(r) = 0.6745\,\frac{1-r^2}{\sqrt{n}}$$

If $r < P.E\,(r)$, correlation co-efficient is definitely not significant. If $r > 6\,P.E\,(r)$, correlation co-efficient is significant.

Example 1 If $r = 0.5$ and $n = 16$ is r significant.

$$P.E\,(r) = 0.6745\,\frac{1-r^2}{\sqrt{n}}$$

$$= 0.6745\,\frac{\left(1-(0.5)^2\right)}{\sqrt{16}}$$

$$= 0.6745\,\frac{(1-0.25)}{4}$$

$$= 0.6745\,\frac{(0.75)}{4}$$

$$= 0.1264$$

$6 \times P.E = 0.7588$

$r < 6\,(P.E)$

$\therefore$ r is not significant.

Workout Problem

1. Find Karl Pearson's co-efficient of correlation of the following data.

Age of husband (x):	20	22	23	25	25	28	29	30	30	34
Age of wife (y):	18	20	22	24	21	26	26	25	27	29

(B. U., September, 2000)

Solution

x	y	$u = x - 25$	$v = y - 24$	uv	u^2	v^2
20	18	–5	–6	30	25	36
22	20	–3	–4	12	9	16
23	22	–2	–2	4	4	4
25	24	0	0	0	0	0

Contd.

x	y	$u = x - 25$	$v = y - 24$	uv	u^2	v^2
25	21	0	–3	0	0	9
28	26	3	2	6	9	4
29	26	4	2	8	16	4
30	25	5	1	5	25	1
30	27	5	3	15	25	9
34	29	9	5	45	81	25
Total 266	238	16	–2	125	194	108

Here $\overline{X} = 26.6, \overline{Y} = 23.8$

It will be tedious to calculate from the actual mean, ∴ we take the deviation from assumed mean.

$$r(x, y) = r(u, v)$$

$$r(u, v) = \frac{n\sum uv - \sum u \sum v}{\sqrt{n\sum u^2 - (\sum u)^2}\sqrt{n\sum v^2 - (\sum v)^2}}$$

$$= \frac{10 \times 125 - (16 \times -2)}{\sqrt{10 \times 194 - (16)^2}\sqrt{10 \times 108 - (-2)^2}}$$

$$= \frac{1250 + 32}{\sqrt{1684}\sqrt{1076}}$$

$$= \frac{+1282}{1346.0996}$$

$$r = 0.9524$$

2. Calculate Karl Pearson's co-efficient of correlation between percentage of pass and failure from the following data.

No. of student (x):	800	600	900	700	500	400
No. passed (y):	480	300	450	560	450	300

(B. U., April 1999)

Solution

X	Y	$u = \frac{X - 650}{10}$	$v = \frac{X - 450}{10}$	uv	u^2	v^2
800	480	15.0	3.0	45	225	9
600	300	–5.0	–15.0	75	25	225
900	450	25.0	0	0	625	0
700	560	5.0	11.0	55	25	121
500	450	–15.0	0	0	225	0
400	300	–25.0	–15.0	375	625	225
Total 3900	2540	0	–16	550	2150	580

$$r(u, v) = \frac{n\sum uv - \sum u \sum v}{\sqrt{n\sum u^2 - (\sum u)^2}\sqrt{n\sum v^2 - (\sum v)^2}}$$

$$= \frac{6 \times 550 - 0}{\sqrt{6 \times 2150}\sqrt{6 \times 580 - (-16)^2}}$$

$$= \frac{3300}{\sqrt{12900}\sqrt{3224}}$$

$$= \frac{3300}{6448.999}$$

$$r(u, v) = 0.5117$$

$$r(x, y) = r(u, v)$$

$$= 0.5117$$

3. Calculate correlation between age and playing habits of students.

Age (years):	15	16	17	18	19	20
No. of students:	250	200	150	120	100	80
Regular players:	200	150	90	48	30	12

(B. U., April, 1998)

Solution

x	%	y	$u = x - 17$	$v = y - 50$	u^2	u^3	uv
15	$\frac{200}{250} \times 100$	80	–2	+30	4	900	–60
16	$\frac{150}{200} \times 100$	75	–1	+25	1	625	–25
17	$\frac{90}{150} \times 100$	60	0	+10	0	100	0
18	$\frac{48}{120} \times 100$	40	1	–10	1	100	–10
19	$\frac{30}{100} \times 100$	30	2	–20	4	400	–40
20	$\frac{12}{80} \times 100$	15	3	–35	9	1225	–105
	Total		3	0	19	3550	–240

$$r(x, y) = r(u, v)$$

$$r(x, y) = r(u, v) = \frac{n\sum uv - \sum u \sum v}{\sqrt{n\sum u^2 - (\sum u)^2}\sqrt{n\sum v^2 - (\sum v)^2}}$$

$$= \frac{n\sum uv}{\sqrt{n\sum u^2 - (\sum u)^2}\sqrt{n\sum v^2}} \quad \because \sum v = 0$$

$$= \frac{-6 \times 240}{\sqrt{6 \times 19 - (3)^2}\sqrt{6 \times 3550}}$$

$$= \frac{-1440}{\sqrt{105}\sqrt{21300}}$$

$$= \frac{-1440}{14955.55}$$

$$= -0.0963$$

4. Calculate Karl Pearson's co-efficient of correlation for the following data.

Price (in Rs) (x):	21	22	23	24	25	26	27	28	29	30
Demand (in 000 units) (y):	18	19	19	16	17	16	16	15	13	11

(B. U., November, 1997)

Solution

	X	Y	$u = \frac{X - 25.5}{0.5}$	$v = y - 16$	uv	u^2	v^2
	21	18	–9	+2	–18	81	4
	22	19	–7	3	–21	49	9
	23	19	–5	3	–15	25	9
	24	16	–3	0	0	9	0
	25	17	–1	1	–1	1	1
	26	16	1	0	0	1	0
	27	16	3	0	0	9	0
	28	15	5	–1	–5	25	1
	29	13	7	–3	–21	49	9
	30	11	9	–5	–45	81	25
Total	255	160	0	0	–126	330	58

$$r(u, v) = \frac{n\sum uv}{\sqrt{n\sum u^2}\sqrt{n\sum v^2}} \quad \because \sum u = 0, \sum v = 0$$

$$= \frac{\sum uv}{\sqrt{\sum u^2}\sqrt{\sum v^2}}$$

$$= \frac{-126}{\sqrt{330}\sqrt{58}}$$

$$= \frac{-126}{138.3474}$$

$$= 0.9108$$

$$r(x, y) = r(u, v)$$

$$= 0.9108$$

5. Calculate Karl Pearson's co-efficient of correlation.

x:	1	2	3	4	5	6	7
y:	3	5	6	8	10	11	13

(B. U., October, 1996)

Solution

	x	y	$u = x - 4$	$v = y - 8$	uv	u^2	v^2
	1	3	–3	–5	15	9	25
	2	5	–2	–3	6	4	9
	3	6	–1	–2	2	1	4
	4	8	0	0	0	0	0
	5	10	+1	2	2	1	4
	6	11	2	3	6	4	9
	7	13	3	5	15	9	25
Total	28	56	0	0	46	28	76

$$r(u, v) = \frac{\sum uv}{\sqrt{\sum u^2}\sqrt{\sum v^2}}$$

$$= \frac{46}{\sqrt{28}\sqrt{76}}$$

$$= \frac{46}{5.2915 \times 8.7178}$$

$$= \frac{46}{46.1302}$$

$$= 0.9972$$

$$r(x, y) = r(u, v)$$

$$= 0.9972$$

6. Calculate co-efficient of correlation using Karl Pearson's method.

Advertisement ('000 Rs) (x):	39	65	62	90	82	75	25	98	36	78
Sale (in Lakh Rs) (y):	47	53	58	86	62	68	60	91	51	84

(B. U., April, 1996)

Solution

x	y	$u = x - 65$	$v = y - 66$	uv	u^2	v^2
39	47	–26	–19	494	676	361
65	53	0	–13	0	0	169
62	58	–3	–8	24	9	64
90	86	25	20	500	625	400
82	62	17	–4	–68	289	16
75	68	10	2	20	100	4
25	60	–40	–6	240	1600	36
98	91	33	25	825	1089	625
36	51	–29	–15	435	841	225
78	84	13	18	234	169	324
Total 650	660	0	0	2704	5398	2224

$$r(u, v) = \frac{\sum uv}{\sqrt{\sum u^2}\sqrt{\sum v^2}}$$

$$= \frac{2704}{\sqrt{5398}\sqrt{2224}}$$

$$= \frac{2704}{73.4711 \times 47.1593}$$

$$= \frac{2704}{3464.8456}$$

$$= 0.7804$$

$$r(x, y) = r(u, v)$$

$$= 0.7804$$

7. A company wanted to assess the impact of R and D expenditure (x) on annual profit (y) following table presents data for past 8 years.

x:	9	7	5	10	4	5	3	2
y:	45	42	41	60	30	34	25	20

Compute the correlation co-efficient. Comment on the value.

(B. U., November, 1993)

Solution

X	Y	$u = \frac{X-5}{1}$	$v = \frac{y-40}{1}$	uv	u^2	v^2
9	45	+4	5	20	16	25
7	42	+2	2	4	4	4
5	41	0	1	0	0	1

Contd.

X	Y	$u = \frac{X-5}{1}$	$v = \frac{y-40}{1}$	uv	u^2	v^2
10	60	5	20	100	25	400
4	30	–1	–10	+10	1	100
5	34	0	–6	0	0	36
3	25	–2	–15	+30	4	225
2	20	–3	–20	60	9	400
45	297	5	–23	224	59	1191

$$r(u, v) = \frac{n\sum uv - \sum u \sum v}{\sqrt{n\sum u^2 - (\sum u)^2}\sqrt{n\sum v^2 - (\sum v)^2}}$$

$$= \frac{8 \times 224 - 5 \times (-23)}{\sqrt{8 \times 59 - (5)^2}\sqrt{8 \times 1191 - (-23)^2}}$$

$$= \frac{1792 + 115}{\sqrt{447}\sqrt{8999}}$$

$$= \frac{1907}{2005.6303}$$

$$= 0.9508$$

$$r(x, y) = r(u, v)$$

$$= 0.9508$$

Comment:

$$P.E(r) = 0.6745\frac{(1 - r^2)}{\sqrt{n}}$$

$$= 0.6745\frac{(1 - (0.9508)^2)}{\sqrt{8}}$$

$$= 0.6745\frac{0.0960}{2.8284}$$

$$= 0.0229$$

$$6\,P.E(r) = 6 \times 0.0229$$

$$= 0.1374$$

$$6\,P.E(r) < r$$

$\therefore r$ **is significant**

8. Given below are the monthly incomes and their net saving of a sample of 10 supervisory staff belonging to a firm. Calculate the correlation co-efficient.

Employee No.: 1 2 3 4 5 6 7 8 9 10

Monthly income (x):	780	360	980	250	750	820	900	620	650	390
Net saving: (y)	84	51	91	60	68	62	86	58	53	47

(B. U., April, 1992)

Solution

	x	y	$u = \dfrac{X - 650}{10}$	$v = y - 66$	uv	u^2	v^2
	780	84	13	18	234	169	324
	360	51	–29	–15	435	841	225
	980	91	33	25	825	1089	625
	250	60	–40	–6	240	1600	36
	750	68	+10	2	20	100	4
	820	62	17	–4	68	289	16
	900	86	25	20	500	625	400
	620	58	–3	–8	24	9	64
	650	53	0	–13	0	0	169
	390	47	–26	–19	494	676	361
Total	6500	660	0	0	2840	5398	2224

$$r(u, v) = \frac{\sum uv}{\sqrt{\sum u^2}\,\sqrt{\sum v^2}}$$

$$= \frac{2840}{\sqrt{5398}\,\sqrt{2224}}$$

$$= \frac{2840}{73.4711 \times 47.1593}$$

$$= \frac{2840}{3464.846}$$

$$= 0.8197$$

$$r(x, y) = r(u, v)$$

$$= 0.8197$$

9. From the following data calculate the correlation co-efficient.

$$\sum x = 50,\ \sum y = 100,\ \sum x^2 = 255,\ \sum y^2 = 4594,\ \sum xy = 630,\ \&\ n = 10.$$

Solution

$$r = \frac{n\sum xy - \sum x \sum y}{\sqrt{n\sum x^2 - (\sum x)^2}\,\sqrt{n\sum y^2 - (\sum y)^2}}$$

$$= \frac{10 \times 630 - 50 \times 100}{\sqrt{10 \times 255 - (50)^2}\,\sqrt{10 \times 4594 - (100)^2}}$$

$$= \frac{6300 - 5000}{\sqrt{2550 - 2500}\ \sqrt{45940 - 10000}}$$

$$= \frac{1300}{1340.522}$$

$$r = 0.9698$$

10. From the following details find the value of σ_Y.

$r = 0.6$, Cov $(X, Y) = 12$ and $\sigma_X = 5$

Solution We know that

$$r = \frac{Cov\ (X, Y)}{\sigma_X\ \sigma_Y}$$

$$0.6 = \frac{12}{5 \times \sigma_Y}$$

$$\therefore\ \sigma_Y = \frac{12}{5 \times 0.6}$$

$$\sigma_Y = 4$$

11. From the following data, compute the co-efficient of correlation between X and Y.

	X Series	Y Series
No. of items	15	15
Arithmetic mean	25	18
Sum of squares of Deviation from mean	136	138

Summation of product of X and Y series from their respective arithmetic means = 122.

Solution

$$r\ (x, y) = \frac{\Sigma\ (X - \overline{X})\ (Y - \overline{Y})}{\sqrt{\Sigma\ (X - \overline{X})^2\ \Sigma\ (Y - \overline{Y})^2}}$$

$$= \frac{122}{\sqrt{136 \times 138}}$$

$$= 0.8905$$

12. Find the correlation co-efficient.

(i) $Cov\ (u, v) = 3,\ \sigma_u^2 = 4.5,\ \sigma_v^2 = 5.5.$

(ii) $\Sigma x = 125,\ \Sigma y = 100,\ \Sigma x^2 = 650,\ \Sigma y^2 = 436,\ \Sigma xy = 520,\ \&\ n = 25.$

Solution

(i) $$r = \frac{Cov\ (u, v)}{\sigma_u\ \sigma_v}$$

$$= \frac{3}{\sqrt{4.5} \times \sqrt{5.5}}$$

$$= \frac{3}{4.9749}$$

$$= 0.603$$

(ii) $$r = \frac{n\sum xy - \sum x \sum y}{\sqrt{n\sum x^2 - (\sum x)^2}\sqrt{n\sum y^2 - (\sum y)^2}}$$

$$= \frac{25 \times 520 - 25 \times 100}{\sqrt{25 \times 650 - (125)^2}\sqrt{25 \times 436 - (100)^2}}$$

$$= \frac{13000 - 12500}{\sqrt{625}\sqrt{900}}$$

$$= \frac{50}{750}$$

$$= 0.0667$$

13. Find the probable error.

(i) $r = 0.05, n = 25$

(ii) $r = +0.526, n = 50$

comment on the value of r.

(i) $$P.E(r) = 0.6745 \frac{(1 - r^2)}{\sqrt{n}}$$

$$= 0.6745 \frac{(1 - (0.05)^2)}{\sqrt{25}}$$

$$= 0.6745 \frac{0.9975}{5}$$

$$= 0.1292$$

$r <$ P.E (r) $\therefore$ r is insignificant

(ii) $$P.E(r) = 0.6745 \frac{(1 - r^2)}{\sqrt{n}}$$

$$= 0.6745 \frac{(1 - (0.526)^2)}{\sqrt{50}}$$

$$= 0.6745 \frac{0.7233}{7.0711}$$

$$= 0.06623$$

$6 \times P.E\,(r) = 6 \times 0.06623$
$= 0.39738$

$r > 6\,P.E\,(r)$ ∴ r is significant.

14. How do you interpret 'r' when it is (a) +1 (b) –1 (c) 0?
    (a) When $r = 1$, it means Y is proportional to X, it means as X increases Y also increases. There is a perfect positive linear association between X and Y.
    (b) When $r = -1$, it means Y is inversely proportional to X, it means as X increases (decreases) y decreases (increases). This ensures perfect negative association.
    (c) When $r = 0$, it means there is no linear association between two variables.

6.6 SPEARMAN'S RANK CORRELATION

In correlation co-efficient we used the actual value to measure the degree of relationship, here we use the rank of the given data. Rank correlation is useful when we study the relationship between quantitative characteristic like beauty, intelligence etc. Ranks are obtained by arranging the data in order of their merits.

Spearman's rank correlation formula is given by

$$\rho = 1 - \frac{6\sum d^2}{n\left(n^2 - 1\right)}$$

where d is the difference between the rank of the two variables.

Example 1 Calculate rank correlation from the following.

Marks is Statistics (x):	15	20	25	18	40	60	80
Marks in Accountancy (y):	40	30	50	36	20	10	60

Solution

x	y	Rank of x Rx	Rank of y Ry	$d = Rx - Ry$	d^2
15	40	7	3	4	16
20	30	5	5	0	0
25	50	4	2	2	4
18	36	6	4	2	4
40	20	3	6	–3	9
60	10	2	7	–5	25
80	60	1	1	0	0
				Total	58

$$\rho = 1 - \frac{6\sum d^2}{n\left(n^2 - 1\right)}$$

$$= 1 - \frac{6 \times 58}{7\left(7^2 - 1\right)}$$

$$= 1 - \frac{348}{7(49-1)}$$

$$= 1 - \frac{348}{336}$$

$$= 1 - 1.0357$$

$$\rho = -0.0357$$

6.7 REPEATED RANKS

If there are more than one item with the same value in the series then common ranks are given to the repeated items. This common rank is the average of the ranks which these items would have attained if they were slightly different from each other and the next item gets the rank next to the ranks (actual rank). As a result there is a small adjustment in the rank correlation formula.

$$\text{Adjustment factor} = \frac{m(m^2-1)}{12}$$

This adjustment factor is added to $\sum d^2$ (where m is the number of items repeated) for each repeated values.

$$\rho = 1 - \frac{6\sum\left[d^2 + \sum \frac{m(m^2-1)}{12}\right]}{n(n^2-1)}$$

Example 1 Calculate rank correlation from the following:

Marks in Statistics (*X*)	15	20	28	12	40	60	20	80
Marks in Accountancy (*Y*)	40	30	50	30	20	10	30	60

(B. U., September, 1998)

Solution

X	Y	Rank X R_X	Rank Y R_Y	$d = R_X - R_Y$	d^2
15	40	7	3	4	16
20	30	5.5	5	0.5	0.25
28	50	4	2	2	0.4
12	30	8	5	3	9
40	20	3	7	–4	16
60	10	2	8	–6	36
20	30	5.5	5	0.5	0.25
80	60	1	1	0	0
				Total	81.5

In X series: 20 is repeated 2 times, the actual ranks will be 5 and 6

$\therefore$ the average is 5.5, the mark 15 gets the seventh position.

Here $m = 2$

$$\therefore \frac{m\left(m^2-1\right)}{12} = \frac{12\,(4-1)}{12} = \frac{2\times 3}{12} = \frac{1}{2}$$

In Y series: 30 is repeated 3 times, it shares 4th, 5th and 6th position $\therefore$ 30 gets the average rank as 5. 20 gets the seventh position.

Here $m = 3$

$$\frac{m\left(m^2-1\right)}{12} = \frac{3\times(9-1)}{12} = 2$$

$$\rho = 1 - \frac{\left[\dfrac{6\Sigma d^2 + \Sigma\left(m\left(m^2-1\right)\right)}{12}\right]}{n\left(n^2-1\right)}$$

$$= 1 - \frac{[6\times 81.5 + 0.5 + 2]}{8(64-1)}$$

$$= 1 - \dot{0}.9752$$

$$\therefore \rho = 0.0248$$

Solved Examples

1. Ten competitors in a beauty contest are ranked by three judges in the following order.

First judge:	1	5	4	8	9	6	10	7	3	2
Second judge:	4	8	7	6	5	9	10	3	2	1
Third judge:	6	7	8	1	5	10	9	2	3	4

Use the rank correlation co-efficient to discuss which pair of judges have the nearest approach to common tastes in beauty.

(B. U., October, 2001)

Solution

First judge R_x	*Second judge* R_y	*Third judge* R_z	$d_{xy}^2=(R_x-R_y)^2$	$d_{xz}^2=(R_x-R_z)^2$	$d_{yz}^2=(R_y-R_z)^2$
1	4	6	9	25	4
5	8	7	9	4	1
4	7	8	9	16	1
8	6	1	4	49	25

Contd.

First judge R_x	Second judge R_y	Third judge R_z	$d_{xy}^2=(R_x-R_y)^2$	$d_{xz}^2=(R_x-R_z)^2$	$d_{yz}^2=(R_y-R_z)^2$
9	5	5	16	16	0
6	9	10	9	16	1
10	10	9	0	1	1
7	3	2	16	25	1
3	2	3	1	0	1
2	1	4	1	4	9
	Total		74	156	44

$$\rho_{xy} = 1 - \frac{6 \times \Sigma d_{xy}^2}{N(n^2-1)} \qquad \rho_{xz} = 1 - \frac{6\Sigma d_{xz}^2}{n(n^2-1)}$$

$$= 1 - \frac{6 \times 74}{10(10^2-1)} \qquad = 1 - \frac{6 \times 156}{10(10^2-1)}$$

$$= 1 - \frac{6 \times 74}{990} \qquad = 1 - \frac{6 \times 56}{990}$$

$$= 0.5515 \qquad = 0.0545$$

$$\rho_{yz} = 1 - \frac{6\Sigma d_{yz}^2}{n(n^2-1)}$$

$$= 1 - \frac{6 \times 44}{10(10^2-1)}$$

$$= 1 - \frac{6 \times 44}{990}$$

$$\rho = 0.7333$$

Rank correlation of 1st and 2nd = 0.5515

Rank correlation of 2nd and 3rd = 0.7333

Rank correlation of 1st and 3rd = 0.0545

∴ 2nd and 3rd judges have the nearest approach to common taste in beauty.

2. Find rank correlation co-efficient for the following data and give your comments.

Marks in Accounts (x):	84	56	89	58	59	67	74	78
Marks in Maths (y):	38	69	56	58	63	78	87	77

(B. U., April, 1997)

Solution

x	y	Rank x Rx	Rank y Ry	$d = Rx - Ry$	d^2
84	38	2	8	–6	36
56	69	8	4	2	4
89	56	1	7	–6	36
58	58	7	6	1	1
59	63	6	5	1	1
67	78	5	2	3	9
74	87	4	1	3	9
78	77	3	3	0	0
				Total	96

$$\rho = 1 - \frac{6\sum d^2}{n\left(n^2 - 1\right)}$$

$$= 1 - \frac{6 \times 96}{8\left(8^2 - 1\right)}$$

$$= 1 - \frac{6 \times 96}{8 \times 63}$$

$$= 1 - 1.1429$$

$$\rho = -0.1429$$

The marks in accountancy and maths are negatively correlated i.e. students who scored good marks in accountancy have not scored good in maths.

3. Find rank correlation co-efficient and comment

Marks in Mathematics (x):	60	70	50	40	80	90	85	96
Marks in Statistics (y):	64	58	72	44	86	80	95	81

(B. U., October, 1996)

Solution

x	y	Rx	Ry	$d = Rx - Ry$	d^2
60	64	6	6	0	0
70	58	5	7	–2	4
50	72	7	5	2	4
40	44	8	8	0	0
80	86	4	2	2	4
90	80	2	4	–2	4
85	95	3	1	+2	4
96	81	1	3	2	4
				Total	24

$$\rho = 1 - \frac{6\sum d^2}{n(n^2 - 1)}$$

$$= 1 - \frac{6 \times 24}{8(8^2 - 1)}$$

$$= 1 - \frac{6 \times 24}{8 \times 63}$$

$$= 1 - 0.2857$$

$$\rho = 0.7143$$

$\rho = 0.7143$, shows that the marks in statistics and mathematics are closely correlated i.e. a student who fairs well in statistics do the same in mathematics.

4. Find rank correlation co-efficient for the following data and give your comments.

Marks in Statistics (x):	84	56	89	58	59	67	74	78
Marks in Law (y):	38	69	56	58	63	78	87	77

(B. U., April, 1995)

Solution

x	y	Rx	Ry	$d = Rx - Ry$	d^2
84	38	2	8	–6	36
56	69	8	4	4	16
89	56	1	7	–6	36
58	58	7	6	1	1
59	63	6	5	2	4
67	78	5	2	3	9
74	87	4	1	3	9
78	77	3	3	0	0
		Total			111

$$\rho = 1 - \frac{6\sum d^2}{n(n^2 - 1)}$$

$$= 1 - \frac{6 \times 111}{8(63)}$$

$$= 1 - 1.3214$$

$$\rho = -0.3214$$

The marks in statistics and law are negatively correlated, i.e. a student who fares well in Statistics need not fare well in law and vice-versa.

5. Ten competitors in a beauty contest are ranked by three judges in the following order.

Judge I:	1	6	5	10	3	2	4	9	7	8

Judge II:	3	5	8	4	7	10	2	1	6	9
Judge III:	6	4	9	8	1	2	3	10	5	7

Use rank correlation co-efficient to determine which pair of judges has the nearest approach to common tastes in beauty.

(B. U., April, 1992)

Solution

R_1	R_2	R_3	$d_1 = R_1 - R_2$	$d_2 = R_1 - R_3$	$d_3 = R_2 - R_3$	d_1^2	d_2^2	d_3^2
1	3	6	–2	–5	–3	4	25	9
6	5	4	1	2	1	1	4	1
5	8	9	–3	–4	–1	9	16	1
10	4	8	6	2	–4	36	4	16
3	7	1	–4	2	6	16	4	36
2	10	2	–8	0	8	64	0	64
4	2	3	2	1	–1	4	1	1
9	1	10	8	–1	–9	64	1	81
7	6	5	1	2	1	1	4	1
8	9	7	–1	1	2	1	1	4
					Total	200	60	214

$$\rho_{12} = 1 - \frac{6\Sigma d_1^2}{n\left(n^2 - 1\right)} = 1 - \frac{6 \times 200}{10\ (100 - 1)} = -0.212$$

$$\rho_{13} = 1 - \frac{6\Sigma d_2^2}{n\left(n^2 - 1\right)} = 1 - \frac{6 \times 60}{10\ (100 - 1)} = 0.636$$

$$\rho_{23} = 1 - \frac{6\Sigma d_3^2}{n\left(n^2 - 1\right)} = 1 - \frac{6 \times 214}{10\ (100 - 1)} = -0.297$$

Since ρ_{13} is highest, we can conclude that the pair I and II judges has more common taste towards beauty.

6. A sample of 5 fathers and their eldest sons gave the following data about their weight in Kgs.

Father (x):	65	60	67	63	74
Son (y):	49	45	57	40	60

Solution

x	y	Rx	Ry	$d = Rx - Ry$	d^2
65	49	3	3	0	0
60	45	5	4	1	1
67	57	2	2	0	0
63	40	4	5	–1	1
74	60	1	1	0	0
				Total	2

$$\rho = 1 - \frac{6\sum d^2}{n(n^2-1)} = 1 - \frac{6 \times 2}{5\,(25-1)} = 0.9$$

7. Calculate the rank correlation.

x:	65	66	67	68	69	70	71	72
y:	67	68	69	72	78	80	82	85

Solution

x	y	Rx	Ry	$d = Rx - Ry$	d^2
65	67	8	8	0	0
66	68	7	7	0	0
67	69	6	6	0	0
68	72	5	5	0	0
69	78	4	4	0	0
70	80	3	3	0	0
71	82	2	2	0	0
72	85	1	1	0	0
				Total	0

$$\rho = 1 - \frac{6\sum d^2}{n(n^2-1)} = 1 - 0 = 1$$

$\therefore$ Perfect correlation (positive)

8. Heights of fathers and sons are given in inches.

Height of father x:	65	66	67	68	69	71	73
Height of son:	67	68	64	68	72	70	69

Solution

x	y	Rx	Ry	$d = Rx - Ry$	d^2
65	67	7	6	1	1
66	68	6	4.5	1.5	2.25
67	64	5	7	−2	4
68	68	4	4.5	−0.5	0.25
69	72	3	1	+2	4
71	70	2	2	0	0
73	69	1	3	−2	4
				Total	15.5

$$\rho = 1 - \frac{6\sum d^2 + \dfrac{\sum m(m^2-1)}{12}}{n(n^2-1)}$$

Here $m = 2$

$$\therefore \quad \frac{m(m^2 - 1)}{12} = \frac{2(4-1)}{12} = \frac{1}{2} = 0.5$$

$$\rho = 1 - \frac{6(15.5 + 0.5)}{7\,(49 - 1)}$$

$= 1 - 0.2857$

$\rho = 0.7143$

9. Compute the rank correlation between IQ's and marks scored in examination.

Personal:	A	B	C	D	E	F
I.Q:	100	110	140	160	120	130
Exam mark:	70	80	81	78	72	75

Solution

Person	I.Q (x)	Exam. marks (y)	Rx	Ry	$d = Rx - Ry$	d^2
A	100	70	6	6	0	0
B	110	80	5	2	3	9
C	140	81	2	1	1	1
D	160	78	1	3	–2	4
E	120	72	4	5	–1	1
F	130	75	3	4	–1	1
					Total	16

$$\rho = 1 - \frac{6\sum d^2}{n\left(n^2 - 1\right)} = 1 - \frac{6 \times 16}{6\,(36 - 1)}$$

$= 1 - 0.4571$

$\rho = 0.5429$

10. The table below shows the respective I.Qs of 7 father and their eldest sons.

Father I.Q (x):	90	98	103	104	105	110	114
Sons I.Q (y):	102	95	107	114	100	98	113

Calculate the rank correlation for the above statistics.

Solution

x	y	Rx	Ry	$d = Rx - Ry$	d^2
90	102	7	4	+3	9
98	95	6	7	–1	1
103	107	5	3	2	4
104	114	4	1	3	9

Contd.

x	y	Rx	Ry	d = Rx – Ry	d²
105	100	3	5	–2	4
110	98	2	6	–4	16
114	113	1	2	–1	1
				Total	44

$$\rho = 1 - \frac{6\sum d^2}{n\left(n^2-1\right)} = 1 - \frac{6 \times 44}{7\,(49-1)}$$

$$= 0.2143$$

11. Answer the following

1. $n = 6$ and 'r' = 0.42857. Find $\sum d^2$

$$r = 1 - \frac{6\sum d^2}{n\left(n^2-1\right)}$$

$$0.42857 = 1 - \frac{6 \times \sum d^2}{6 \times 35}$$

$$210 \times 0.42857 = 210 - 6 \times \sum d^2$$

$$89.9997 = 210 - 6 \times \sum d^2$$

$$\therefore\ 6\sum d^2 = 210 - 89.9997$$

$$= 120.0003$$

$$\sum d^2 = 20$$

2. $n = 10$ and $\sum d^2$ Find r

$$r = 1 - \frac{6\sum d^2}{n\left(n^2-1\right)} = 1 - \frac{6 \times 200}{10\,(100-1)} = -0.2121$$

3. $n = 8$ and $\sum d^2 = 4$, find r

$$r = 1 - \frac{6\sum d^2}{n\left(n^2-1\right)} = 1 - \frac{6 \times 4}{8\,(64-1)} = 0.9524$$

4. $r = -0.2969$ and $n = 10$ Find $\sum d^2$

$$r = 1 - \frac{6\sum d^2}{n\left(n^2-1\right)}$$

$$-0.2969\ r = 1 - \frac{6\Sigma d^2}{10\,(100 - 1)}$$

$$990 \times 0.2969 = 990 \times 6\Sigma d^2$$

$$-293.9310 = 990 - 6\Sigma d^2$$

$$6\Sigma d^2 = 1283.9310$$

$$\Sigma d^2 = 214$$

Exercises

1. Define co-efficient of correlation.
2. State different types of correlation.
3. Define probable error and state its significance.
4. How do you interpret when it is (a) $r = 0.925$ (b) $r = -0.361$ (c) $r =$ zero
5. What is the formula for co-efficient of correlation when the deviation are taken from actual mean.
6. In a distribution $n = 10$, $r = 0.06$ find $P.E\,(r)$.
7. What is correlation.
8. What is the formula to find a correlation co-efficient. What does it measure.
9. Write the formula for rank correlation.
10. When $n = 10$, $\Sigma d^2 = 60$ find r.
11. Calculate the correlation co-efficient for the following heights of father (x) and their sons (y).

x:	65	66	67	67	68	69	70	72
y:	67	68	65	68	72	72	69	71

(Ans. r = .603)

12. Calculate the co-efficient of correlation between x and y for the following.

x:	1	3	4	5	7	8	10
y:	2	6	8	10	14	16	20

(Ans. r (x, y) = 1)

13. The following table gives the marks scored by 6 students in statistics (x) and commerce (y) in an examination. Calculate Karl Pearson's correlation co-efficient.

x:	52	56	58	62	66	68
y:	33	36	42	42	45	51

(Ans. r (x, y) = 0.95)

14. Calculate the co-efficient of correlation between x and y for the following.

x:	2	4	6	8	10
y:	5	7	9	8	11

(Ans. r (x, y) = 0.92)

15. Calculate Karl Pearson's correlation co-efficient.

x:	40	38	35	42	30
y:	30	35	40	36	29

(*Ans.* $r = 0.259$)

16. Calculate Karl Pearson's correlation co-efficient.

x:	1	2	3	4	5
y:	6	10	14	18	22

(*Ans.* $r(x, y) = 1$)

17. Calculate Karl Pearson's correlation co-efficient.

x:	1	5	3	2	1	1	7	3
y:	6	1	0	0	1	2	1	5

(*Ans.* $r(x, y) = 0.29$)

18. If $P.E(r) = 0.1264$, $r = 0.5$, find the value of n.
19. Find 'r' if

$$n = 10,\ \Sigma xy = 120,\ \Sigma x^2 = 900,\ \sigma_y = 8.$$

20. Find r when $n = 9$, $P.E(r) = 0.13274$.
21. From the following data find the Karl Pearson's co-efficient.

$$\Sigma x = 490, \Sigma y = 294, \Sigma xy = 3030$$

$$\Sigma x^2 = 5350, \Sigma y^2 = 1964, n = 49.$$

22. Obtain Spearman's rank correlation for the following data.

x:	68	64	75	50	64	80	75	40	55	64
y:	62	58	68	45	81	60	68	48	50	70

(*Ans.* $\rho = 0.545$)

23. For the given data find rank correlation.

x:	1	2	3	4	5	6	7	8	9	10	11	12	13	14	15	16
y:	1	10	3	4	5	7	2	6	8	11	15	9	14	12	16	13

(*Ans.* $\rho = 0.8$)

24. Ten competitors in a musical test were ranked by three judges x, y and z in the following order.

Judges x:	1	6	5	10	3	2	4	9	7	8
Judges y:	3	5	8	4	7	10	2	1	6	9
Judges z:	6	4	9	8	1	2	3	10	5	7

Using rank correlation, discuss which pair of judges has the nearest approach to common likings in music.

(*Ans.* x & z)

25. Obtain Spearman's rank correlation for the following data and also obtain $r(x, y)$.

x:	110	100	140	120	80	90
y:	70	60	80	60	10	20

(*Ans.* $\rho = 0.9$)
(*Ans.* $r = 0.882$)

26. Calculate rank correlation co-efficient between height (in inches) of 12 fathers (x) and their sons (y).

x:	65	63	67	64	68	62	70	66	68	67	69	71
y:	68	66	68	65	69	66	68	65	71	67	68	70

(***Ans.*** ρ ***= 0.7220***)

27. Calculate rank correlation co-efficient given ranks of x and y.

x:	6	4	5	3	1	2
y:	5	6	3	4	1	2

(***Ans.*** ρ ***= 0.714***)

28. Calculate rank correlation co-efficient gives ranks of x and y.

x:	1	2	3	4	5	6	7	8	9	10
y:	10	7	2	6	4	8	3	1	9	5

(***Ans.*** ρ ***= 0.6464***)

29. Calculate Spearman's co-efficient of rank correlation for the following data.

x:	53	98	95	81	75	61	59	55
y:	47	25	32	37	30	40	39	45

(***Ans.*** ρ ***= –0.905***)

30. Answer the following

1. $\rho = 0.8, \ \Sigma d^2 = 33$ find n.

(***Ans.*** n ***= 10***)

2. $n = 10 \ \rho = 0.2576$ find Σd^2.

(***Ans.*** $\Sigma d^2 = 122.5$)

7

Regression Analysis

Content

- Regression
- Line of Regression
- Least Square method
- Regression Co-efficient
- Relationship between regression co-efficient 'b' and correlation co-efficient *r*.
- Properties of regression co-efficient
- Distinction between correlation and regression analysis.

Formulae

1. Regression co-efficient

(a) $$b_{yx} = \frac{n\sum xy - \sum x \sum y}{n\sum x^2 - (\sum x)^2}$$

(b) $$b_{xy} = \frac{n\sum xy - \sum x \sum y}{n\sum y^2 - (\sum y)^2}$$

(c) When deviations are taken from mean

1. $$b_{yx} = \frac{n\sum xy}{n\sum x^2}$$

2. $$b_{xy} = \frac{n\sum xy}{n\sum y^2}$$

2. Relationship between *r* and *b*.

(a) $$b_{xy} = r \cdot \frac{\sigma_x}{\sigma_y}$$

(b) $b_{yx} = r \cdot \dfrac{\sigma_x}{\sigma_y}$

(c) $r = \sqrt{b_{xy}\, b_{yx}}$

3. Regression equation

(a) Y on X

$$\left(Y - \overline{Y}\right) = b_{yx}\left(X - \overline{X}\right)$$

(b) X on Y

$$\left(X - \overline{X}\right) = b_{xy}\left(Y - \overline{Y}\right)$$

INTRODUCTION

In the previous chapter we discussed about the nature of relation-ship between the variable (i.e. correlation) and the extend to which they are correlated. Correlation analysis just gave us an idea about the association of the variable, but we would like to know how far they are related to each other. i.e., we would like to know the functional relationship between two variables. Regression methods are meant to determine these functional relationship.

7.1 REGRESSION

Regression analysis is a mathematical measure of the average relationship between two or more variables in term of the original units of the data.

In regression analysis there are two types of variable; namely the dependent and independent variable. A dependent variable is the one whose value is to be predicted and independent variable is the one which influences the value of the variable (dependent).

7.2 LINE OF REGRESSION

In scatter diagram (discussed in previous chapter), we find that the points cluster around some curve called curve regression. If curve is a straight line, it is called the line of regression.

"The line of regression is the line which gives the best estimate to the value of one variable for any specific value of the other variable".

Let us consider the relationship between two variables X and Y. Here there are two lines of regression (i) Y on X and (ii) X on Y.

(i) Line of regression Y on X: Here Y is the dependent variable and X is independent variable and this line gives the best estimate for the value of Y for any specified value of X. The regress equation for Y on X is given by

$$Y = a + bX \qquad \text{...(7.1)}$$

(ii) Line of regression X on Y: This line is used to estimate the value of X for any specified value of Y. The regression equation for X on Y is given by

$$X = a + bY \qquad \ldots(7.2)$$

In equations (7.1) and (7.2), a and b are constants and are obtained by least square principle.

7.3 LEAST SQUARE METHOD

Legender's principle of least square is "To minimize the sum of squares of the deviations of the actual values Y (or X) from its estimated values as given by the line of best fit."

Regression Equation of *Y* on *X*:

The regression equation of Y on X is $Y = a + bX$.

Then according to least square principle, we have to minimize

$$E_y = \Sigma(\underset{\substack{\downarrow \\ \text{Actual} \\ \text{Value}}}{Y} - \underset{\substack{\downarrow \\ \text{Estimated} \\ \text{Value}}}{(a + bX)})^2$$

E_y is minimized by partially differentiating it *w.r.t* a and b respectively and equating them to zero. In this way we get two equations.

$$\Sigma Y = na + b\Sigma X \qquad \ldots(7.3)$$

$$\Sigma XY = a\Sigma X + b\Sigma X^2$$

These equations are known as normal equations. Solving these equation simultaneously, we can obtain the values of the constants a and b.

Regression equation of *X* on *Y*:

The regression equation of X on Y is $X = a + bY$.

Here we have to minimize $E_x = \Sigma(X - (a + bY))^2$

Similarly here we get

$$\Sigma X = na + b\Sigma Y \qquad \ldots(7.4)$$

$$\Sigma XY = a\Sigma Y + b\Sigma Y^2$$

as the normal equations and the values of a and b can be obtained from them.

Example Find the regression equation Y on X for the following data.

X:	2	4	6	8
Y:	5	15	20	25

Solution

	X	Y	X^2	Y^2	XY
	2	5	4	25	10
	4	15	16	225	60
	6	20	36	400	120
	8	25	64	625	200
Total	20	65	120	1275	390

The normal equation for Y on X.

$$\Sigma Y = na + b\Sigma X$$

$$\Sigma XY = a\Sigma X + b\Sigma X^2$$

$$65 = 4a + 20b \qquad (1)$$

$$390 = 20a + 120b \qquad (2)$$

$$9 \times 5: \qquad 325 = 20a + 100b \quad (3)$$

$$390 = 20a + 120b \quad (2)$$

$$(3) - (2): \qquad -65 = 0 - 20b$$

$$b = \frac{65}{20} = 3.25$$

Substituting the value of b in (1) we get

$$65 = 4a + 20 \times 3.25$$

$$65 = 4a + 65$$

$$4a = 0$$

$$a = 0$$

$$Y = 3.25X$$

7.4 REGRESSION CO-EFFICIENT

In a regression equation 'a' is the intercept which the line cuts in the axis Y and b is the slope of the line and is called the regression co-efficient. The regression co-efficient is a measure of change in dependent variable corresponding to an unit change in independent variable.

Regression co-efficient of X on Y is denoted by b_{xy} and Y on X is denoted by b_{yx}.

7.5 RELATIONSHIP BETWEEN REGRESSION CO-EFFICIENT '*b*' AND CORRELATION CO-EFFICIENT '*r*'

By definition we know that

$$r = \frac{Cov\,(X, Y)}{\sigma_x\,\sigma_y} = \frac{\sigma_{xy}}{\sigma_x\,\sigma_y}$$

By solving the normal equation 6.3 we get

$$b_{xy} = \frac{\Sigma(X-\overline{X})(Y-\overline{Y})}{\Sigma(X-\overline{X})^2}$$

$$= \frac{\dfrac{\Sigma(X-\overline{X})(Y-\overline{Y})}{n}}{\dfrac{\Sigma(X-\overline{X})^2}{n}}$$

$$= \frac{\sigma_{xy}}{\sigma_x^{\ 2}}$$

Multiplying and dividing r by σ_x we get

$$r = \frac{\sigma_{xy}\ \sigma_x}{\sigma_x^{\ 2}\ \sigma_y}$$

$$= b_{yx}\frac{\sigma_x}{\sigma_y} \quad \text{for } Y \text{ on } X$$

Similarly $r = b_{xy}\dfrac{\sigma_y}{\sigma_x}$ for X on Y.

$\therefore$ we get

$$b_{yx} = \frac{r\sigma_y}{\sigma_x} \quad \text{for } Y \text{ on } X$$

$$b_{xy} = \frac{r\sigma_x}{\sigma_y} \quad \text{for } X \text{ on } Y$$

Now the two regression equations are

$$(Y-\overline{Y}) = b_{yx}(X-\overline{X}) \quad \text{for } Y \text{ on } X$$

$$(X-\overline{X}) = b_{xy}(Y-\overline{Y}) \quad \text{for } X \text{ on } Y$$

Computation of regression co-efficient

Regression Co-efficient for Y on X is given by

$$b_{yx} = \frac{\Sigma(X-\overline{X})(Y-\overline{Y})}{\Sigma(X-\overline{X})^2}$$

$$b_{yx} = \frac{n\Sigma xy - \Sigma x \Sigma y}{n\Sigma x^2 - (\Sigma x)^2}$$

Let $d_x = X - A,\ d_y = Y - A$
Be the deviation from the original units then

$$b_{yx} = \frac{n\Sigma d_{xy} - \Sigma d_x \Sigma d_y}{n\Sigma d_x^{\,2} - (\Sigma d_x)^2}$$

if the deviations are taken from actual means then

$$\Sigma d_x = \Sigma d_y = 0$$

$$\therefore\ b_{yx} = \frac{n\Sigma d_{xy}}{n\Sigma d_x^{\,2}}$$

$$= \frac{\Sigma d_{xy}}{\Sigma d_x^{\,2}}$$

Similarly for X on Y we have

$$b_{xy} = \frac{\Sigma (X - \overline{X})(Y - \overline{Y})}{\Sigma (Y - \overline{Y})^2}$$

$$b_{xy} = \frac{n\Sigma xy - \Sigma x \Sigma y}{n\Sigma Y^2 - (\Sigma Y)^2}$$

Let $d_x = X - A,\ d_y = Y - A$
Be the deviation from the original units then

$$b_{xy} = \frac{n\Sigma d_{xy} - \Sigma d_x d_y}{n\Sigma d_y^{\,2} - (\Sigma d_y)^2}$$

If the deviations are taken from actual means then

$$\Sigma d_x = \Sigma d_y = 0$$

$$\therefore\ b_{xy} = \frac{\Sigma d_{xy}}{\Sigma d_y^{\,2}}$$

Example 1 You are given the following

	Y	X
Mean	10	15
S.D	2	4

$r = 0.8$

Find the regression equation:

Solution Regression equation Y on X

$$b_{xy} = r \cdot \frac{\sigma_y}{\sigma_x} = 0.8 \times \frac{2}{4} = 0.4$$

$$(Y - \overline{Y}) = b_{yx}\,(X - \overline{X})$$

$$(Y - 10) = 0.4\,(X - 15)$$

$$Y - 10 = 0.4X - 6$$

$$Y = 0.4X + 10 - 6$$

$$Y = 0.4X + 4$$

Regression equation X on Y.

$$b_{xy} = r \cdot \frac{\sigma_y}{\sigma_x} = 0.8 \times \frac{4}{2} = 1.6$$

$$(X - \overline{X}) = b_{xy}\,(Y - \overline{Y})$$

$$(X - 15) = 1.6\,(Y - 10)$$

$$X = 1.6Y - 16 + 15$$

$$X = 1.6Y - 1$$

Example Find the regression equation for the following.

X:	2	4	6	8
Y:	10	20	25	30

Solution Regression equation

	X	Y	$d_x = X - 5$	d_y	d_x^2	d_y^2	d_{xy}
	2	10	–3	–5	9	25	–15
	4	20	–1	5	1	25	–5
	6	25	1	10	1	100	10
	8	30	3	15	9	225	45
Total	20	65	0	25	20	375	35

$d_x = X - 5$, $d_y = Y - 15$

Regression equation of y on X.

$$b_{yx} = \frac{n\sum d_{xy} - \sum d_x \sum d_y}{n\sum d_x^2 - (\sum d_x)^2}$$

$$= \frac{4 \times 35}{4 \times 20} \qquad \therefore \sum d_x = 0$$

$$= 1.75$$

$$(Y - \overline{Y}) = b_{yx}\,(X - \overline{X})$$

$(Y - 16.25) = 1.75\,(X - 5)$

$Y = 1.75X - 8.75 + 16.25$

$Y = 1.75X + 7.5$

Regression equation of X on y.

$$b_{xy} = \frac{n\Sigma d_{xy} - \Sigma d_x\, \Sigma d_y}{n\Sigma d_y^{\,2} - (\Sigma d_y)^2}$$

$$= \frac{4 \times 35 - 0}{4 \times 20 - (20)^2}$$

$$= 0.054$$

$$(X - \overline{X}) = b_{xy}\,(Y - \overline{Y})$$

$(X - 5) = 0.054\,(Y - 16.25)$

$X = 0.054 + 4.1225$

7.6 PROPERTIES OF REGRESSION CO-EFFICIENT

1. The range of regression co-efficient is $-\infty$ to ∞
2. The correlation co-efficient between two variable X and Y is given by

   $$r = \sqrt{b_{xy} \times b_{yx}}$$

3. If $b_{xy} > 0 \Leftrightarrow r > 0$

   $b_{xy} < 0 \Leftrightarrow r < 0$

   Property hold the same for regression co-efficient Y on X.
4. If one of the regression co-efficient is greater than unity the other one is less than unity.

   i.e., if $b_{xy} > 1$ then $b_{yx} < 1$
5. If the variable X and Y are independent the regression co-efficient is zero.

7.7 DISTINCTION BETWEEN CORRELATION AND REGRESSION ANALYSIS

As we have discussed about these two analysis separately, now let us distinguish between them.

	Correlation	*Regression*
1.	It studies the relationship between two variables.	It estimates the functional relationship between X and Y variables.
2.	Correlation need not imply cause and effect relationship between the variables under study.	It studies the cause and effect of relationship between the variables.
3.	Correlation analysis is confined only to the study of linear relationship.	Regression analysis deals both linear and non linear relationship of the variables under study.
4.	It measures the degree of Covariance.	It study nature of covariance.
5.	The relationship may be purely a chance and it may not have practical relevance.	There is a perfect relationship and it has practical relevance.
6.	Correlation co-efficient between X and Y is symmetric i.e. $r(x, y) = r(y, x)$	Regression co-efficient is not symmetric i.e. $b_{xy} \neq b_{yx}$
7.	Correlation co-efficient is a relative measure and its range is –1 to 1.	Regression co-efficient is an absolute measure and its range is $-\infty$ to ∞.

Solved Examples

1. Obtain the regression of Y on X and X on Y from the following table and estimate the blood pressure when age is 50.

Age X	*Blood pressure Y*
56	147
42	125
72	160
36	118
63	149
47	128
55	150
49	145
38	115
42	140
68	152
60	155

(B. U., April, 2001)

Solution

X	Y	$d_x (X-55)$	$d_y (Y-140)$	d_x^2	d_y^2	d_{xy}
56	147	1	7	1	49	7
42	125	–13	–15	169	225	195
72	160	17	20	289	400	340
36	118	–19	–22	361	484	418
63	149	8	9	64	81	72
47	128	–8	–12	64	124	96
55	150	0	10	0	100	0
49	145	–6	5	36	25	–30
38	115	–17	–25	289	625	425
42	140	–13	0	169	0	0
68	152	13	12	169	144	156
60	155	5	15	25	225	75
Total 628	1684	–32	4	1636	2482	1754

(a) Regression Y on X

$$b_{yx} = \frac{n\sum d_{xy} - \sum d_x \sum d_y}{n\sum d_x^2 - (\sum d_x)^2}$$

$$= \frac{12 \times 1754 - (-32)(4)}{12 \times 1636 - (-32)^2}$$

$$= \frac{21048 + 128}{19632 - 1024}$$

$$= \frac{21176}{18608}$$

$$= 1.138$$

$$(Y - \overline{Y}) = b_{yx}(X - \overline{X})$$

$$\overline{Y} = \frac{\sum Y}{n} = 140.333$$

$$\overline{X} = \frac{\sum X}{n} = 52.333$$

$$(Y - 140.33) = 1.138(X - 52.33)$$

$$Y = 1.138X - 59.552 + 140.333$$

$$Y = 1.138X + 80.781$$

(b) Regression X on Y

$$b_{xy} = \frac{n\sum d_{xy} - \sum d_x \sum d_y}{n\sum d_y^2 - \left(\sum d_y\right)^2}$$

$$= \frac{12 \times 1754 - (-32)(4)}{12 \times 2482 - (-32)^2}$$

$$= \frac{21176}{29768}$$

$$= 0.711$$

$$(X - \overline{X}) = b_{xy} (Y - \overline{Y})$$

$$(X - 52.333) = 0.711 (Y - 140.333)$$

$$X = 0.711Y - 97.777 + 52.333$$

$$X = 0.711Y - 47.444$$

Blood pressure at the age 50 is given by

$$Y = 1.138 \times 50 + 80.781$$

$$= 137.681$$

$$= 138$$

2. A company wanted to assess the impact of *R* and *D* expenditure (*X*) on annual profit (*Y*). Following table presents data for past 8 years.

X	Y
9	45
7	42
5	41
10	60
4	30
5	34
3	25
2	20

Find the regression equation *Y* on *X*. Estimate the profit for an *R* and *D* expenditure of Rs 8 lakhs.

(B. U., November, 1993)

Solution

	X	Y	$d_x = X - 5$	$d_y = Y - 40$	d_x^2	d_y^2	d_{xy}
	9	45	4	5	16	25	20
	7	42	2	2	4	4	4
	5	41	0	1	0	1	0
	10	60	5	20	25	400	100
	4	30	−1	−10	1	100	10
	5	34	0	−6	0	36	0
	3	25	−2	−15	4	225	30
	2	20	−3	−20	9	400	60
Total	45	297	5	−23	59	1191	224

Regression on y on X

$$b_{yx} = \frac{n\Sigma d_{xy} - \Sigma d_x \Sigma d_y}{n\Sigma d_x^2 - (\Sigma d_x)^2}$$

$$= \frac{8 \times 224 - (5)(-23)}{12 \times 59 - (5)^2}$$

$$= \frac{1792 + 115}{708 - 25}$$

$$= \frac{1907}{683}$$

$$= 2.792$$

$$\overline{X} = \frac{45}{8}, \overline{Y} = \frac{297}{8}$$

$$\overline{X} = 5.625, \overline{Y} = 37.125$$

equation Y on X is $\left(Y - \overline{Y}\right) = b_{yx}\left(X - \overline{X}\right)$

$$(Y - 37.125) = 2.792\,(X - 5.625)$$

$$Y = 2.792X - 15.705 + 37.125$$

$$= 2.792X + 21.42$$

∴ the expected profit when R and D expenditure is 8 lakh is

$$Y = 2.792 \times 8 + 21.42$$

$$= 43.756$$

$$= 44 \text{ lakhs}$$

3. Calculate the two line of regression for the following data.

X:	1	2	3	4	5
Y:	3	6	9	12	15

Solution

	X	Y	$d_x = X - 3$	$d_y = Y - 9$	d_x^2	d_y^2	d_{xy}
	1	3	–2	–6	4	36	12
	2	6	–1	–3	1	9	3
	3	9	0	0	0	0	0
	4	12	1	3	1	9	3
	5	15	2	6	4	36	12
Total	15	45	0	0	10	90	30

Regression Y on X

$$b_{yx} = \frac{\Sigma d_{xy}}{\Sigma d_x^2}$$

Regression X on Y

$$b_{xy} = \frac{\Sigma d_{xy}}{\Sigma d_y^2}$$

$$= \frac{30}{10}$$

$$= 3$$

$\overline{X} = 3, \overline{Y} = 9$

Equation: $(Y - \overline{Y}) = b_{yx}(X - \overline{X})$

$$(Y-9) = 3(X-3)$$

$$Y = 3X$$

$$= \frac{30}{90}$$

$$= \frac{1}{3}$$

$\overline{X} = 3, \overline{Y} = 9$

Equation: $(X - \overline{X}) = b_{xy}(Y - \overline{Y})$

$$(X - 3) = \frac{1}{3}(Y - 9)$$

$$X = \frac{1}{3}Y$$

4. From the data below, find the two regression equation.

Marks in Economics (X)	Marks in Statistics (Y)
25	43
28	46
35	49
32	41
31	36
36	32
29	31
38	30
34	33
32	39

Compute co-efficient correlation between X and Y.

(B. U., November, 1994)

X	Y	$d_x = X - 32$	$d_y = Y - 38$	d_x^2	d_y^2	d_{xy}
25	43	–7	5	49	25	–35
28	46	–4	8	16	64	–32
35	49	3	11	9	121	33
32	41	0	3	0	9	0
31	36	–1	–2	1	4	2
36	32	4	–6	16	36	–24
29	31	–3	–7	9	49	+21
38	30	6	–8	36	64	–48
34	33	2	–5	4	25	–10
32	39	0	1	0	1	0
Total 320	380	0	0	137	398	–93

$$\overline{X} = \frac{320}{10} = 32$$

$$\overline{Y} = \frac{380}{10} = 38$$

Regression Y on X

$$b_{yx} = \frac{\Sigma d_{xy}}{\Sigma d_x^2}$$

$$= \frac{-93}{137}$$

$$= -0.6788$$

Equation: $(Y - \overline{Y}) = b_{yx}\ (X - \overline{X})$

$$(Y - 38) = -0.6788\ (X - 32)$$

$$= -0.6788X + 21.7216X + 38$$

$$Y = -0.6788X + 59.7216$$

Regression X on Y

$$b_{xy} = \frac{\Sigma d_{xy}}{\Sigma d_y^2}$$

$$= \frac{-93}{398}$$

$$= -0.2336$$

Equation: $(X - \overline{X}) = b_{xy}\ (Y - \overline{Y})$

$$(X - 32) = -0.2336\ (Y - 38)$$

$$= -\ 0.2336Y + 8.8768 + 32$$

$$X = -0.2336Y + 40.8768$$

The co-efficient of correlation is

$$r = \sqrt{b_{xy}\ b_{yx}}$$

$$= \sqrt{(-0.6788)\ (-0.2336)}$$

$$= 0.3982$$

5. Calculate the regression equation for the following data and also computes Karl Pearson's co-efficient of correlation.

No. of Students (X):	800	600	900	700	500	400
No. of passed (Y):	480	300	450	560	450	310

Solution

	X	Y	$d_x = X - 650$	$d_y = Y - 425$	d_x^2	d_y^2	d_{xy}
	800	480	150	55	22500	3025	8250
	600	300	−50	−125	2500	15625	6250
	900	450	250	25	62500	625	6250
	700	560	50	135	2500	18225	6750
	500	450	−150	25	22500	625	−3750
	400	310	−250	−115	625000	13225	28750
Total	3900	2550	0	0	175000	51350	52500

$$\overline{X} = \frac{3900}{6} = 650 \quad \overline{Y} = \frac{2550}{6} = 425$$

Regression Y on X

$$b_{yx} = \frac{\sum d_{xy}}{\sum d_x^2}$$

$$= \frac{52500}{175000}$$

$$= 0.3$$

Equation: $(Y - \overline{Y}) = b_{yx}(X - \overline{X})$

$$(Y - 425) = 0.3\,(X - 650)$$

$$Y = 0.3X - 195 + 425$$

$$= 0.3X + 230$$

Regression X on Y

$$b_{xy} = \frac{\sum d_{xy}}{\sum d_y^2}$$

$$= \frac{52500}{51350}$$

$$= 1.022$$

Equation: $(X - \overline{X}) = b_{xy}(Y - \overline{Y})$

$$(X - 650) = 1.022\,(Y - 425)$$

$$X = 1.022Y - 434.35 + 650$$

$$= 1.022Y + 215.65$$

$$X = 1.022Y + 215.65$$

The correlation co-efficient

$$r = \sqrt{b_{xy}\, b_{yx}}$$

$$= \sqrt{1.022 \times 0.3}$$

$$= 0.5537$$

6. You are given below the following in formation about advertisement and sales.

	Adv. Exp. (X) (Rs Crores)	Sales (Y) (Rs Crores)
Mean	20	120
S.D	5	25

Correlation co-efficient = 0.8

1. Calculate the two regression equations.
2. Find the likely sales when advertisement expenditure is Rs 25 crores.

(B. U., October, 2001)

(B. U., October, 2000)

Solution

Regression equation Y on X

$$(Y - \overline{Y}) = b_{yx}(X - \overline{X})$$

$$= r \cdot \frac{\sigma_y}{\sigma_x}(X - \overline{X})$$

$$(Y - 120) = 0.8 \times \frac{25}{5}(X - 20)$$

$$Y = 4X - 80 + 120$$

$$Y = 4X + 40$$

Regression equation X on Y

$$(X - \overline{X}) = b_{xy}(Y - \overline{Y})$$

$$= r \cdot \frac{\sigma_x}{\sigma_y}(Y - \overline{Y})$$

$$(X - 20) = 0.8 \times \frac{5}{25}(Y - 120)$$

$$X = 0.16Y - 19.2 + 20$$

$$X = 0.16Y + 0.8$$

1. The two regression equations are

(a) $Y = 4X + 40$

(b) $X = 0.16Y + 0.8$

2. When $X = \text{Rs } 25$

$$Y = 4 \times 25 + 40$$
$$= 140$$

The likely sales is Rs 140 crores.

7. Given No. of pairs = 12

	X	Y
A M	74.50	125.50
Std. Deviation	13.07	15.85

Summation of products of corresponding deviations from respective means = 2176. Calculate the co-efficient of correlation between X and Y and find the regressions equations.

(B. U., November, 1999)

Solution

$$\text{Correlation co-efficient } r = \frac{Cov(X, Y)}{\sigma_x \sigma_y}$$

$$Cov(X, Y) = \frac{2176}{12} = 181.333$$

$$r = \frac{181.333}{13.07 \times 15.85}$$

$$= 0.875$$

Regression equation Y on X

$$(Y - \overline{Y}) = r \cdot \frac{\sigma_y}{\sigma_x} (X - \overline{X})$$

$$(Y - 125.5) = 0.875 \times \frac{15.85}{13.07} (X - 74.5)$$

$$Y = 1.0611X - 79.053 + 125.5$$

$$Y = 1.0611X - 46.447$$

Regression equation X on Y

$$(X - \overline{X}) = r \cdot \frac{\sigma_x}{\sigma_y} (Y - \overline{Y})$$

$$(X - 74.5) = 0.875 \times \frac{13.07}{15.85} (Y - 125.5)$$

$$X = 0.722Y - 90.611 + 74.5$$

$$X = 0.722Y - 16.111$$

8. From the following details estimate the value of A when B = 55.

Mean of A = 39.5, Mean of B = 47.5

Std. Dev. of A = 10.8, Std. Dev. of B = 16.8

Co-efficient of correlation between A and B = 0.42.

(B. U., April, 1999)

Solution Regression equation of A on B is

$$(A - \overline{A}) = b_{AB} (B - \overline{B})$$

$$(A - \overline{A}) = r \cdot \frac{\sigma_A}{\sigma_B} (B - \overline{B})$$

$$(A - 39.5) = 0.42 \times \frac{10.8}{16.8}(B - 47.5)$$

$$A = 0.27B - 12.825 + 39.5$$
$$= 0.27B + 26.675$$

Value of A when B = 55 is

$$A = 0.27 \times 55 + 26.675$$
$$= 14.85 + 26.675$$
$$= 41.525$$

9. From the following data find the marks in English. If the marks in Hindi is 50.

	Hindi (X)	English (Y)
Mean	40	50
S.D	10	16

$r = +0.5$

(B. U., April, 1998)

Solution: Regression equation Y on X

$$(Y - \overline{Y}) = r \cdot \frac{\sigma_y}{\sigma_x}(X - \overline{X})$$

$$(Y - 50) = 0.5 \times \frac{10}{16}(X - 40)$$

$$Y = 0.3125X - 12.5 + 50$$
$$Y = 0.3125X + 37.5$$

Marks in English when marks scored in Hindi is 50.

$$Y = 0.3125 \times 50 + 37.5$$
$$= 15.625 + 37.5$$
$$= 53.125$$
$$= 53 \text{ marks}$$

10. The information given below relates to the advertisement and sales of a company.

	Adv. Exp. (Rs lakhs) (X)	Sale (Rs lakhs) (Y)
Arithmetic mean	20	100
S.D	3	12

Correlation co-efficient between X and $Y = 0.8$

1. Find two regression equations.

2. What should be the advertisement expenditure, if the company wants to attain a sales target of Rs 120 lakhs?

(B. U., November, 1997)

Solution: Regression equation y on X

$$(Y - \overline{Y}) = r \cdot \frac{\sigma_y}{\sigma_x}(X - \overline{X})$$

$$(Y - 100) = 0.8 \times \frac{\sigma_y}{\sigma_x}(X - 20)$$

$$Y = 3.2X - 64 + 100$$

$$Y = 3.2X + 36$$

Regression equation X on y

$$(X - \overline{X}) = r \cdot \frac{\sigma_x}{\sigma_y}(Y - \overline{Y})$$

$$(X - 20) = 0.8 \times \frac{3}{12}(Y - 100)$$

$$X = 0.2Y - 20 + 20$$

$$X = 0.2Y$$

1. The two regression equations are

$$Y = 3.2X + 36$$

$$X = 0.2Y$$

2. Sales is Rs 120 lakhs

$$X = 0.2 \times 12.$$

$$X = 24 \text{ lakhs}$$

$\therefore$ the expect expenditure is 24 lakhs

11. Following data are given for marks in English (X) and marks in Maths (Y) at a certain examination.

	X	Y
Mean marks	39.5	47.5
S.D marks	10.8	16.8

Co-efficient of correlation is $r = 0.42$. Find the regression equation of Y and X and estimate the marks in Maths when marks in English is 50.

(B. U., April, 1997)

Solution: Regression equation Y on X

$$(Y - \overline{Y}) = r \cdot \frac{b_y}{b_x}(X - \overline{X})$$

$$(Y - 47.5) = 0.42 \times \frac{16.8}{10.8}(X - 39.5)$$

$$Y = 0.653X - 25.806 + 47.5$$

$$= 0.653X + 21.694$$

Regression equation X on Y

$$(X - \overline{X}) = r \cdot \frac{b_x}{b_y}(Y - \overline{Y})$$

$$(X - 39.5) = 0.42 \times \frac{10.8}{16.8}(Y - 47.5)$$

$$X = 0.27Y - 12.825 + 39.5$$

$$= 0.27Y + 26.675$$

Marks in Maths when English = 50

$$Y = 0.653 \times 50 + 21.694$$

$$= 0.653 \times 50 + 21.694$$

$$= 32.65 + 21.694$$

$$= 54.344$$

$$= 54 \text{ marks}$$

12. You are given the following data.

	X	Y
Mean	20	10
S.D	3	4

$r = +0.9$

Estimate the value of Y when $X = 25$

(B. U., October, 1996)

Solution Regression equation Y on X

$$(Y - \overline{Y}) = r \cdot \frac{\sigma_y}{\sigma_x}(X - \overline{X})$$

$$(Y - 10) = 0.9 \times \frac{4}{3}(X - 20)$$

$$Y = 1.2X - 24 + 10$$
$$= 1.2X - 14$$

value of Y when $X = 25$ is

$$Y = 1.2 \times 25 - 14$$
$$= 30 - 14$$
$$= 16$$

13. To study the effect of rain on yield of wheat the following results were obtained.

	Mean	S.D
Yield (X)	800	12
Rainfall (Y)	50	02

$r = 0.80$

estimate the yield when the rainfall is 80 inches.

(B. U., April, 1995)

Solution Regression equation of X on Y

$$(X - \overline{X}) = r \cdot \frac{\sigma_x}{\sigma_y}(Y - \overline{Y})$$

$$(X - 800) = 0.80 \times \frac{12}{2}(Y - 50)$$

$$X = 4.8Y - 240 + 800$$
$$= 4.8Y + 560$$

when rainfall is 80 inches yield is

$$y = 4.8 \times 80 + 560$$
$$= 384 + 560$$
$$= 944$$

yield = 944

14. Find the two regression equations, regression coefficient and co-efficient of correlation from the following figures.

$$\Sigma X = 120,\ \Sigma Y = 432,\ \Sigma XY = 4992,\ \Sigma X^2 = 1394,\ \Sigma Y^2 = 18{,}252 \text{ and } n = 12.$$

Solution

$$\overline{X} = \frac{\Sigma X}{n} = \frac{120}{12} = 10$$

$$\overline{Y} = \frac{\Sigma Y}{n} = \frac{432}{12} = 36$$

$$b_{xy} = \frac{n\Sigma XY - \Sigma X \Sigma Y}{n\Sigma X^2 - (\Sigma X)^2}$$

$$= \frac{12 \times 4992 - 120 \times 432}{12 \times 1394 - (120)^2}$$

$$= \frac{8064}{2325}$$

$$= 3.5$$

$$b_{yx} = \frac{n\Sigma XY - \Sigma X \Sigma Y}{n\Sigma Y^2 - (\Sigma Y)^2}$$

$$= \frac{12 \times 4992 - 120 \times 432}{12 \times 18252 - (432)^2}$$

$$= \frac{8064}{32400}$$

$$= 0.25$$

Regression equation Y on X is

$$(Y - \overline{Y}) = b_{yx}\,(X - \overline{X})$$

$$(Y - 36) = 0.25\,(X - 10)$$

$$Y = 0.25X + 33.5$$

Regression equation X on Y is

$$(X - \overline{X}) = b_{xy}\,(Y - \overline{Y})$$

$$(X - 10) = 3.5\,(Y - 36)$$

$$X = 3.5Y - 116$$

15. Find the two regression equations.

Given	X	Y
Mean	25	22
S.D	4	5

$r = +\,0.8$.

Solutions:

Y on X	X on Y
$(Y-\overline{Y}) = r \cdot \frac{\sigma_y}{\sigma_x}(X-\overline{X})$	$(X-\overline{X}) = r \cdot \frac{\sigma_x}{\sigma_y}(Y-\overline{Y})$
$(Y-22) = 0.8 \times \frac{5}{4}(X-25)$	$(X-25) = 0.8 \times \frac{4}{5}(Y-22)$
$Y-22 = (1)(X-25)$	$X-25 = 0.64Y - 14.08$
$Y = X-25+22$	$X = 0.64Y - 14.08 + 25$
$Y = X-3$	$X = 0.64Y + 10.92$

16. If $b_{xy} = 0.268$ and $b_{yx} = 0.25$, find r.

Solution:

$$r = \sqrt{b_{xy}\, b_{yx}}$$

$$r = \sqrt{0.268 \times 0.25}$$

$$= 0.259$$

Exercises

1. Define the term regression.
2. What are lines of regression?
3. What do you mean by regression coefficient?
4. What is the relation ship between correlation co-efficients and regression co-efficient.
5. State any two differences between regression and correlation analysis.
6. Find n when $r = 0.5$, $\Sigma Y^2 = 3.6$, $\Sigma XY = 27$, $\sigma_x = 3$.

   (Ans. n = 9)
7. If $b_{xy} = 0.4915$, $b_{yx} = 1.219$, find r

   (Ans. r = 0.774)
8. Obtain the two regression equation from the following data.

   $n = 20$, $\Sigma X = 80$, $\Sigma Y = 40$, $\Sigma X^2 = 1680$, $\Sigma Y^2 = 320$, $\Sigma XY = 480$

   (Ans. Y = 0.2857X + 0.8571, X = 1.5Y + 1)
9. Obtain the two regression equation for the following data.

X:	2	4	6	8	10
Y:	5	7	9	8	11

(Ans. Y = 0.65X + 4.1, X = 1.3Y – 4.4)

10. Calculate the two regression equations from the data given below.

X:	10	12	13	12	16	15
Y:	40	38	43	45	37	43

(***Ans.*** $Y = 0.25X + 44.25$, $X = 0.12Y + 17.92$)

11. Find the regression equation from the following.

	X	*Y*
Mean	10	90
S.D	3	12
Correlation	0.8	

(a) Find the value of Y when $X = 15$

(b) Find the value of X when $Y = 190$

(***Ans.*** $Y = 3.2X + 58$, $X = 0.2Y - 8$, $a = 106$, $b = 30$)

12. Find the regression equation for the following.

	X	*Y*
Mean	18	100
S.D	14	20

$r = +0.8$

(***Ans.*** $X = 0.56Y - 38$, $Y = 1.1429X + 79.43$)

13. Given $n = 450$, $\Sigma X = 18{,}000$, $\Sigma Y = 21600$, $\Sigma d_{xy} = 42{,}075$ Find r.

(***Ans.*** $r = 0.487$)

14. Find the regression equation for Y on X.

	A	*B*
Mean	170	192
S.D	60	20

$r = +0.6$

(***Ans.*** $Y = 0.1999X - 33.99$)

15. Find r, when $b_{xy} = 0.4999$, $b_{yx} = 1.125$.

(***Ans.*** $r = 0.75$)

8

Time Series

Content

- Time Series
- Variations in time series
- Least square method for fitting the linear trend

Formula

1. Trend equation $Y_i = a + bx$

$$\text{where } a = \frac{\Sigma y}{n}$$

$$b = \frac{\Sigma x y}{\Sigma x^2}$$

Key Point

Note whether the given series is odd or even in number, accordingly choose the year of origin.

INTRODUCTION

Forecasting or predicting is an essential tool in any decision making process. The quality of the forecast is dependent on the information that are extracted from past data. Time series analysis is one of the quantitative methods, that we use for forecasting.

8.1 TIME SERIES

Time series refer to any group of statistical information collected at regular intervals of times. Time series analysis is used to detect the change in patterns in these collected data. Based on such changes in pattern that occurs in the group of information a business man can easily make decision about his future and accordingly can plan his future business.

8.2 VARIATIONS IN TIME SERIES (COMPONENTS OF TIME SERIES)

The changes or variation in the values of any phenomenon over a period of time are due to various forces or factors and their cumulative effects. These interacting forces due to which the value of a variable change over a period of time is known as component of time series.

The following are the four major components of a given time series data.

1. Secular trend
2. Seasonal fluctuations
3. Cyclical fluctuations
4. Irregular variation

1. Secular Trend

In secular trend the value of the variable tends to increase or decrease over a long period of time. The steady increase in the cost of living recorded is an example of secular trend. Though there seems a great fluctuation in the cost of living from year to year, the fluctuation that happens in a long-term period is a trend towards a steady increase.

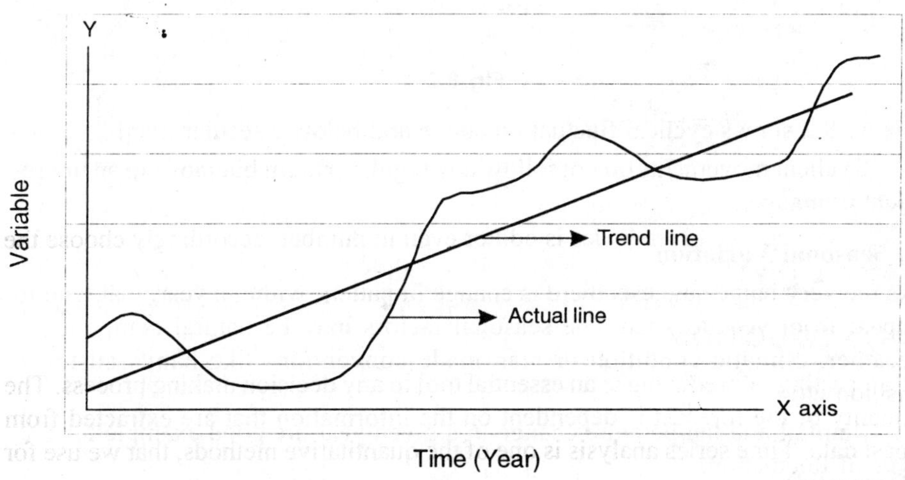

Fig. 8.1

Figure 8.1 shows a secular trend is an increasing but fluctuating time series.

eg: (i) Increase in rate of unemployment year to year.

(ii) Increase of pollution year to year.

(iii) Increase in population (birth rate) year to year.

2. Cyclical Fluctuation

In cyclical fluctuation the value of the variable can have both rising and falling phases. The most common example of cyclical fluctuation is business cycle. In business we can see both peak and slump period. During peak time the value of the variable is above the trend line and during the slump period it is below the trend line. The four phases in the oscillatory movement are (i) prosperity (ii) recession (iii) depression (iv) recovery.

Thus we can see a cyclical pattern.

The successive phase between peak and slump period is atlest 1 year and it can be as many as 5 or 10 years.

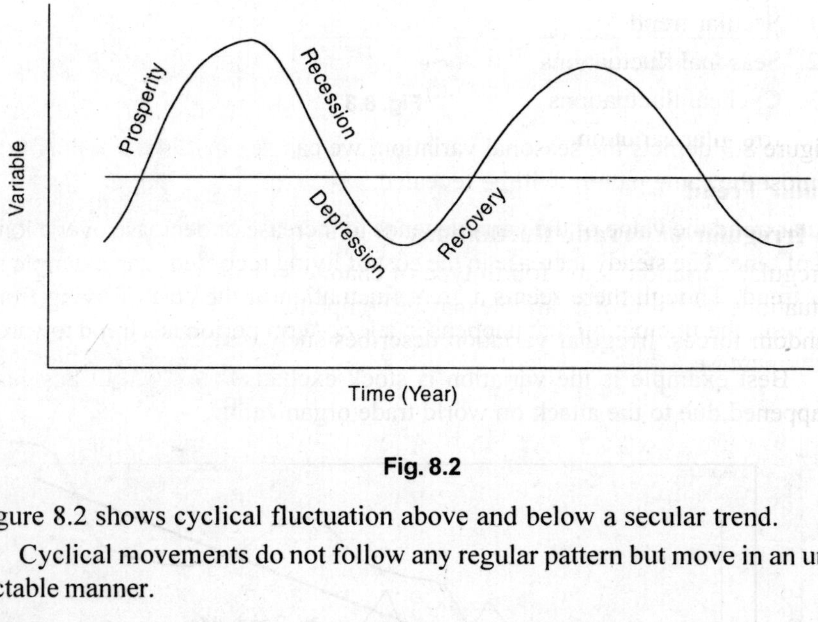

Fig. 8.2

Figure 8.2 shows cyclical fluctuation above and below a secular trend.

Cyclical movements do not follow any regular pattern but move in an unpredictable manner.

3. Seasonal Variation

As the very name suggests there is change in pattern within a year, that tend to repeat from year to year. The seasonal factors may be natural components-weather, climatic condition or man made conventions like habits customs, fashion etc.

Rainfall and temperature change in a given year are good example for this type of fluctuation.

Other examples are:

(i) Sales of sweaters increase during winter season.

(ii) Increase in sales of cold drinks and ice creams during summer season.

(iii) Sales and profits in departmental stores increases during festivals like Diwali, Christmas, etc.

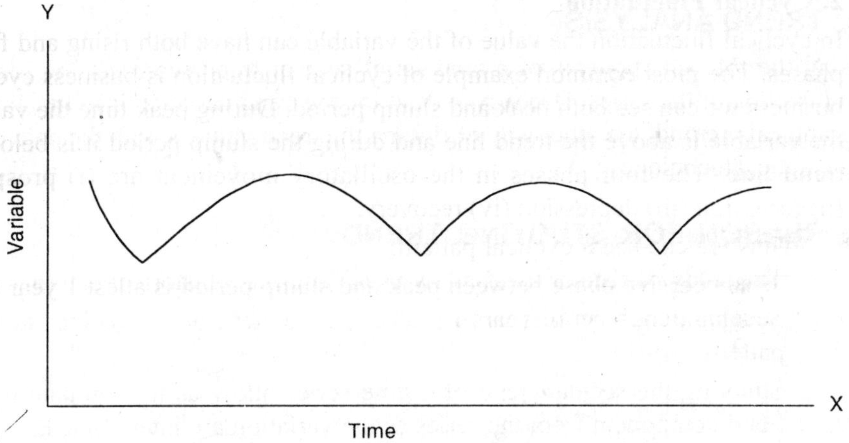

Fig. 8.3

Figure 8.3 depicts the seasonal variation, we can see that in the following years almost the same pattern will be repeated.

4. Irregular or erratic fluctuations

Irregular variation is the fourth type of change in time series analysis. In many situations the value of a variable may be completely unpredictably changed due to random forces. Irregular variation describes such manner.

Best example is the variation is stock exchanges and world business that happened due to the attack on world trade organization.

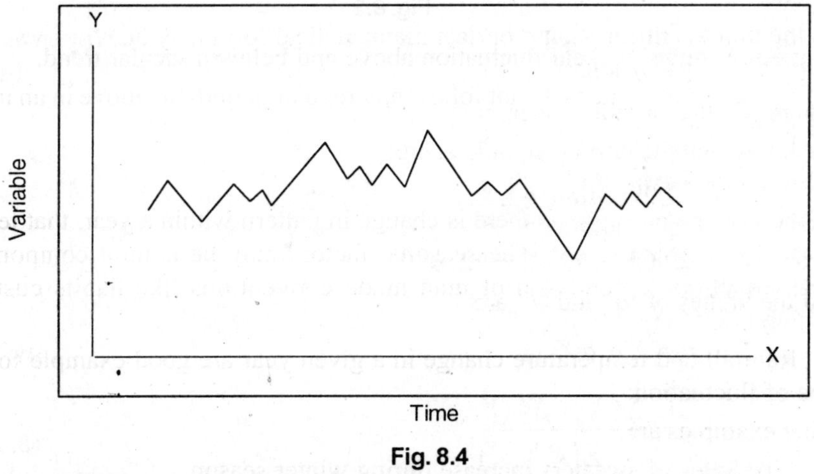

Fig. 8.4

Figure 8.4 Depicts irregular variation.

Other examples are:

(i) Rise in price of vegetables due to lorry strikes.

(ii) Rise in death rate in Gujarat due to an earthquake.

8.3 TREND ANALYSIS

Among the four components of a time series secular trend represent the long term direction of the series. One way to describe the trend variation is to fit a line by graphical method, the other way of finding the trend value is with the help of mathematical technique.

8.4 REASON FOR STUDYING TREND

1. The study of secular trend allow us to describe a historical pattern.
2. Secular trends permits us to project past pattern or to predict future pattern.
3. Studying the secular trend of a time series allow us to eliminate the trend component from the series as the variation are inter related. i.e., if we have to study about the short term trend (cyclical trend), we have to eliminate the long term fluctuation.

8.5 FITTING THE LINEAR TREND BY THE LEAST SQUARE METHOD

Least square method is the most accurate method of finding the trend values by using mathematical techniques. We have already discussed about this method in the previous chapter.

Linear Trend: If the values of the variable are plotted on a graph and if the plotted points fall more or less in the pattern around a straight line, it is called a linear trend.

As the trend is linear its appropriate mathematical form may be written as

$$Y = a + bX \qquad \ldots (8.1)$$

Where y is the variable under study.

We know that the normal equations are

$$\Sigma Y = na + b\Sigma Y$$

$$\Sigma XY = a\Sigma X + b\Sigma X^2$$

and the values of 'a' and 'b' are

$$a = \overline{Y} - b\overline{X} \qquad \ldots (8.2)$$

$$b = \frac{\Sigma(X - \overline{X})(Y - \overline{Y})}{\Sigma(X - \overline{X})^2} \qquad \ldots (8.3)$$

now the trend equation is

$$Y_c = a + bX$$

1. When there is an odd number of element in the time series.

Year (Y)	$(Y - \overline{Y})$ = (Year – Year of origin)	Time (X)
1995	1995–1998	–3
1996	1996–1998	–2
1997	1997–1998	–1
1998	1998–1998	0
1999	1999–1998	1
2000	2000–1998	2
2001	2001–1998	3

Where year of origin is exactly the middle year.

Now by choice $\Sigma X = 0$

Therefore the equation (8.2) and (8.3) reduces to

$$a = \overline{Y} = \frac{\Sigma Y}{n}$$

$$b = \frac{\Sigma XY}{\Sigma X^2}$$

Hint

$$b = \frac{\Sigma(X - \overline{X})(Y - \overline{Y})}{\Sigma(X - \overline{X})^2}$$

$$= \frac{n\Sigma XY - \Sigma X\,\Sigma Y}{n\Sigma X^2 - (\Sigma X)^2}$$

2. Method of trend when there is a even number of elements in the time series.

Year (Y)	$(Y - \overline{Y})$	$(Y - \overline{Y}) \times 2$	Time (X)
1995	1995–1998.5	–3.5 × 2	–7
1996	1996–1998.5	–2.5 × 2	–5
1997	1997–1998.5	–1.5 × 2	–3
1998	1998–1998.5	–0.5	–1
1999	1999–1998.5	0.5	+1
2000	2000–1998.5	1.5	3
2001	2001–1998.5	2.5	5
2002	2002–1998.5	3.5	7

Where $\overline{Y}$ = average of the middle years.

Here again by choice $\sum X = 0$

Therefore the equation (8.2) and (8.3) reduce

$$a = \overline{Y} = \frac{\sum Y}{n}$$

$$b = \frac{\sum XY}{\sum X^2}$$

Aliter Method:

1. When there is an odd number of element in the time series.

In this case we choose the middle most value as $X = 0$ and the remaining value are accordingly written so that $\sum X = 0$ i.e. the middle most year will have the value zero and the points above that value will be decreased by one and the value below zero will be increased by one and that yields $\sum X = 0$.

For example:

	Year Y	*Time X*
	1995	–3
	1996	–2
	1997	–1
Middle year	1998	0
	1999	1
	2000	2
	2001	3
		$\sum X = 0$

Thereby using the formulae

$$a = \overline{Y} = \frac{\sum Y}{n}$$

$$b = \frac{\sum XY}{\sum X^2}$$

we can find the constants

2. When there is an even number of element in the time series.

In this case the middle two value will be assigned with $X = -1$ and $X = +1$ and the value above –1 will be decreased by 2 and the value below 1 will be increased by 2 so that $\sum X = 0$.

For example:

	Year (Y)	*Time (X)*
	1995	−7
	1996	−5
	1997	−3
Middle year	1998	−1
	1999	+1
	2000	3
	2001	5
	2002	7
		$\sum X = 0$

And there by applying the formulae.

$$a = \overline{Y} = \frac{\sum Y}{n}$$

$$b = \frac{\sum XY}{\sum X^2}$$

we can find the constants.

Solved Examples

1. Give below is the data on production of a company (in lakhs).

Year:	1983	1984	1985	1986	1987	1988	1989
Production:	15	14	18	20	17	24	27

Fit the linear trend. Compute the trend values for each year. Estimate the production for year 1991.

(B. U., November, 1993)

Solution

Year (1)	Output (2)	X ((1) − 1986)	X^2	XY	Trend value Y_c
1983	15	−3	9	−45	13.3928
1984	14	−2	4	−28	15.3571
1985	18	−1	1	−18	17.3214
1986	20	0	0	0	19.2857
1987	17	1	1	17	21.25
1988	24	2	4	48	23.2143
1989	27	3	9	81	25.1786
Total	135	0	28	55	

Now from the table

$$a = \frac{\sum Y}{n} = \frac{135}{7} = 19.2857$$

$$b = \frac{\sum XY}{\sum X^2} = \frac{55}{28} = 1.9643$$

The equation for the best fit is given by $Y_c = 19.2857 + 1.9643X$.(1)

The trend Y_c for each year can be calculate by substituting the value for X in equation (1). To estimate the production for year 1991.

Year of origin is 1986

$\therefore \quad X = 1991 - 1986 = 5$

$\therefore \quad Y_c = 19.2857 + 1.9643 \times 5$

$= 29.1072$

$\therefore$ The production in year 1991 is estimated to 29.1072 lakhs.

2. Fit a trend equation of the form $Y = a + bX$ bY the method of least squares to the following data and find by which year the production will reach 63 million tons. Also obtain trend values.

Year:	1966	1967	1968	1969	1970	1971	1972	1973
Production: (in millions)	50.3	52.7	49.8	57.3	56.6	60.7	62.1	56.0

(B. U., April, 1992)

Solution

Year (1)	Production (Y)	X	X^2	XY
1966	50.3	–7	49	–352.1
1967	52.7	–5	25	–263.5
1968	49.8	–3	9	–149.4
1969	57.3	–1	1	–57.3
1970	56.6	1	1	56.6
1971	60.7	3	9	182.1
1972	62.1	5	25	310.5
1973	56.0	7	49	392.0
Total	445.5	0	168	118.9

Year of origin = 1969.5 (even data)

$\therefore \quad X = (\text{year (1)} - 1969.5) \times 2$

$$a = \frac{\sum Y}{n} = \frac{445.5}{8} = 55.6875$$

$$b = \frac{\sum XY}{\sum X^2} = \frac{118.9}{168} = 0.7077$$

The equation for the best fit is given by

$Y_c = 55.6875 + 0.7077X$

To estimate the year when the production is 63 million.

$$63 = 55.6875 + .7077X$$

$$\therefore \quad 0.7077X = 63 - 55.6875 = 7.3125$$

$$\therefore \quad X = \frac{7.3125}{0.7077}$$

$$= 10.33$$

$$= 11 \text{(we are rounding off to the next year)}$$

we know that

$$X = (\text{year} - 1969.5) \times 2$$

$$\therefore \quad \frac{11}{2} = (\text{year} - 1969.5)$$

$$5.5 = \text{Year} - 1969.5$$

$$\therefore \quad \text{Year} = 1969.5 + 5.5$$

$$= 1975.0$$

In the year 1975, the production is expected to be 63 million tons.

3. Fit a straight line trend using the method of least squares and estimate the profits for the year 1990.

Year:	1981	1982	1983	1984	1985	1986	1987
Profits (Rs '000):	57	65	63	72	69	78	82

(B. U., November, 1994)

Year	Profits Y	X	X^2	XY	Y_c
1981	57	–3	9	–171	57.96
1982	65	–2	4	–130	61.785
1983	63	–1	1	–63	65.607
1984	72	0	0	0	69.4286
1985	69	1	1	69	73.25
1986	78	2	4	156	77.071
1987	82	3	9	246	80.893
Total	486		28	107	

Year of origin = 1984

Now from the table

$$a = \frac{\Sigma Y}{n} \qquad b = \frac{\Sigma XY}{\Sigma X^2}$$

$$= \frac{486}{7} \qquad = \frac{107}{28}$$

$$= 69.4286 \qquad = 3.8214$$

The equation for the best fit is given by

$$Y_c = 69.4286 + 3.8214X$$

Profit for the year 1990 is

$$Y_c = 69.4286 + 3.8214\ (1990\text{–}1984)$$
$$= 69.4286 + 3.8214 \times 6$$
$$= 92.357$$

Estimated profit for the year 1990 is Rs 92.357

4. Fit by the method of least squares a linear equation $Y = mX + c$ to the following data.

Year:	1982	1983	1984	1985	1986	1987	1988	1989
Production: (in tons)	9.7	10.1	10.2	10.7	11.9	12.9	12.4	14.8

obtain trend values for the years 1996 and 1997.

(B. U., November, 1995)

Solution

Year (1)	Production (Y)	X	X^2	XY	Trend value Y_c
1982	9.7	–7	49	–67.9	9.2334
1983	10.1	–5	25	–50.5	9.906
1984	10.2	–3	9	–30.6	10.5786
1985	10.7	–1	1	–10.7	11.2512
1986	11.9	1	1	11.9	11.9238
1987	12.9	3	9	38.7	12.5964
1988	12.4	5	25	62	13.269
1989	14.8	7	49	103.6	13.9416
Total	92.7	0	168	56.5	

Year of origin = 1985.5

$$\therefore X = (\text{year (1)} - 1985.5) \times 2$$

Now from the table

$$m = \frac{\Sigma Y}{n} \qquad c = \frac{\Sigma XY}{\Sigma X^2}$$

$$m = \frac{92.7}{8} \qquad c = \frac{56.5}{168}$$

$$= 11.5875 \qquad = .3363$$

The equation for the best fit is given by

$$Y_c = 11.5875 + 0.3363X$$

Trend value for the years

1996 is

$$Y_c = 11.5875 + 0.3363\ ((1996 - 1985.5) \times 2)$$
$$= 11.5875 + 0.3363\ (21)$$
$$= 18.6498$$
$$Y_c = 18.6498 \text{ tons}$$

1997 is

$$X = (1997 - 198.5) \times 2 = 11.5 \times 2 = 23$$

$$\therefore\ Y_c = 11.5875 + 0.3363 \times 23$$

$$Y_c = 19.3224 \text{ tons}$$

5. From the following details fit a straight line trend.

Year:	1989	1990	1991	1992	1993	1994	1995
Production:	160	180	184	166	188	198	184

(B. U., October, 1996)

Solution

Year (1)	Production (Y)	X	X^2	XY	Trend value Y_c
1989	160	–3	9	–480	168
1990	180	–2	4	–360	172
1991	184	–1	1	–184	176
1992	166	0	0	0	180
1993	188	1	9	188	184
1994	198	2	4	396	188
1995	184	3	1	552	192
Total	1260	0	28	112	

Now from the table

$$a = \frac{\Sigma Y}{n} = \frac{1260}{7} = 180 \qquad b = \frac{\Sigma XY}{\Sigma X^2} = \frac{112}{28} = 4$$

The best fit equation is

$$Y_c = 180 + 4X$$

Graphical method

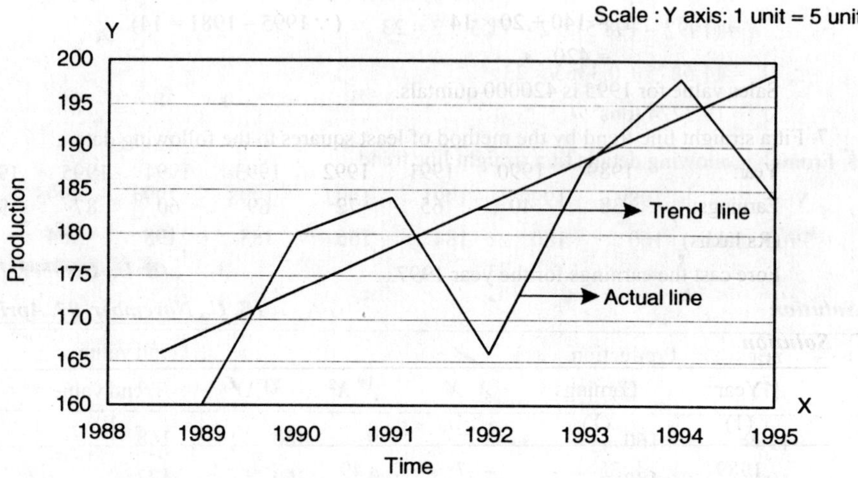

6. Estimate the value of sales for the year 1985 by using the method of least square for trend analysis.

Year:	1979	1980	1981	1982	1983
Sales: ('000 quintals)	100	120	140	160	180

(B. U., April, 1997)

Solution

Year (1)	Sales (Y)	X	X^2	XY	Trend value Y_c
1979	100	–2	4	–200	100
1980	120	–1	1	–120	120
1981	140	0	0	0	140
1982	160	1	1	160	160
1983	180	2	4	360	180
Total	700	0	10	200	

Now from table

$$a = \frac{\sum Y}{n} \qquad b = \frac{\sum XY}{\sum X^2}$$

$$= \frac{700}{5} \qquad = \frac{200}{10}$$

$$a = 140 \qquad b = 20$$

The equation for the best fit is given by

$$Y_c = 140 + 20X$$

The estimated sales value for 1995

$Y_c = 140 + 20 \times 14$ $\quad (\because 1995 - 1981 = 14)$

$= 420$

Sales value for 1995 is 420000 quintals.

7. Fit a straight line trend by the method of least squares to the following data.

Year:	1989	1990	1991	1992	1993	1994	1995	1996
Earnings: (Rs lakhs)	38	40	65	72	69	60	87	95

Fore cast the earnings for the year 1997.

(B. U., November, 97, April, 2001)

Solution

Year (1)	Earnings (Y)	X	X^2	XY	Trend value Y_c
1989	38	–7	49	–266	40.081
1990	40	–5	25	–200	47.415
1991	65	–3	9	–195	54.749
1992	72	–1	1	–72	62.083
1993	69	1	1	69	69.417
1994	60	3	9	180	76.751
1995	87	5	25	435	84.085
1996	95	7	49	665	91.419
Total	526	0	168	616	

$X = ((1) - 1992.5) \times 2$

from the table we get

$$a = \frac{\Sigma Y}{n} \qquad b = \frac{\Sigma XY}{\Sigma X^2}$$

$$= \frac{526}{8} \qquad = \frac{616}{168}$$

$$= 65.75 \qquad = 3.667$$

The straight line trend is given by

$Y_c = 65.75 + 3.667X$

The earnings for the year 1997.

$Y_c = 65.75 + 3.667\,(9)$ $\quad (\because X = 2 \times (1997 - 1992 = 5))$

$Y_c = 98.753$

The earnings for the year 1997 is estimated to be Rs 98.753 lakhs.

8. Fit an equation of the type $Y = a + bX$ to be following data.

Year:	1991	1992	1993	1994	1995	1996
Production: (000's tons)	70	72	88	80	90	92

(B. U., April, 1998)

Solution

Year (1)	Production (Y)	X	X^2	XY
1991	70	–5	25	–350
1992	72	–3	9	–216
1993	88	–1	1	–88
1994	80	1	1	80
1995	90	3	9	270
1996	92	5	25	460
Total	492	0	70	156

Now from the table

$$a = \frac{\sum Y}{n} = \frac{492}{6} = 82 \qquad b = \frac{\sum XY}{\sum X^2} = \frac{156}{70} = 2.2286$$

∴ linear trend is given by

$$Y_c = 82 + 2.2286X$$

9. Fit a straight line trend by the method of least squares.

Year:	1990	1991	1992	1993	1994	1995
Production: (000's tons)	70	72	88	80	90	92

(*B. U., October, 1998*)

Solution

Year (1)	Production (Y)	X	X^2	XY	Y_c
1990	70	–5	25	–350	70.557
1991	72	–3	9	–216	75.134
1992	88	–1	1	–88	79.711
1993	80	1	1	80	84.289
1994	90	3	9	270	88.866
1995	92	5	25	460	93.443
Total	492	0	70	156	

(X is calculated by 2nd method (even)).

Trend line is $Y_c = 82 + 2.2886X$

Graphical representation of trend line.

Scale: Y axis: 1 unit: 5000 tons

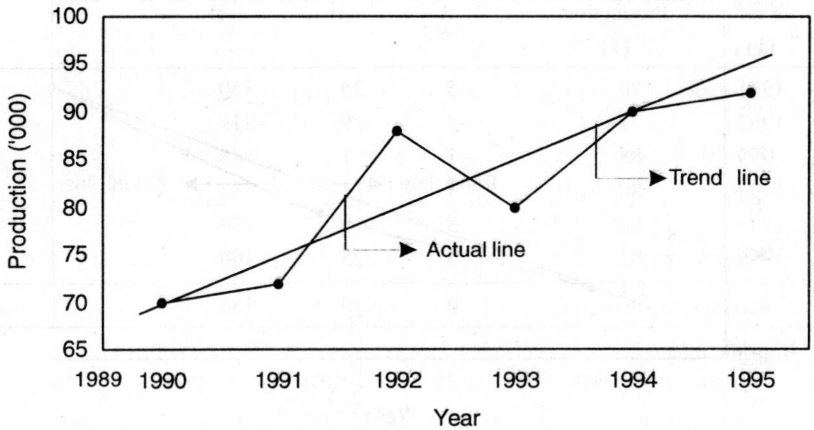

10. Fit a straight line trend to the data given below by the method of least squares.

Year:	1989	1990	1991	1992	1993	1994
Output: (Rs Crore)	672	824	968	1205	1464	1758

(B. U., April, 1999)

Year	Output (Y)	X	X^2	XY	Trend value Y_c
1989	672	–5	25	–3360	606.57
1990	824	–3	9	–2472	823.34
1991	968	–1	1	–968	1040.11
1992	1205	1	1	1205	1256.89
1993	1464	3	9	4392	1473.66
1994	1758	5	25	8790	1690.43
Total	6891	0	70	7587	

Now from the table

$$a = \frac{\Sigma Y}{n} = \frac{6891}{6} = 1148.5$$

$$b = \frac{\Sigma XY}{\Sigma X^2} = \frac{7587}{70} = 108.3857$$

$\therefore$ The linear trend equation is $Y_c = 1148.5 + 108.3857X$ graphical representation of trend line.

Scale: Y axis: 1 unit = 200 units

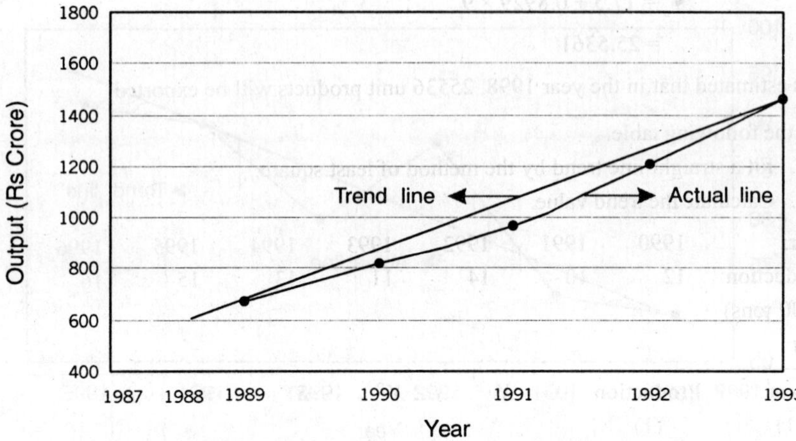

11. The number of units of a product exported during 1990–97 is given below. Fit a straight line trend and estimate the export for the year 1998.

Year:	1990	1991	1992	1993	1994	1995	1996	1997
No. of units exported: (in '000)	12	13	13	16	19	23	21	23

(B. U., Nov./Dec., 1999)

Solution

Year (1)	No. of units exported (Y)	X	X^2	XY	Trend value Y_c
1990	12	–7	49	–84	11.2497
1991	13	–5	25	–65	13.0355
1992	13	–3	9	–39	14.8213
1993	16	–1	1	–16	16.6071
1994	19	1	1	19	18.3928
1995	23	3	9	69	20.1786
1996	21	5	25	105	21.9644
1997	23	7	49	161	23.7502
Total	140	0	168	150	

Year of origin: 1993.5 (Use even No. year table)

Now from the table

$$a = \frac{\Sigma Y}{n} = \frac{140}{8} = 17.5$$

$$b = \frac{\Sigma XY}{\Sigma X^2} = \frac{150}{168} = 0.8929$$

The equation for the best fit is given by $Y_c = 17.5 + 0.8929X$.

Export for the year 1998 is

$$Y_c = 17.5 + 0.8929 \times 9$$

$$= 25.5361$$

it is estimated that in the year 1998, 25536 unit products will be exported.

12. Find the following table.

1. Fit a straight line trend by the method of least square.
2. Calculate the trend value.

Year:	1990	1991	1992	1993	1994	1995	1996
Production: ('000 tons)	12	10	14	11	13	15	16

Solution

Year (1)	Production (Y)	X	X^2	XY	Trend value Y_c
1990	12	–3	9	–36	10.75
1991	10	–2	4	–20	11.5
1992	14	–1	1	–14	12.25
1993	11	0	0	0	13
1994	13	1	1	13	13.75
1995	15	2	4	30	14.50
1996	16	3	9	48	15.25
Total	91	0	28	21	

From the table

$$a = \frac{\Sigma Y}{n} \qquad b = \frac{\Sigma XY}{\Sigma X^2}$$

$$= \frac{91}{7} \qquad = \frac{21}{28}$$

$$= 13 \qquad = 0.75$$

The trend line is $Y_c = 13 + 0.75X$

Graphical representation of the linear trend:

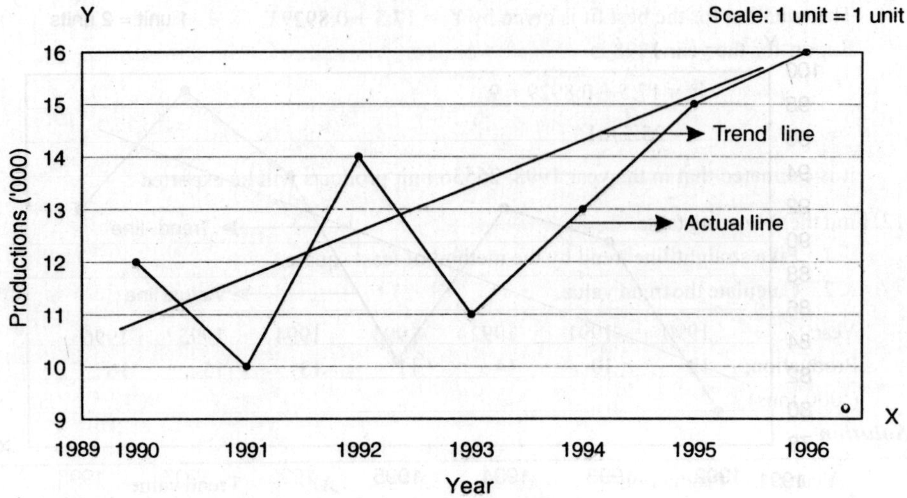

13. Given below are the figures of production of a factory.

Year:	1992	1993	1994	1995	1996	1997	1998
Production: ('000 tons)	80	90	92	83	94	99	92

Fit a straight line trend.

Solution:

Year (1)	Production (Y)	X	X^2	XY	Trend value Y_c
1992	80	–3	9	–240	84
1993	90	–2	4	–180	86
1994	92	–1	1	–92	88
1995	83	0	0	0	90
1996	94	1	1	94	92
1997	99	2	4	198	94
1998	92	3	9	276	96
Total	630	0	28	56	

Now from the table

$$a = \frac{\sum Y}{n} = \frac{630}{7} = 90$$

$$b = \frac{\sum XY}{\sum X^2} = \frac{56}{28} = 2$$

The best fit is given by

$$Y_c = 90 + 2X$$

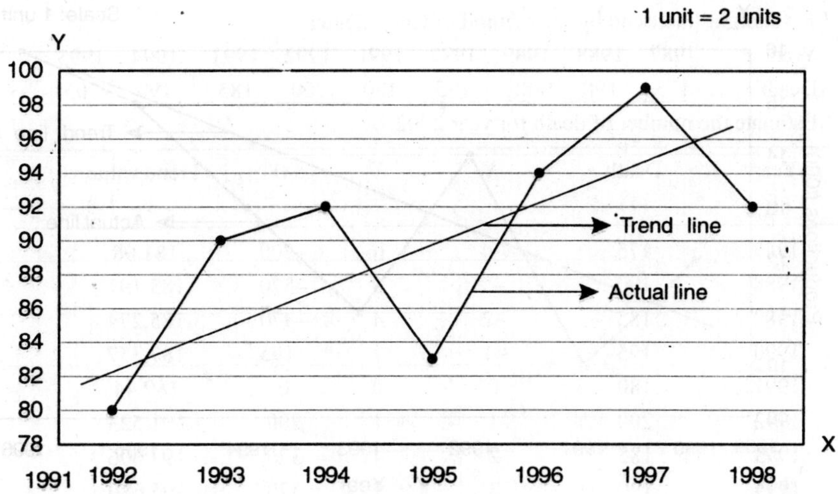

14. Here are data describing the air pollution rate (in pp) in a western city.

Year:	1980	1985	1990	1995
Pollution rate:	220	350	800	2450

Fit the linear trend and estimate the pollution rate for the year 2002.

Solution

Year (1)	Pollution rate (Y)	X	X^2	XY
1980	220	–15	225	–3300
1985	350	–5	25	–1750
1990	800	5	25	4000
1995	2450	15	225	36750
Total	3850	0	500	35700

Year of origin: 1987.5

$$X = (\text{year (1)} - 1987.5) \times 2$$

Now from the table

$$a = \frac{\Sigma Y}{n} \qquad b = \frac{\Sigma XY}{\Sigma X^2}$$

$$= \frac{3820}{4} \qquad = \frac{35700}{500}$$

$$= 955 \qquad = 71.4$$

$\therefore$ The linear trend equation is given by $Y_c = 955 + 71.4X$

Estimated pollution rate in the year 2002 is $Y_c = 955 + 71.4 \times 29$

$= 3025.6$ *pp*.

15. Fit a straight line trend by the method of least squares.

Year:	1987	1988	1989	1990	1991	1992	1993	1994	1995
Death:	175	190	185	195	180	200	185	190	205

Estimate the number of death for year 2002.

Year (1)	Death (Y)	X	X^2	XY	Trend value Y_c
1987	175	–4	16	–700	181.08
1988	190	–3	9	–570	183.191
1989	185	–2	4	–370	185.274
1990	195	–1	1	–195	187.357
1991	180	0	0	0	189.44
1992	200	1	1	200	191.523
1993	185	2	4	370	193.606
1994	190	3	9	570	195.689
1995	205	4	16	820	197.772
Total	1705	0	60	125	

Now from the table

$$a = \frac{\sum Y}{n} = \frac{1705}{9} = 189.44 \qquad b = \frac{\sum XY}{\sum X^2} = \frac{125}{60} = 2.083$$

The equation for the best fit is $Y_c = 189.44 + 2.083X$

$$Y_c = 189.44 + 2.083\,(2002 - 1991)$$
$$= 189 + 2.083\,(11)$$
$$= 212 \text{ death}$$

16. The following table describes speed post rates from 1975 to 1987

Year:	1975	1976	1977	1978	1979	1980	
Rate:	5	5	8	8	10	13	
Year:	1981	1982	1983	1984	1985	1986	1987
Rate:	15	18	20	22	25	25	29

Fit a trend line equation.

Solution:

Year	Rate (Y)	X	X^2	XY
1975	5	–6	36	–30
1976	5	–5	25	–25
1977	8	–4	16	–40

Contd

Year	Rate (Y)	X	X^2	XY
1978	8	–3	9	–24
1979	10	–2	4	–20
1980	13	–1	1	–13
1981	15	0	0	0
1982	18	1	1	18
1983	20	2	4	40
1984	22	3	9	66
1985	25	4	16	100
1986	25	5	25	125
1987	29	6	36	174
Total	203	0	182	371

Year of origin - 1981

From the table

$$a = \frac{\Sigma Y}{n} \qquad b = \frac{\Sigma XY}{\Sigma X^2}$$

$$= \frac{203}{13} \qquad = \frac{371}{182}$$

$$= 15.615 \qquad = 2.0385$$

∴ The trend line equation is $Y_c = 15.615 + 2.0385X$

17. Fit a straight line trend by the method of least square.

Year:	1990	1991	1992	1993	1994	1995	1996	1997
Y:	16	14	20	18	12	23	28	25

Solution:

Year (1)	Production (Y)	X	X^2	XY
1990	16	–7	49	–112
1991	14	–5	25	–70
1992	20	–3	9	–60
1993	18	–1	1	–18
1994	12	1	1	12
1995	23	3	9	69
1996	28	5	25	140
1997	25	7	49	175
Total	156	0	168	136

$X = (\text{year} - 1993.5) \times 2$

Now from the table

$$a = \frac{\Sigma Y}{n} \qquad b = \frac{\Sigma XY}{\Sigma X^2}$$

$$= \frac{156}{8} \qquad = \frac{136}{168}$$

$$= 19.5 \qquad = 0.8095$$

The total equation is given by

$$Y_c = 19.5 + 0.8095X$$

18. Fit a straight line trend for the following data.

Year:	1996	1997	1998	1999	2000
Per capita:	25	30	32	37	40

(Consumption of cotton in Kgs)

Estimate the per capita consumption for the year 2002. Represent the trend line graphically.

Solution

Year	Consumption (Y)	X	XY	X^2	Trend value Y_c
1996	25	–2	–50	4	25.4
1997	30	–1	–30	1	29.1
1998	32	0	0	0	32.8
1999	37	1	37	1	36.5
2000	40	2	80	4	40.2
Total	164	0	37	10	

Year of origin: 1998

Form the table

$$a = \frac{\Sigma Y}{n} \qquad b = \frac{\Sigma XY}{\Sigma X^2}$$

$$= \frac{164}{5} \qquad = \frac{37}{10}$$

$$= 32.8 \qquad = 3.7$$

The straight line trend equation for the given data is given by

$$Y_c = 32.8 + 3.7X$$

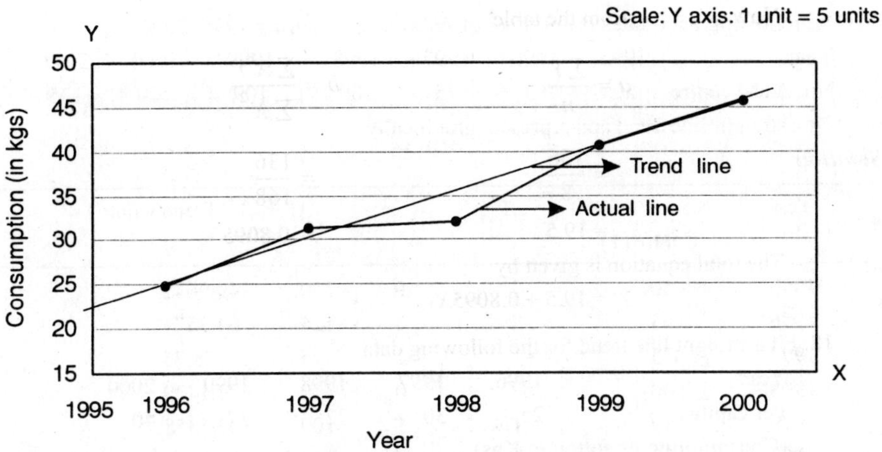

19. The following data describes the export quantity of a company.

Year:	1995	1996	1997	1998	1999	2000	2001
Export:	12	13	13	16	16	19	23

Fit a straight line trend and estimate the export for the year 2002.

Solution

Year	Export (Y)	X	X^2	XY	Trend value Y_c
1995	12	–3	9	–36	
1996	13	–2	4	–26	
1997	13	–1	1	–13	
1998	16	0	0	0	
1999	16	1	1	16	
2000	19	2	4	38	
2001	23	3	9	69	
Total	112	0	28	48	

Year of origin: 1998

From the table

$$a = \frac{\Sigma Y}{n} = \frac{112}{7} = 16$$

$$b = \frac{\Sigma XY}{\Sigma X^2} = \frac{48}{28} = 1.7143$$

The straight line that best fits the data is given by

$$Y_c = 16 + 1.7143X$$

The export for the year 2002.

$$Y_c = 16 + 1.7143\ (4)$$
$$= 22.8572$$
$$= 23 \text{ units}$$

20. The following data states the number of ATM centres during 1995–2001.

Year:	1995	1996	1997	1998	1999	2000	2001
No. ATM centre:	50	63	75	100	109	120	135

Fit a straight line trend and represent graphically.

Solution

Year	No. of ATM Centre (Y)	X	X^2	XY	Trend value Y_c
1995	50	–3	9	–150	49.9642
1996	63	–2	4	–126	64.3571
1997	75	–1	1	–75	78.75
1998	100	0	0	0	93.1429
1999	109	1	1	109	107.5358
2000	120	2	4	240	121.9287
2001	135	3	9	405	136.3216
Total	652	0	28	403	

From the table

$$a = \frac{\sum Y}{n} = \frac{652}{7} = 93.1429,$$

$$b = \frac{\sum XY}{\sum X^2} = \frac{403}{28} = 14.3929$$

∴ The equation of straight line trend is given by $Y_c = 93.1429 + 14.3929X$

Questions

1. What are the components of time series.
2. What are cyclical fluctuation.
3. What do you mean by seasonal variation.
4. What do you mean by secular trend.
5. What do you mean by analysis of time series.

Exercises

1. The following table shows the disposable income per family.

Year:	1980	1981	1982	1983	1984	1985
D.I:	6100	6500	6700	6900	7000	7200

Compute the straight line by least square method.

[*Ans:* $Y_c = 6733.333 + 92.308X$]

2. Fit a straight line trend and estimate the sales for the year 2003.

Year:	1997	1998	1999	2000	2001
Sales: ('000)	150	179	189	200	205

[Ans: $Y_c = 184.6 + 13.1X$
Sales for the year 2003 is 237000]

3. Fit a straight line trend and estimate the production for the year 2003.

Year:	1995	1996	1997	1998	1999	2000
Production: (tons)	200	210	217	223	237	250

[Ans: $Y_c = 222.833 + 4.815X$
Sales for 2003: 276 tons]

4. From the following details fit a straight line trend.

Year:	1990	1991	1992	1993	1994	1995
Output:	300	310	317	400	417	425

[Ans: $Y_c = 361.5 + 14.7X$]

5. Calculate the trend value for the following data.

Year:	1997	1998	1999	2000	2001
Table sold:	92	111	120	127	140

[Ans: $Y_c = 118 + 11.2X$]

6. Following are the profits of a concern from 1993 to 1998. Fit a straight line trend.

Year:	1993	1994	1995	1996	1997	1998
Profit: ('000)	72	77	79	83	87	90

[Ans: $Y_c = 81.333 + 1.771X$]

7. Fit a straight line trend and show the data graphically with trend values form the following data.

Year:	1990	1991	1992	1993	1994	1995
No. of pcs sold:	50	110	350	1100	2000	3800

Estimate the number of pcs that will sold in the year 2003.

[Ans: $Y_c = 1235 + 359.27X$
Sales 2003: 8778 pcs]

8. Fit by the method of least squares a linear equation $y = mx + c$ to the following data.

Year:	1990	1991	1992	1993	1994
Y:	4	6	7	8	10

Obtain trend values for the years 2003 and 2005. ***[Ans: $Y_c = 3.5 + 14X$***
$Y_{2005} = 22$, $Y_{2003} = 19$]

9. Estimate the value of sales for the year 2003 by using the method of least square for trend analysis.

Year:	1980	1981	1982	1983	1984	1985
Sale:	35	42	44	47	49	51

[Ans: $Y_c = 44.667 + 1.486X$
$Y_{2003} = 105$]

10. Find the following table.

    1. Fit a straight line trend by the method of least square.
    2. Calculate the trend value.

Year:	1990	1991	1992	1993	1995
No. of telephone:	2000	2050	2250	2500	2700

[*Ans:* $Y_c = 2300 + 185X$]

9

Interpolation and Extrapolation

Content

- Interpolation and extrapolation
- Differentiation
- Assumption and significance
- Methods of interpolation
  Graphical, Newton's method
  Lagrange's method.

Formulae

1. Newton's advancing difference method

$$y_x = y_0 + x\,\Delta^1 y_0 + \frac{x(x-1)}{2!}\Delta^2 y_0 + \frac{x(x-1)(x-2)}{3!}\Delta^3 y_0$$

$$+ \ldots + \frac{x(x-1)(x-2)\ldots(x-(n-1))}{n!}\Delta^n y_0$$

Where $x = \dfrac{x_I - x_0}{h}$

2. Lagrange's formula

$$y_x = y_o\,\frac{(x-x_1)(x-x_2)\ldots(x-x_n)}{(x_0-x_1)(x_0-x_2)\ldots(x_0-x_n)}$$

$$+\, y_1\,\frac{(x-x_0)(x-x_2)\ldots(x-x_n)}{(x_1-x_0)(x_1-x_2)\ldots(x_1-x_n)}$$

$$+ \quad \vdots$$

$$+\, y_n\,\frac{(x-x_0)(x-x_2)\ldots(x-x_{n-1})}{(x_n-x_1)(x_n-x_2)\ldots(x_n-x_{n-1})}$$

Key points

1. Newton's advancing difference method is used when the arguments are equally spaced.
2. Lagrange's method can be used when the arguments are unequally spaced.

INTRODUCTION

Interpolation and extrapolation comes under mathematical approach known as numerical analysis. Numerical analysis is the process of estimating a value of the dependent variable for given value of the independent variable without establishing functional relationship through the data set. In the previous chapter the method of estimating a value of the dependent variable Y corresponding to a given value of X has been explained. There we found out the functional relationship of Y and X. But here we obtain the most likely estimate of a certain quantity from the relevant facts.

9.1 INTERPOLATION

"Insertion of the most likely estimate under certain assumption" which means to estimate the value of Y for a given value of X with in the range of X values. Interpolation helps in estimating missing or unrecorded information within the given facts.

Extrapolation: "Estimation of past or future figure" which means to estimate the value of Y for a given value of X outside the range of X.

9.2 DISTINCTION BETWEEN INTERPOLATION AND EXTRAPOLATION

	Interpolation	*Extrapolation*
1.	Estimates the value within the range of given data.	Estimates the value outside the range of the given value.
2.	This technique cannot be extended to find the values of future data. It is used to find the missing values within the given range.	This technique can be used to find the future date and missing past data.

9.3 ASSUMPTION

Interpolation and extrapolation are based on the following assumptions.

1. To get more accurate values, it is assumed that there is no violent fluctuation within the given set of data.
2. There should be a regularity in variation of variate values.
3. There should be a definite mathematical relation between the dependent (Y) and the independent (X) variables (i.e., $y = f(x)$).

9.4 SIGNIFICANCE OF INTERPOLATION AND EXTRAPOLATION

There are various uses of interpolation and extrapolation, a few are mentioned below.

1. The data are mostly collected at regular intervals of time but at times there arise a need for intermediate values, at those moment interpolation helps in estimating such values.
2. When one wants to predict the future data or wants to know about the past value extrapolation helps.
3. Interpolation helps in estimating the missing values of the dependent variable within the given range.
4. For comparison of data, it is necessary that the data should belong to the same reference period. Interpolation and extrapolation helps in the transformation to the same period, if the data are not so.

9.5 METHODS OF INTERPOLATION

There are two approaches.

1. Graphical Method
2. Algebraic Method

Graphical Method

In the graphical method the values of paired variate X and Y ($=f(x)$) are plotted on the graph paper after appropriately choosing the scales along abscissa and ordinates. The plotted points are jointed by a smooth line sequencely. Now the value of Y for the given value of $X = x_0$ is estimated by drawing a line at the given point $X = x_0$ parallel to Y-axis, so as to touch the line or smooth curve. From the intersection point, draw a line perpendicular to Y-axis. The reading corresponding to the point that touches the Y-axis is the estimated value of Y for the given value of X.

This method provides a good estimate if the relation between X and Y $(=f(x)$ is linear.

Algebraic Method

There a number of methods available in algebraic method of interpolation which gives more accurate value than graphical method. Here we will deal with two important methods.

1. Newton's formula for advancing differences
2. Lagrange's method

Newton's Formula for Advancing Differences

This is also known as Newton's Gregory forwarded formula method. This method is used for interpolation and it is based on the difference between consecutive item. Process of obtaining the difference between the value of entry (y) will continue till only one difference remain. This method is applicable when the arguments advance with equal intervals such as x, $x + h$, $x + 2h$ …

Difference Table

Argument x	Entry y	I difference $\Delta^1 y$	II difference $\Delta^2 y$	III difference $\Delta^3 y$	IV difference $\Delta^4 y$	V difference $\Delta^5 y$
x_0	y_0					
		$y_1 - y_0 = \Delta^1 y_0$				
x_1	y_1		$\Delta^1 y_0 - \Delta^1 y_1 = \Delta^2 y_0$			
		$y_2 - y_1 = \Delta^1 y_1$		$\Delta^2 y_1 - \Delta^2 y_0 = \Delta^3 y_0$		
x_2	y_2				$\Delta^3 y_1 - \Delta^3 y_0 = \Delta^4 y_0$	
		$y_3 - y_2 = \Delta^1 y_2$	$\Delta^1 y_2 - \Delta^1 y_1 = \Delta^2 y_1$	$\Delta^2 y_2 - \Delta^2 y_1 = \Delta^3 y_1$		$\Delta^4 y_1 - \Delta^4 y_0 = \Delta^5 y_0$
x_3	y_3				$\Delta^3 y_2 - \Delta^3 y_1 = \Delta^4 y_1$	
		$y_4 - y_3 = \Delta^1 y_3$	$\Delta^1 y_3 - \Delta^1 y_2 = \Delta^2 y_2$	$\Delta^2 y_3 - \Delta^2 y_2 = \Delta^3 y_2$		
x_4	y_4					
		$y_5 - y_4 = \Delta^1 y_4$	$\Delta^1 y_4 - \Delta^1 y_3 = \Delta^3 y_3$			
x_5	y_5					

Where $\Delta \equiv$ difference

Formula To estimate the value of Y for the given value of $X = x$ we use the following formula.

$$y_x = y_0 + x\Delta^1 y_0 + \frac{x(x-1)}{2!}\Delta^2 y_0 + \frac{x(x-1)(x-2)}{3!}\Delta^3 y_0$$

$$+ \ldots + \frac{x(x-1)(x-2)\ldots(x-(n-1))}{n!}\Delta^n y_0 \qquad \ldots (9.1)$$

Where $x = \dfrac{x_l - x_0}{h}$

Where x_l is the value of X for which the value of Y is to be estimated.

x_0 = The first value of the X series.

h = Interval between the consecutive items of the argument (x).

Example 1 From the following table of yearly premium for polices maturing at different years estimate the premium for the policies maturing at the age of 47 years.

Agė (in years):	45	50	55	60	65
Premium (in Rs):	287	240	208	186	171

Argument x	Argument y	$\Delta^1 y$	$\Delta^2 y$	$\Delta^3 y$	$\Delta^4 y$
45	287				
50	240	240 – 287 = –47	–32 –(– 47) = 15		
55	208	208 – 240 = –32	–22 –(– 32) = 10	10 – 15 = –5	–3 –(–5) = 2
60	186	186 – 208 = –22	–15 (–22) = 7	7 – 10 = –3	
65	171	171 – 186 = –15			

Here $y_0 = 287$, $\Delta^1 y_0 = -47$, $\Delta^2 y_0 = 15$,

$\Delta^3 y_0 = -5$, $\Delta^4 y_0 = 2$, $x_l = 47$.

$$x = \frac{x_l - x_o}{h} = \frac{47 - 45}{5} = 0.4$$

Using formula 8.1

$$y_x = y_0 + x\Delta^1 y_0 + \frac{x(x-1)}{2!}\Delta^2 y_0 + \frac{x(x-1)(x-2)}{3!}\Delta^3 y_0$$

$$+ \frac{x(x-1)(x-2)\ldots(x-3)}{4!}\Delta^4 y_0$$

$$y_x = 287 + 0.4(-47) + \frac{0.4(0.4-1)}{2}(15) + \frac{0.4(0.4-1)(0.4-2)}{3}(-5)$$

$$+ \frac{0.4(0.4-1)(0.4-2)\ldots(0.4-3)}{24}(2)$$

$Y_x = 287 - 18.8 - 1.8 - 0.032 - 0.0832$

$= 266.285$

$=$ Rs 266.

Lagrange's Method

Lagrange's method is applicable for both interpolation and extrapolation. Lagrange's method is used when the argument are unequally spaced.

Let X: $x_0, x_1, \ldots, x_n$

Y: $y_0, y_1, \ldots, y_n$

be the values of the argument x and $y = f(x)$.

Formula The value of y that has to be estimated for the given argument x is given by

$$y_x = \frac{(x-x_1)(x-x_2)\ldots(x-x_n)}{(x_0-x_1)(x_0-x_2)\ldots(x_0-x_n)} y_o$$

$$+ \frac{(x-x_0)(x-x_2)\ldots(x-x_n)}{(x_1-x_0)(x_1-x_2)\ldots(x_1-x_n)} y_1$$

$$\vdots$$

$$+ \frac{(x-x_0)(x-x_2)\ldots(x-x_{n-1})}{(x_n-x_1)(x_n-x_2)\ldots(x_n-x_{n-1})} y_n \qquad \ldots (9.2)$$

Where x is the value of the argument for which y_x has to be estimated.

Example 1 Using Lagrange's formula interpolation the value of y when $x = 6$

X:	2	4	8
Y:	7	21	73

Solution

X:	x_0	x_1	x_2
	2	4	8
Y:	y_0	y_1	y_2
	7	21	73

Using the formula (9.2)

$$y_x = y_o \frac{(x-x_1)(x-x_2)}{(x_0-x_1)(x_0-x_2)} + y_1 \frac{(x-x_0)(x-x_2)}{(x_1-x_0)(x_1-x_2)}$$

$$+ y_2 \frac{(x - x_0)(x - x_1)}{(x_2 - x_0)(x_2 - x_1)}$$

Here $x = 6$

$$y_x = 7 \frac{(6-4)(6-8)}{(2-4)(2-8)} + 21 \frac{(6-2)(6-8)}{(4-2)(4-8)}$$

$$+ 73 \frac{(6-2)(6-4)}{(8-2)(8-4)}$$

$$= -2.333 + 21 + 24.333$$
$$= 42.997$$
$$= 43$$

Solved Examples

1. Following are the annual premiums charged by an insurance company for a policy of Rs 1000. Calculate the premium payable at age 26.

Age in years:	20	25	30	35	40
Premium Rs:	23	26	30	35	42

(B. U., November, 1993)

Solution Newton's forward interpolation method

Age in year	*Premium Rs*	$\overset{1}{\Delta} y$	$\overset{2}{\Delta} y$	$\overset{3}{\Delta} y$	$\overset{4}{\Delta} y$
20	23 y_0				
25	26 y_1	3 $\Delta y_0 = y_1 - y_0$			
30	30 y_2	4 $\Delta y_1 = y_2 - y_1$	1 $\overset{2}{\Delta} y_0 - \Delta y_1 = \Delta y_0$		
35	35 y_3	5 $\Delta y_2 = y_3 - y_2$	1 $\overset{2}{\Delta} y_1 - \Delta y_2 = \Delta y_1$	0 $\overset{3}{\Delta} y_0$	1 $\overset{4}{\Delta} y_0$
40	42 y_4	7 $\Delta y_3 = y_4 - y_3$	2 $\overset{3}{\Delta} y_2 - \Delta y_3 = \Delta y_2$	1 $\overset{3}{\Delta} y_1$	

Formula

$$Y_0 = y_0 + x \Delta^1 y_0 + \frac{x(x-1)}{2!} \Delta^2 y_0 + \frac{x(x-1)(x-2)}{3!} \Delta^3 y_0$$

$$+ \ldots + \frac{x(x-1)(x-2)\ldots(x-(n-1))}{n!} \Delta^n y_0$$

$$x = \frac{x_I - x_0}{h}$$

Premium payable at age 26.

$x = (26 - 20)/5 = 1.2$

$$y = 23 + 1.2 \times 3 + \frac{1.2(1.2-1)}{2!} \times 1 + \frac{1.2(1.2-1)(1.2-2)}{3!} \times 0 + \frac{1.2(1.2-1)(1.2-2)\dots(1.2-3)}{4!} \times 1$$

$= 23 + 3.6 + .12 + 0 + .0144$

$= 26.73$

= Rs 26.73

Thus the premium payable at age 26 is Rs 26.73.

2. The following table gives the expectation of life $C(x)$ at age x. Calculate expectation of life at age 12 by using Newton's forward interpolation formula.

x:	10	15	20	25	30	35
$C(x)$:	35.4	32.2	29.1	26.0	23.1	20.4

(B. U., April, 1992)

Solution Newton's forward interpolation method.

x	$C(x)$	$\Delta^1 y$	$\Delta^2 y$	$\Delta^3 y$	$\Delta^4 y$	$\Delta^5 y$
10	35.4					
15	32.2	–3.2				
20	29.1	–3.1	0.1	–0.1		
25	26.0	–3.1	0		0.3	–0.5
30	23.1	–2.9	0.2	0.2	–0.2	
35	20.4	–2.7	0.2	0		

Expectation of life at age 12:

$$x = \frac{12-10}{5} = 0.4$$

$$y = y_0 + x\Delta^1 y_0 + \frac{x(x-1)}{2!}\Delta^2 y_0 + \frac{x(x-1)(x-2)}{3!}\Delta^3 y_0 + \dots$$

$$= 35.4 + 0.4 \times (-3.2) + \frac{0.4(0.4-1)}{2!} \times 0.1 + \frac{0.4(0.4-1)(0.4-2)}{3!} \times (-0.1) + \frac{0.4(0.4-1)(0.4-2)(0.4-3)}{4!} \times (0.3) + \frac{0.4(0.4-1)(0.4-2)(0.4-3)(0.4-4)}{5!} \times (-0.5)$$

$$= 35.4 - 1.28 + \frac{(-0.024)}{2!} + \frac{(-0.0384)}{3!} + \frac{(-0.29952)}{4!} + \frac{(-1.79712)}{5!}$$

$= 34.07$

Thus the expected life at age 12 is 34.00.

3. Find by suitable method of interpolation the value of y when $x = 7$.

x:	2	4	6	8
y:	7	21	43	73

(B. U., November, 1995)

Solution Newton's forward interpolation method

x	y	$\Delta^1 y$	$\Delta^2 y$	$\Delta^3 y$
2	7			
4	21	14	8	
6	43	22	8	0
8	73	30		

$$x = \frac{7-2}{2} = \frac{5}{2} = 2.5$$

$$y = y_0 + x\Delta^1 y_0 + \frac{x(x-1)}{2!}\Delta^2 y_0 + \frac{x(x-1)(x-2)}{3!}\Delta^3 y_0$$

$$= 7 + 2.5 \times 14 + \frac{2.5(2.5-1)}{2!} \times 8 + \frac{2.5(2.5-1)(2.5-2)}{3!} \times 0$$

$$= 7 + 35 + 14 = 56$$

Thus the value of y when $x = 7$ is 56.

4. Calculate the expectation of life at the age of 22 using the following data.

Age:	10	15	20	25	30
Expectation of life:	36	30	28	26	24

(B. U., April, 1996)

Solution Newton's forward interpolation method

Age	*Expectation of Life y*	$\Delta^1 y$	$\Delta^2 y$	$\Delta^3 y$	$\Delta^4 y$
10	36				
15	30	–6			
20	28	–2	4	–4	
25	26	–2	0	0	4
30	24	–2	0		

$$x = \frac{22-10}{5} = 2.4$$

$$y = y_0 + x\Delta^1 y_0 + \frac{x(x-1)}{2!}\Delta^2 y_0 + \frac{x(x-1)(x-2)}{3!}\Delta^3 y_0 + \frac{x(x-1)(x-2)(x-3)}{4!}\Delta^4 y_0$$

$$= 36 + 2.4 \times (-6) + \frac{2.4(2.4-1)}{2} \times 4 + \frac{2.4(2.4-1)(2.4-2)}{6} \times (-4)$$

$$+ \frac{2.4(2.4-1)(2.4-2)(2.4-3)}{24} \times 4$$

$= 36 - 14.4 + 6.72 + -0.896 - 0.1344$

$= 27.28$

$= 27$ years.

$\therefore$ the expectation of life at 22 is 27 years.

5. Calculate the premium at the age of 22 from the following data.

Age (x):	20	25	30	35	40
Premium 1000 sum (y):	30	40	55	75	100

(B. U., October, 1996)

Solution Newton's forward interpolation method

x	y	$\Delta^1 y$	$\Delta^2 y$	$\Delta^3 y$	$\Delta^4 y$
20	30				
		10			
25	40		5		
		15		0	
30	55		5		0
		20		0	
35	75		5		
		25			
40	100				

$$x = \frac{22-20}{5} = 0.4$$

$$y = y_0 + x\Delta^1 y_0 + \frac{x(x-1)}{2!}\Delta^2 y_0$$

$$y = 30 + 0.4 \times 10 + \frac{0.4(0.4-1)}{2} \times 5$$

$= 30 + 4 + (-0.06)$

$= 33.94$

$= 34$

The premium sum assured at the age of 22 is 34.

6. Form a difference table and find the value of $f(x)$ when $x = 9$.

x:	3	7	11	15	19
$f(x)$:	42	43	47	53	60

(B. U., April, 1997)

Solution Newton's advance difference method.

x	$f(x) = y$	$\Delta^1 y$	$\Delta^2 y$	$\Delta^3 y$	$\Delta^4 y$
3	42				
		1			
7	43		3		
		4		–1	
11	47		2		0
		6		–1	
15	53		1		
		7			
19	60				

$$X = \frac{9-3}{4} = 1.5$$

$$y_x = y_0 + x\Delta^1 y_0 + \frac{x(x-1)}{2!}\Delta^2 y_0 + \frac{x(x-1)(x-2)}{3!}\Delta^3 y_0 + \frac{x(x-1)(x-2)(x-3)}{4!}\Delta^4 y_0$$

$$= 42 + 1.5 \times 1 + \frac{1.5(1.5-1)}{2} \times 3 + \frac{1.5(1.5-1)(1.5-2)}{6} \times -1 + 0$$

$$= 42 + 1.5 + 1.125 + 0.0625$$

$$= 44.6875$$

$$= 45$$

This the value of $f(x)$ when $x = 9$ is 45.

7. Estimate the number of students who get more than 48 but not more than 50 marks from the following data.

Mark upto:	45	50	55	60	65
No. of student:	447	484	505	511	514

(B. U., November, 2001)

Solution Newton's advance difference method.

Mark upto x	y	$\Delta^1 y$	$\Delta^2 y$	$\Delta^3 y$	$\Delta^4 y$
45	447				
		37			
50	484		–16		
		21		1	
55	505		–15		11
		6		12	
60	511		–3		
		3			
65	514				

$$x = \frac{48-45}{5} = \frac{3}{5} = 0.6$$

$$y_x = y_0 + x\Delta^1 y_0 + \frac{x(x-1)}{2!}\Delta^2 y_0 + \frac{x(x-1)(x-2)}{3!}\Delta^3 y_0$$

$$+ \frac{x(x-1)(x-2)\dots(x-3)}{4!}\Delta^4 y_0$$

$$= 447 + 0.6 \times 37 + \frac{0.6(0.6-1)}{2}(-16) + \frac{0.6(0.6-1)(0.6-2)}{6} \times 1$$

$$+ \frac{0.6(0.6-1)(0.6-2)}{24} \times 11$$

$= 447 + 22.2 + 1.92 + 0.056 - 0.3696$

$= 470.80$

$= 471$

No. of students scoring upto 50 = 484

No. of students scoring up to 48 = 471

∴ No. of students scoring marks between 48 and 50 is 484 – 471 = 13

8. Estimate the probable number of persons earning between 450 and 480 from the following table.

Income (Rs):	350–400	400–450	450–500	500–550	550–600
No. of person:	38	82	54	16	10

(B. U., April, 1999)

Solution Let us form the less than cumulative frequency table.

Income less than	*No. of person* y	$\Delta^1 y$	$\Delta^2 y$	$\Delta^3 y$	$\Delta^4 y$
400	38				
		82			
450	120		–28		
		54		–10	
500	174		–38		42
		16		32	
550	190		–6		
		10			
600	200				

Solution

$$x = \frac{480-400}{50} = \frac{80}{50} = 1.6$$

$$y_x = y_0 + x\Delta^1 y_0 + \frac{x(x-1)}{2!}\Delta^2 y_0 + \frac{x(x-1)(x-2)}{3!}\Delta^3 y_0$$

$$+ \frac{x(x-1)(x-2)\dots(x-3)}{4!}\Delta^4 y_0$$

$$= 38 + 1.6 \times 82 + \frac{1.6(1.6-1)}{2}(-28) + \frac{1.6(1.6-1)(1.6-2)}{6} \times (-10)$$

$$+ \frac{1.6(1.6-1)(1.6-2)(1.6-3)}{24} \times 42$$

$Y_x = 38 + 131.2 - 13.44 + 0.64 + 0.9408$
$= 157.3408$
$= 157$ person.

No. of person earning income upto 450 = 120
No. of person earning income upto 480 = 157
∴ No. of person earning between 450 and 480 is 157 – 120 = 37 persons.

9. From the following table of yearly premium for policies maturing at different ages, estimate the premium for policies maturity at the age of 47 years.

Age (in years):	45	50	55	60	65
Premium (in Rs):	287	240	208	186	171

(B. U., October, 2001)

Solution Newton's difference method

Age (in years)	*Premium (y)*	$\Delta^1 y$	$\Delta^2 y$	$\Delta^3 y$	$\Delta^4 y$
45	287				
		–47			
50	240		15		
		–32		–5	
55	208		10		2
		–22		–3	
60	186		7		
		–15			
65	171.				

$$x = \frac{47-45}{5} = 0.4$$

$$y_x = y_0 + x\Delta^1 y_0 + \frac{x(x-1)}{2!}\Delta^2 y_0 + \frac{x(x-1)(x-2)}{3!}\Delta^3 y_0$$

$$+ \frac{x(x-1)(x-2)\ldots(x-3)}{4!}\Delta^4 y_0$$

$$= 287 + 0.4 \times (-47) + \frac{0.4(0.4-1)}{2} \times 15 + \frac{0.4(0.4-1)(0.4-2)}{6} \times (-5)$$

$$+ \frac{0.4(0.4-1)(0.4-2)(0.4-3)}{24} \times 2$$

$= 287 + (-18.8) - 1.8 - .32 - 0.0832$
$= 265.9968$
$= 266$

10. Given below are the marks obtained by 200 students of a university in statistics paper.

Marks:	40–50	50–60	60–70	70–80	80–90
No. of students:	32	42	55	40	31

Interpolation the number of students who have obtained less than 65 marks.

(B. U., November, 1999)

Solution Newton's forward interpolation method.

Marks upto	y	$\Delta^1 y$	$\Delta^2 y$	$\Delta^3 y$	$\Delta^4 y$
50	32				
		42			
60	74		13		
		55		–28	
70	129		–15		34
		40		6	
80	169		–9		
		31			
90	200				

$$x = \frac{x_1 - x_0}{h} = \frac{65 - 50}{10} = 1.5$$

$$y_x = y_0 + x\Delta^1 y_0 + \frac{x(x-1)}{2!}\Delta^2 y_0 + \frac{x(x-1)(x-2)}{3!}\Delta^3 y_0 + \frac{x(x-1)(x-2)\ldots(x-3)}{4!}\Delta^4 y_0$$

$$= 32 + 1.5 \times 42 + \frac{1.5(1.5-1)}{2} \times 13 + \frac{1.5(1.5-1)(1.5-2)}{6} \times (-28) + \frac{1.5(1.5-1)(1.5-2)(1.5-3)}{24} \times 31$$

$$= 32 + 63 + 4.875 + 1.75 - 0.727$$

$$= 102.352$$

$$= 102 \text{ students.}$$

11. Below are given the wages earned per month in a certain factory. Calculate the number of workers earning more than Rs 75 per month.

Monthly income (Rs):	upto Rs 50	60	70	80	90	100
No. of workers:	50	150	300	500	700	800

(B. U., November, 1994)

Solution Newton's forward interpolation method

Monthly income upto	y	$\Delta^1 y$	$\Delta^2 y$	$\Delta^3 y$	$\Delta^4 y$	$\Delta^5 y$
50	50					
		100				
60	150		50			
		150		0		
70	300		50		–50	
		200		–50		0
80	500		0		–50	
		200		–100		
90	700		–100			
		100				
100	800					

$$x = \frac{x_l - x_0}{h} = \frac{75 - 50}{10} = 1.5$$

$$y_x = y_0 + x\Delta^1 y_0 + \frac{x(x-1)}{2!}\Delta^2 y_0 + \frac{x(x-1)(x-2)}{3!}\Delta^3 y_0$$

$$+ \frac{x(x-1)(x-2)\ldots(x-3)}{4!}\Delta^4 y_0$$

$$+ \frac{x(x-1)(x-2)(x-3)(x-4)}{5!}\Delta^5 y_0$$

$$= 50 + 1.5 \times 100 + \frac{1.5(1.5-1)}{2} \times 50 + \frac{1.5(1.5-1)(1.5-2)}{6} \times 0$$

$$+ \frac{1.5(1.5-1)(1.5-2)(1.5-3)}{24} \times (-50) + 0$$

$= 50 + 150 + 18.8 + 0 - 1.2$

$= 200.6$

$= 201$

Workers earning up 75 = 201

∴ Worker earning more than 75 is total – 201

= 800 – 201

= 599 workers

12. Calculate the expected life of refrigerators at the age of 7 using the following data.

Age:	up 2	2–4	4–6	6–8	8–10	10–12
Girls:	5	16	13	7	5	4

Solution

Age	y	$\Delta^1 y$	$\Delta^2 y$	$\Delta^3 y$	$\Delta^4 y$	$\Delta^5 y$
2	5					
		16				
4	21		−3			
		13		−3		
6	34		−6		7	
		7		4		−10
8	41		−2		−3	
		5		0		
10	46		−1			
		4				
12	50					

$$x = \frac{7-2}{2}\ \frac{5}{2} = 2.5$$

$$y_x = y_0 + x\Delta^1 y_0 + \frac{x(x-1)}{2!}\Delta^2 y_0 + \frac{x(x-1)(x-2)}{3!}\Delta^3 y_0$$

$$+ \frac{x(x-1)(x-2)(x-3)}{4!}\Delta^4 y_0 + \frac{x(x-1)(x-2)(x-3)(x-4)}{5!}\Delta^5 y_0$$

$$= 5 + 2.5 \times 16 + \frac{2.5(2.5-1)}{2} \times (-3) + \frac{2.5(2.5-1)(2.5-2)}{6} \times (-3)$$

$$+ \frac{2.5(2.5-1)(2.5-2)(2.5-3)}{24} \times 7$$

$$+ \frac{2.5(2.5-1)(2.5-2)(2.5-3)(2.5-4)}{120} \times (-10)$$

$$= 5 + 40 + (-5.6) - 0.9 - 0.3 - 0.1$$

$$= 38.1$$

$$= 38$$

13. Estimate the number of employees earning at least 63 but less than Rs 75 from the following data.

Wage:	50–60	60–70	70–80	80–90	90–100	100–110
No. of employees (y):	8	10	16	14	10	7

Solution Newton's forward interpolation method

Wage upto	y	$\Delta^1 y$	$\Delta^2 y$	$\Delta^3 y$	$\Delta^4 y$	$\Delta^5 y$
60	8					
		10				
70	18		6			
		16		–8		
80	34		–2			
		14		–2		
90	48		–4		6	–3
		10		1		
100	58		–33			
		7				
110	65					

(i) When income is Rs 63

$$x = \frac{x_I - x_0}{h} = \frac{63 - 60}{10} = 0.3$$

$$y_x = y_0 + x\Delta^1 y_0 + \frac{x(x-1)}{2!}\Delta^2 y_0 + \frac{x(x-1)(x-2)}{3!}\Delta^3 y_0$$

$$+ \frac{x(x-1)(x-2)(x-3)}{4!}\Delta^4 y_0 + \frac{x(x-1)(x-2)(x-3)(x-4)}{5!}\Delta^5 y_0$$

$$= 8 + 0.3 \times 10 + \frac{0.3(0.3-1)}{2} \times 6 + \frac{0.3(0.3-1)(0.3-2)}{6} \times (-8)$$

$$+ \frac{0.3(0.3-1)(0.3-2)(0.3-3)}{24} \times 6$$

$$+ \frac{0.3(0.3-1)(0.3-2)(0.3-3)(0.3-4)}{120} \times -3$$

$$= 10$$

(ii) When income is Rs 75

$$x = \frac{x_I - x_0}{h} = \frac{75 - 60}{10} = 1.5$$

Applying the same formula (1) with $x = 1.5$ we get

$$y_x = 8 + 1.5 \times 10 + \frac{1.5(1.5-1)}{2} \times 6 + \frac{1.5(1.5-1)(1.5-2)}{6} \times (-8)$$

$$+ \frac{1.5(1.5-1)(1.5-2)(1.5-3)}{24} \times 6$$

$$+ \frac{1.5(1.5-1)(1.5-2)(1.5-3)(1.5-4)}{120} \times -3$$

$$= 26.24$$

$$= 26$$

The No. of employee between 63 and 75 = 26 – 10 = 16.

14. Interpolate the number of students who have obtained marks less than 77 from the following data.

Marks:	50–55	55–60	60–65	65–70	70–75	75–80
No. of students:	6	10	22	30	16	12

Solution

Marks upto	y	$\Delta^1 y$	$\Delta^2 y$	$\Delta^3 y$	$\Delta^4 y$	$\Delta^5 y$
55	6					
		10				
60	16		12			
		22		–4		
65	38		8		–18	
		30		–22		50
70	68		–14		32	
		16		+10		
75	84		–4			
		12				
80	96					

$$x = \frac{x_l - x_0}{h} = \frac{77-55}{5} = 4.4$$

$$y_x = y_0 + x\Delta^1 y_0 + \frac{x(x-1)}{2!}\Delta^2 y_0 + \frac{x(x-1)(x-2)}{3!}\Delta^3 y_0$$

$$+ \frac{x(x-1)(x-2)(x-3)}{4!}\Delta^4 y_0 + \frac{x(x-1)(x-2)(x-3)(x-4)}{5!}\Delta^5 y_0$$

$$= 6 + 4.4 \times 10 + \frac{4.4(4.4-1)}{2} \times 12 + \frac{4.4(4.4-1)(4.4-2)}{6} \times (-4)$$

$$+ \frac{4.4(4.4-1)(4.4-2)(4.4-3)}{24} \times (-18)$$

$$+ \frac{4.4(4.4-1)(4.4-2)(4.4-3)(4.4-4)}{120} \times 50$$

$$Y_x = 6 + 44 + 89.76 + (-23.936) + (-37.699) + 8.378$$
$$= 86.503$$
$$= 87$$

15. Use Lagrange's formula and estimate from the following data the number of workers getting income not exceeding Rs 26 per month.

Income not exceeding (Rs):	15	25	30	35
No. of workers:	36	40	45	48

(B. U., November, 1997)

Solution Lagrange's method

Income not exceeding	*No. of workers*
15 (x_0)	36 (y_0)
25 (x_1)	40 (y_1)
30 (x_2)	45 (y_2)
35 (x_3)	48 (y_3)

$x = 26$

$$y_x = y_0 \frac{(x-x_1)(x-x_2)(x-x_3)}{(x_0-x_1)(x_0-x_2)(x_0-x_3)} + y_1 \frac{(x-x_0)(x-x_2)(x-x_3)}{(x_1-x_0)(x_1-x_2)(x_1-x_3)}$$

$$+ y_2 \frac{(x-x_0)(x-x_1)(x-x_3)}{(x_2-x_0)(x_2-x_1)(x_2-x_3)} + y_3 \frac{(x-x_0)(x-x_1)(x-x_2)}{(x_3-x_0)(x_3-x_1)(x_3-x_2)}$$

$$= 36 \frac{(26-25)(26-30)(26-35)}{(15-25)(15-30)(15-35)} + 40 \frac{(26-15)(26-30)(26-35)}{(25-15)(25-30)(25-35)}$$

$$+ 45 \frac{(26-15)(26-25)(26-35)}{(30-15)(30-25)(30-35)} + 48 \frac{(26-15)(26-25)(26-30)}{(35-15)(35-25)(35-30)}$$

$$= 36 \frac{(1)(-4)(-9)}{(-10)(-15)(-20)} + 40 \frac{(11)(-4)(-9)}{(10)(-5)(-10)} + 45 \frac{(11)(1)(-9)}{(15)(5)(-5)}$$

$$+ 48 \frac{(11)(1)(-4)}{(20)(10)(5)}$$

$= 41.88$

$= 42$ person.

16. Using interpolation estimate the business done in 1985 from the following data.

Year:	1982	1983	1984	1986	1987
Business done (lakhs):	150	235	365	525	780

(*B. U., April, 2001*)

Solution Lagrange's method

x:	x_0 1982	x_1 1983	x_2 1984	x_3 1986	x_4 1987
y:	y_0 150	y_1 235	y_2 365	y_3 525	y_4 780

$x = 1985$

$$y_x = y_0 \frac{(x-x_1)(x-x_2)(x-x_3)(x-x_4)}{(x_0-x_1)(x_0-x_2)(x_0-x_3)(x_0-x_4)}$$

$$+ y_1 \frac{(x-x_0)(x-x_2)(x-x_3)(x-x_4)}{(x_1-x_0)(x_1-x_2)(x_1-x_3)(x_1-x_4)}$$

$$+ y_2 \frac{(x - x_0)(x - x_1)(x - x_3)(x - x_4)}{(x_2 - x_0)(x_2 - x_1)(x_2 - x_3)(x_2 - x_4)}$$

$$+ y_3 \frac{(x - x_0)(x - x_1)(x - x_2)(x - x_4)}{(x_3 - x_0)(x_3 - x_1)(x_3 - x_2)(x_3 - x_4)}$$

$$+ y_4 \frac{(x - x_0)(x - x_1)(x - x_2)(x - x_3)}{(x_4 - x_0)(x_4 - x_1)(x_4 - x_2)(x_4 - x_3)}$$

$$= 150 \frac{(1985 - 1983)(1985 - 1984)(1985 - 1986)(1985 - 1987)}{(1982 - 1983)(1982 - 1984)(1982 - 1986)(1982 - 1987)}$$

$$+ 235 \frac{(1985 - 1982)(1985 - 1984)(1985 - 1986)(1985 - 1987)}{(1983 - 1982)(1983 - 1984)(1983 - 1986)(1983 - 1987)}$$

$$+ 365 \frac{(1985 - 1982)(1985 - 1983)(1985 - 1986)(1985 - 1987)}{(1984 - 1982)(1984 - 1983)(1984 - 1986)(1984 - 1987)}$$

$$+ 525 \frac{(1985 - 1982)(1985 - 1983)(1985 - 1984)(1985 - 1987)}{(1986 - 1982)(1986 - 1983)(1986 - 1984)(1986 - 1987)}$$

$$+ 780 \frac{(1985 - 1982)(1985 - 1983)(1985 - 1984)(1985 - 1986)}{(1987 - 1982)(1987 - 1983)(1987 - 1984)(1987 - 1986)}$$

$$= 150 \frac{(2) \times (1) \times (-1)(-2)}{(-1)(-2)(-4)(-5)} + 235 \frac{(3) \times (1) \times (-1)(-2)}{(1)(-1)(-3)(-4)}$$

$$+ 365 \frac{(3) \times (2) \times (-1)(-2)}{(2)(1)(-2)(-3)} + 525 \frac{(3) \times (2) \times (1)(-2)}{(4)(3)(2)(-1)}$$

$$+ 780 \frac{(3) \times (2) \times (1)(-1)}{(5)(4)(3)(1)}$$

$$Y_x = 15 - 117.5 + 365 + 262.5 - 78$$
$$= 447$$

17. Following are the annual premium charged by an insurance company for policy of Rs 1000. Calculate the premium payable at age 30.

Age in years:	20	25	35	40
Premium Rs:	23	26	35	42

Solution Lagrange's method

x:	x_0 20	x_1 25	x_2 35	x_3 40
y:	y_0 23	y_1 26	y_2 35	y_3 42

$x = 30$

$$y_x = y_o \frac{(x-x_1)(x-x_2)(x-x_3)}{(x_0-x_1)(x_0-x_2)(x_0-x_3)} + y_1 \frac{(x-x_0)(x-x_2)(x-x_3)}{(x_1-x_0)(x_1-x_2)(x_1-x_3)}$$

$$+ y_3 \frac{(x-x_0)(x-x_1)(x-x_2)}{(x_3-x_0)(x_3-x_1)(x_3-x_2)}$$

$$= 23 \frac{(30-25)(30-35)(30-40)}{(20-25)(20-35)(20-40)} + 26 \frac{(30-20)(30-35)(30-40)}{(25-20)(25-35)(25-40)}$$

$$+ 35 \frac{(30-20)(30-25)(30-40)}{(35-20)(35-25)(35-40)} + 42 \frac{(30-20)(30-25)(30-35)}{(40-20)(40-25)(40-35)}$$

$$= 23 \frac{(+5)(-5)(-10)}{(-5)(-15)(-20)} + 26 \frac{(10)(-5)(-10)}{(5)(-10)(-15)} + 35 \frac{(10)(5)(-10)}{(15)(10)(-5)}$$

$$+ 42 \frac{(10)(5)(-5)}{(20)(15)(5)}$$

$= -3.833 + 17.333 + 23.333 - 7$

$= 29.833$

$= 30$

$\therefore$ the premium payable at age 30 is Rs 30.

18. Using interpolation find the value of $f(x)$ when $x = 15$.

x:	3	7	11	19
$f(x)$:	42	43	47	60

Solution Lagrange's method

x:	x_0	x_1	x_2	x_3
	3	7	11	19
y:	y_0	y_1	y_2	y_3
	42	43	47	60

Here $x = 15$

Using Lagrange's formula we get

$x = 30$.

$$y_x = y_o \frac{(x-x_1)(x-x_2)(x-x_3)}{(x_0-x_1)(x_0-x_2)(x_0-x_3)} + y_1 \frac{(x-x_0)(x-x_2)(x-x_3)}{(x_1-x_0)(x_1-x_2)(x_1-x_3)}$$

$$+ y_2 \frac{(x-x_0)(x-x_1)(x-x_3)}{(x_2-x_0)(x_2-x_1)(x_2-x_3)} + y_3 \frac{(x-x_0)(x-x_1)(x-x_2)}{(x_3-x_0)(x_3-x_1)(x_3-x_2)}$$

$$= 42 \frac{(15-7)(15-11)(15-19)}{(3-7)(3-11)(3-19)} + 43 \frac{(15-3)(15-11)(15-19)}{(7-3)(7-11)(7-19)}$$

$$+ 47 \frac{(15-3)(15-7)(15-19)}{(11-3)(11-7)(11-19)} + 60 \frac{(15-3)(15-7)(15-11)}{(19-3)(19-7)(19-11)}$$

$$= 42 \frac{(8 \times 4 \times -4)}{(-4 \times -8 \times -16)} + 43 \frac{(12 \times 4 \times -4)}{(4 \times -4 \times -12)} + 47 \frac{(12 \times 8 \times -4)}{(8 \times 4 \times -8)}$$

$$+ 60 \frac{(12 \times 8 \times 4)}{(16 \times 12 \times 8)}$$

$$= 10.5 - 43 + 70.5 + 15$$

$$= 53$$

Exercises

1. What is interpolation?
2. Explain the term extrapolation.
3. Distinguish between interpolation and extrapolation.
4. What are the assumption of interpolation?
5. Interpolate the number of workers earning between Rs 20 to Rs 25.

Income (Rs):	0–10	10–20	20–30	30–40	40–50
No. of person:	20	25	70	95	115

(*Ans. 31*)

6. Below are given the wages earned by workers per month in a certain factory. Interpolation the number of workers earning upto Rs 25.

Wages:	0–10	10–20	20–30	30–40	40–50	50–60
No. of workers:	20	25	70	95	115	100

(*Ans. 396*)

7. From the data given below estimate the number of persons living between the age of 35 and 42.

Age in years:	20	30	40	50
No. of persons:	513	439	346	243

(*Ans. 68*)

8. From the following table find the number of students who have scored less than 45 marks.

Marks upto:	40	50	60	70	80
No. of students:	31	73	124	159	190

(*Ans. 48*)

9. Calculate the premium at the age of 17 from the following data.

Age of years:	15	25	35	45	55
Premium (Rs 1000):	11.1	12.6	14.3	16.1	18.3

(*Ans. Rs 11.36*)

10. Using a suitable method of interpolation calculate the number of workers earning more than Rs 90 per month.

Monthly income (Rs):	upto Rs 50	60	70	80	100
No. of workers:	50	150	300	500	800

(***Hint***: Lagrange's method)

(***Ans. 700***)

11. Use Lagrange's formula and estimate the expectation of life at the age of 20 using the following data.

Age:	10	15	25	30
Expectation of life:	36	30	26	24

(***Ans. 28***)

12. Use suitable method of interpolation, calculate the premium at the age of 30 from the following data.

Age:	20	25	35	40
Premium: (Rs 1000 policy)	30	40	75	100

(***Ans. 55***)

13. Estimate the number of students who get more than 55 from the following data.

Mark upto:	45	50	60	65
No. of students:	447	484	511	514

(***Ans. 505***)

14. Using interpolation estimate the output of a factory in 1986 from the following data.

Year:	1974	1978	1982	1990
Output: (in 000 tones)	25	60	80	170

(***Ans. 100, thousand tones***)

15. Estimate the value of y for $x = 6$ from the following data.

x:	2	4	7	8
y:	7	21	51	73

(***Ans. 43***)

Part III

Previous Years Question Papers

First B. B. M Examination, April 1995
(New Scheme)

MANAGEMENT
Business Mathematics and Statistics (Paper – IV)

Time: 3 Hours **Max. Marks: 90**

Instructions:

1. Neatness in work carries weightage.
2. Necessary calculations to be shown in the body of answers.

Part – I
(Business Mathematics)
SECTION – A

Answer any five of the following. Each question carries 3 marks. **(5 × 3 = 15)**

1. (a) Give the meaning of
   1. a diagonal matrix.
   2. a scalar matrix.
   3. a unit matrix.

   (b) If $\vec{a} = (2, 2, 1)$ $\vec{b} = (1, 4, 5)$

   Find $\vec{a} + \vec{b}$ and $2\vec{a} - 5\vec{b}$.

   (c) Evaluate $\lim_{x \to 3} \dfrac{x^2 - 5x + 6}{x^2 - 9}$.

   (d) Differentiate *w.r.t. x.* $y = 4\sqrt{x} + \dfrac{3}{x} + 4e^x$.

   (e) If $u = (x - y)(x + y)$ find $\dfrac{\partial u}{\partial x}$ and $\dfrac{\partial u}{\partial y}$.

   (f) Evaluate $\int (3x^3 + 7x^2 + 1)\, dx$.

   (g) Find the value of ${}^{10}P_3$ and ${}^{14}C_{12}$.

   (h) The total cost in Rs of output x is given by $C = \dfrac{2x}{3} + \dfrac{35}{2}$.

   (i) Find the marginal cost and average cost when output is 3 units.

SECTION – B

Answer any two of the following. Each questions carries 15 marks. **(5 + 5 + 5)**

2. (a) Given $A = \begin{bmatrix} 2 & 1 & 5 \\ 4 & 1 & 2 \\ 2 & 2 & 3 \end{bmatrix}$ $B = \begin{bmatrix} 4 & 7 & -2 \\ 2 & 3 & -4 \\ 1 & 1 & 3 \end{bmatrix}$

verify whether $(AB)' = B'A'$.

(b) If $A = \begin{bmatrix} 2 & 3 \\ 3 & 5 \end{bmatrix}$ verify that $A^2 - 7A + 1 = 0$ and hence obtain A^{-1}.

(c) Find the number of permutations of the letters of the word TOMORROW. In how many of these are the 3 0's not together.

3. (a) Solve by matrix method **(5 + 5 + 5)**

$x + y + z = 6$

$x - y + z = 2$

$2x + y - z = 1$

(b) If $\vec{a} = (3, 2)$ $\vec{b} = (5, -1)$ $\vec{c} = (-4, 2)$

find the vectors $2\vec{a} + \vec{b} - 3\vec{c}$ and $3\vec{a} + 2\vec{b} - \vec{c}$.

(c) Evaluate

(i) $\lim_{x \to -2} \dfrac{3x^2 + 5x - 2}{x^2 - 3x - 10}$.

(ii) $\lim_{x \to \infty} \dfrac{(x + 5)(2x + 5)}{(3x + 1)(x - 2)}$.

4. (a) Find **(5 + 5 + 5)**

$\dfrac{dy}{dx}$ if $y = \dfrac{2x + 1}{(1 - x)^2}$.

(b) The cost function for producing x units of product is $C(x) = x^3 - 12x^2 + 48x + 11$ in Rupees and the revenue function is $R = 83x - 4x^2 - 21$.

Find the output for which profit is maximum.

(c) Evaluate $\displaystyle\int_1^2 \left(x^2 + \frac{1}{x^2} - x \right) dx$.

5. (a) If $A = \begin{bmatrix} 2 & 4 \\ 5 & 6 \end{bmatrix}$ $B = \begin{bmatrix} 4 & 2 \\ 2 & 4 \end{bmatrix}$ **(5 + 5 + 5)**

Find a matrix X such that $2A + B + 4x = 0$.

(b) Evaluate $\int (x^n \log n)\, dx$.

(c) Define marginal cost and marginal revenue. If the total revenue 'R' and total cost 'C' of a firm are given by $R = 30x - x^2$ and $C = 20 + 6x$. Find the marginal revenue and marginal cost.

PART – II
(Statistics)
SECTION – A

Answer any five of the following questions. Each question carries 3 marks. **(5 × 3 = 15)**

6. (a) Define "Statistics".
   (b) What is secondary data?
   (c) What are the types of classification?
   (d) What is Tabulation?
   (e) What is range? What is co-efficient of range?
   (f) What is a component bar diagram?
   (g) What is an Ogive?
   (h) Give the regression equation of x on y.

SECTION – B

Answer any two of the following. Each question carries 15 marks.

7. (a) Classify the given paragraph in terms of the number of letters in each word by taking the class interval of 1–3, 4–6, 7–9 etc.

   The extra-ordinary general meeting may be convened by the Board of Directors on the written requisition by members holding not less than one tenth of the total voting rights in the company, where it has no share capital or one tenth of the paid up capital of the company. The Board of Directors must arrange to convene a meeting to consider the matters referred to in the requisition within a period of twenty one days. In any case the meeting must be convened by the Directors on a date not later than forty five days from the date of deposit of the requisition.

   (b) Show the following data of expenditure of an average working class family by a pie diagram.

Item of expenditure	*% of total expenditure*
Food	65
Clothing	10
Housing	12
Fuel and lighting	05
Miscellaneous	08

8. (a) Calculate the mean and median from the following data:

Marks	*No. of Students*
0–10	5
10–20	10
20–30	25
30–40	30
40–50	20
50–60	10

(b) Draw an ogive for the following distribution. Read the values of Q_1 and Q_3 and verify the result by calculation.

Marks	*No. of Students*
50–55	6
55–60	10
60–65	22
65–70	30
70–75	16
75–80	12
80–100	15

9. (a) Calculate the mean deviation (from median) from the following data:

Wages per week	*No. of Workers*
10–20	4
20–30	6
30–40	10
40–50	20
50–60	10
60–70	6
70–80	4

(b) The number of runs scored by Cricketers *A* and *B* during a test series consisting of 5 matches shown below for each of the 10 innings.

Cricketer A	*Cricketer B*
5	51
26	47
97	36
76	60
112	58
89	39
6	44
108	42
24	71
16	50

Find which Cricketer may be considered more consistent.

10. (a) Find rank correlation coefficient for the following data and given your comments.

Marks in Statistics (X)	*Marks in Law (Y)*
84	38
56	69
89	56
58	58
59	63
67	78
74	87
78	77

(b) To study the effect of rain on yield of wheat the following results were obtained.

	Mean	*Standard Deviation*
Yield (X)	800	12
Rainfall (Y)	50	02

$r = 0.80$

Estimate the yield when the rainfall is 80 inches.

First B. B. M Examination, Oct./Nov. 1995
(New Scheme)

BUSINESS MANAGEMENT (Paper – IV)
Business Mathematics and Statistics

Time: 3 Hours **Max. Marks: 90**

Instructions:

1. Neatness in work carries weightage.
2. Necessary calculations should be shown in the body of answers.

Part – I
(Business Mathematics)
SECTION – A

Answer any five of the following. Each question carries 3 marks. **(5 × 3 = 15)**

1. (a) What is "Break Even Point"?

(b) If $\vec{a} = (2, 3, -1)$ $\vec{b} = (1, 0, 4)$ find $2\vec{a} + 3\vec{b}$ and $3\vec{a} - 2\vec{b}$.

(c) Evaluate $\lim_{x \to 2} \frac{3x^2 - 4x + 1}{2x^2 + 4x - 1}$.

(d) Find $\frac{dy}{dx}$ if $y = \left(x + \frac{1}{x}\right) \log x$.

(e) If $u = 2x^2 + 3xy + 4y^2$ find $\frac{\partial u}{\partial x}$ and $\frac{\partial u}{\partial y}$.

(f) Evaluate

$$\int \left(\frac{x^2 + 2x - 1}{\sqrt{x}} \right) dx.$$

(g) If ${}^nC_8 = {}^nC_{12}$ find ${}^nC_{18}$.

(h) If $A = \begin{bmatrix} 1 & 3 & -2 & 5 \\ 3 & 1 & 2 & 6 \end{bmatrix}$ and $B = \begin{bmatrix} 4 & 7 & 1 & 3 \\ 2 & 5 & 2 & -3 \end{bmatrix}$ find $3A + 4B$ and $4A - 3B$.

SECTION – B

Answer any two of the following questions. Each question carries 15 marks.

2. (a) If $A = \begin{bmatrix} 1 & 1 & -1 \\ 2 & -3 & 4 \\ 3 & -2 & 3 \end{bmatrix}$ $B = \begin{bmatrix} -1 & -2 & -1 \\ 6 & 12 & 6 \\ 5 & 10 & 5 \end{bmatrix}$ and $C = \begin{bmatrix} -1 & -1 & 1 \\ 2 & 2 & -2 \\ -3 & -3 & 3 \end{bmatrix}$

Show that AB is a null matrix but AC is not a null matrix.

(b) Evaluate

(i) $$\lim_{x \to \infty} \frac{x^2 + 3x + 10}{x^2 + 5x + 2}$$

(ii) $$\lim_{x \to 3} \frac{x^3 - 27}{x - 3}$$

(c) For a manufacturer of electrical appliances, the daily cost of production C for x units is given by $C(x) = \text{Rs } 10.04x + \text{Rs } 446$. If the price of a unit is Rs 17 determine the minimum number of units that must be produced and sold daily to ensure no loss.

(5 + 5 + 5)

3. (a) Find the inverse of

$$A = \begin{bmatrix} 1 & 1 & 2 \\ 2 & -1 & 1 \\ 2 & 2 & 1 \end{bmatrix}.$$

(b) Solve

$$|A| = \begin{bmatrix} 2 & x & 4 \\ 3 & 2 & 1 \\ 1 & 2 & 3 \end{bmatrix} = -3.$$

(c) If $f(x) = 2x^2 - 4x + 4$ find x such that $2f(x) = f(2x)$. (5 + 5 + 5)

4. (a) Find $\frac{dy}{dx}$ if

(i) $$y = \frac{e^x + 1}{e^x - 1}$$

(ii) $$y = \sqrt{x} - 4x^{\frac{3}{2}} + \frac{5}{x}.$$

(b) The total revenue R and the total cost C functions of a firm are given by $R = 30x - x^2$ and $C = 20 + 6x$

(i) Find the marginal revenue and marginal cost.

(ii) Find the average cost and average revenue.

(iii) Find the output at which marginal revenue equals marginal cost.

(c) Solve using matrix method

$8x + 5y = 3$

$5x + 3y = 2$

(5 + 5 + 5)

5. (a) If $A = \begin{bmatrix} 2 & 1 & 3 \\ 3 & 4 & 5 \end{bmatrix}$ find $B = \begin{bmatrix} 4 & 3 \\ 0 & 6 \\ 5 & 7 \end{bmatrix}$. Find A′ + B and A′ + B. Verify that $(A + B')' = A' + B$.

(b) Evaluate

(i) $\int xe^x \, dx$

(ii) $\int_1^2 \left(6x^{-5} + 2x^4 + 4\right) dx$

(c) Find the number of permutations of the letters of the word MANAGEMENT. In how many of these do the M's come together? **(5 + 5 + 5)**

PART – II
(Statistics)
SECTION – A

Answer any five of the following questions. Each question carries 3 marks. **(5 × 3 = 15)**

6. (a) What is primary data?
(b) Explain any three types of classification.
(c) Define a statistical table.
(d) Give the singular definition of statistics.
(e) What is correlation?
(f) Give the assumptions under interpolation.
(g) What is regression coefficient?
(h) What are the components of Time Series?

SECTION – B

Answer any two of the following. Each question carries 15 marks.

7. (a) Find Quartile Deviation and its co-efficient from the following data:

Marks (X)	*No. of Students (F)*
23	4
29	9
34	11
38	18
47	23
53	31

(b) Draw a less than ogive from the following data and estimate the number of employees earning atleast Rs 63 but less than Rs 75.

Wages	*No. of employees*
50–60	8
60–70	10
70–80	16

80–90	14
90–100	10
100–110	5
110–120	2

8. (a) Draw a blank table to present the following information regarding students of Bangalore University.

(1) Faculty : Arts, Science, Commerce

(2) Class : Degree, Pre-University.

(3) Students : Boys, Girls

(4) Age : Below 20 years, More than 20 Years.

(b) Construct a sub-divided bar diagram from the following data, regarding the number of students in 2 colleges.

College	*Arts*	*Science*	*Commerce*	*Agriculture*	*Total*
A	1,200	800	600	400	3,000
B	750	500	300	450	2,000

9. (a) Fit by the method of least squares a linear equation $y = mx + c$ to the following data:

Year	*Production (in tons)*
1982	9.7
1983	10.1
1984	10.2
1985	10.7
1986	11.9
1987	12.9
1988	12.4
1989	14.8

Obtain trend values for the years 1996 and 1997.

(b) Find by suitable method of interpolation the value of y when $x = 7$.

x	y
2	7
4	21
6	43
8	73

10. (a) Calculate the mean deviation (from median) and its coefficient from the following data:

Wages per week (Rs)	*No. of workers*
10–20	4
20–30	6
30–40	10
40–50	20
50–60	10
60–70	6
70–80	4

(b) From the data given in part (a) above, calculate standard deviation and variance.

First B. B. M Examination, April 1996
(New Scheme)

MANAGEMENT (Paper – IV)
Business Mathematics and Statistics

Time: 3 Hours **Max. Marks: 90**

Instructions:

1. Neatness in work carries weightage.
2. Necessary working to be shown in the body of answers.

Part – I
(Business Mathematics)
SECTION – A

Answer any five of the following. Each question carries 3 marks. **(5 × 3 = 15)**

1. (a) Find the value ${}^8P_3$ and ${}^8C_3$.

(b) If $A = \begin{pmatrix} 5 & 3 \\ 4 & 6 \end{pmatrix}$ and $B = \begin{pmatrix} 6 & 8 \\ 9 & 1 \end{pmatrix}$ find $2A + 3B$.

(c) If $\bar{a} = (2, 2, 1)$ and $\bar{b} = (1, 4, 5)$ find $2\bar{a} + \bar{b}$ and $3\bar{a} - 4\bar{b}$.

(d) Evaluate $\lim_{x \to 3} \dfrac{x^2 - 9}{x^2 - 5x + 6}$.

(e) Differentiate *w.r.t. x*. $Y = x^2 e^x$.

(f) $\int (x^3 + 3x^2 - 13)\, dx$.

(g) What do you mean by a diagonal matrix?

(h) For the cost function $Y = 1000x - 18x^2 + 3x^3$ for x units, find the Marginal Cost.

SECTION – B

Answer any two of the following. Each question carries 15 marks. **(2 × 15 = 30)**

2. (a) If $A = \begin{bmatrix} 2 & 3 & 4 \\ 1 & 2 & 3 \\ 1 & 1 & 2 \end{bmatrix}$ and $B = \begin{bmatrix} 1 & 3 & 0 \\ 1 & 2 & 1 \\ 0 & 0 & 2 \end{bmatrix}$.

Find AB and BA.

(b) If $A = \begin{bmatrix} 1 & 0 & 4 \\ 2 & 4 & 3 \end{bmatrix}$ and $= \begin{bmatrix} 4 & 0 & 1 \\ 3 & 4 & 2 \end{bmatrix}$

Find Matrix Y such that $A + 2Y = 4B$.

(c) How many different arrangements can be made by using all the letters in the Word MATHEMATICS? How many of them begin with *C*?

3. (a) Evaluate $\begin{vmatrix} 3 & 8 & 6 \\ 9 & 8 & 2 \\ 6 & 12 & 2 \end{vmatrix}$.

(b) Solve by Matrix Method

$6x + 5y = 2$

$4x - 3y = 14$

(c) If $\vec{a} = (3, 2, 1)$, $\vec{b} = (2, 0, 3)$ and $\vec{c} = (1, 2, 3)$ find the vectors $3\vec{a} + \vec{b} + \vec{c}$ and $\vec{a} + \vec{b} + \vec{c}$.

4. (a) If $A = \begin{bmatrix} 1 & 2 & 3 \\ 2 & 4 & 6 \\ 1 & 2 & 3 \end{bmatrix}$

Verify $(5A)' = 5A'$.

(b) $Y = x\left(x^2 + 1\right)$ find $\frac{dy}{dx}$.

(c) $\int_1^2 \left(x^3 + 3x^2 + x\right) dx$.

5. (a) The total profit in rupees of a drug company from the manufacture and sale of *x* drug bottles is given by

$$Y = \frac{-x^2}{400} + 2x - 80$$

How many bottles must the company sell to achieve the maximum profit?

(b) If $A = \begin{bmatrix} 5 & 6 & 7 \\ 8 & 9 & 0 \\ 1 & 2 & 3 \end{bmatrix}$, $B = \begin{bmatrix} 3 & 2 & 1 \\ 9 & 0 & 8 \\ 7 & 6 & 5 \end{bmatrix}$ find

(i) $3A + 5B$

(ii) $5A + 3B$

(c) Differentiate the following with respect to *x*.

$Y = (3x^2 + 4x - 5)^3$

PART – II
(Statistics)
SECTION – A

Answer any five of the following questions. Each questions carries 3 marks. **(5 × 3 = 15)**

6. (a) What do you mean by a 'Statistical Table'?

(b) Define Mode.
(c) What do you mean by relative measures of dispersion?
(d) What do you mean by Seasonal Variation?
(e) What are regression equations?
(f) What are objectives of averaging?
(g) Define standard Deviation?
(h) What do you mean by correlation?

SECTION – B

Answer any two of the following. Each questions carries 15 marks.

7. (a) Draw an ogive of the following distribution. Read the values of Q_1 and Q_3.

Wages (in Rs)	*No. of workers*
70–80	4
80–90	6
90–100	10
100–110	16
110–120	12
120–130	7
130–140	3

(b) Find the median of the following

Weekly Wages (in Rs)	*No. of workers*
60–70	5
70–80	15
80–90	20
90–100	30
100–110	20
110–120	8

8. (a) Calculate the expectation of life at the age of 22 using the following data.

Age	10	15	20	25	30
Expectation of life	36	30	28	26	24

(b) The prices of two shares *X* and *Y* are given below:

X Co. Ltd.	*Y Co. Ltd.*
55	108
54	107
52	105
53	106
56	105

Find which of the two is more stable?

9. (a) Calculate coefficient of correlation using Karl Pearson's Method:

Advertisement Cost (in '000 Rs)	*Sales (in lakh Rs)*
39	47
65	53
62	58
90	86
82	62
75	68
25	60
98	91
36	51
78	84

(b) You are given the following

	Y	*X*
Mean	10	15
Standard Deviation	2	4

$r = +0.8$

Estimate the value of Y when $X = 20$.

First B. B. M Examination, October 1996
(New Scheme)

BUSINESS MATHEMATICS AND STATISTICS
(Paper – IV)

Time: 3 Hours **Max. Marks: 90**

Instructions:

1. Neatness carries due weightage.
2. Necessary workings to be shown in the body of answers.

Part – I
(Business Mathematics)
SECTION – A

Answer any five of the following. Each question carries 3 marks. **(5 × 3 = 15)**

1. (a) Find the value of ${}^{10}P_6$ and ${}^{12}C_6$.

(b) If $A = \begin{bmatrix} 3 & 2 \\ 1 & 5 \end{bmatrix}$ and $B = \begin{bmatrix} 6 & 7 \\ 8 & 9 \end{bmatrix}$ find $5A + 3B$.

(c) If $\vec{a} = (2, 1, 2)$ and $\vec{b} = (4, 5, 1)$ find $\vec{a} + \vec{b}$ and $\vec{a} - \vec{b}$.

(d) Evaluate $\lim_{x \to 1} \dfrac{x^2 - 1}{x^2 - 3x + 2}$.

(e) Differentiate with respect to x
$y = 4x^2 + 3x + 12$.

(f) If $Y = \sqrt{x}\left(x^2 + 2\right)$ find $\dfrac{dy}{dx}$.

(g) $\int \left(4x^3 + 3x^2 + 12\right) dx$.

(h) The cost function for producing x units of a product is $C(x) = x^3 - 15x^2 + 24x + 15$. Find the marginal cost function.

SECTION – B

Answer any two of the following. Each question carries 15 marks.

2. (a) If $A = \begin{bmatrix} 1 & 2 & 3 \\ 4 & 5 & 6 \\ 7 & 8 & 9 \end{bmatrix}$ and $B = \begin{bmatrix} 9 & 8 & 7 \\ 6 & 5 & 4 \\ 3 & 2 & 1 \end{bmatrix}$

Find (i) AB (ii) BA.

(b) If $A = \begin{bmatrix} 8 & 7 \\ 6 & 5 \end{bmatrix}$ and $B = \begin{bmatrix} 5 & 6 \\ 7 & 8 \end{bmatrix}$

Find Matrix X such that $3A + 2B + X = 0$.

(c) In how many ways can the letters of the word MANAGEMENT be arranged? How many of these begin with M?

3. (a) Solve by Matrix Method

$3x - y = 6$

$2x + 3y = 15$

(b) If $\vec{a} = (2, 1, 3)$, $\vec{b} = (2, 1, 4)$ and $\vec{c} = (0, 1, 2)$

Find (i) $\vec{a} - \vec{b}$, (ii) $4(\vec{c} - \vec{a})$.

(c) Find the output for which profit is maximum

$C(x) = x^3 - 12x^2 + 48x + 11$

$R = 83x - 4x^2 - 21$

4. (a) Find the value

$$\begin{vmatrix} 1 & 2 & 3 \\ 4 & 5 & 6 \\ 7 & 8 & 9 \end{vmatrix}$$

(b) Find $\frac{dy}{dx}$ if $y = (3x - 5)(x^2 + 2x)$.

(c) Find $\frac{dy}{dx}$ if $y = (x^2 + 1)^4$.

5. (a) $\int_1^5 (5x^4 + 4x^3 + 3x^2)\, dx$.

(b) If $A = \begin{bmatrix} 2 & 6 \\ 4 & 5 \end{bmatrix}$ and $B = \begin{bmatrix} 4 & 3 \\ 2 & 4 \end{bmatrix}$

Find Matrix X such that $2A + B + 4X = 0$.

(c) A company has a total cost function of $Y = 2x^3 - 3x^2 + 12x$ where Y represents cost and x represents quantity.

(i) What is the Marginal cost function?

(ii) What is the average cost function?

PART – II
(Statistics)
SECTION – A

Answer any five of the following question. Each question carries 3 marks. **(5 × 3 = 15)**

6. (a) What do you mean by Tabulation of data?

(b) Give the meaning of 'mean deviation'

(c) What do you mean by cyclical variation?
(d) What do you mean by interpolation?
(e) What are the merits of standard deviation?
(f) If $bxy = -0.8$ and $byx = -0.9$. What is value of r?
(g) Give the regression equation of X on Y.
(h) What is an Ogive?

SECTION – B

Answer any two of the following. Each question carries 15 marks.

7. (a) Show the following data by a pie diagram:

Items of expenditure	*% of total expenditure*
Food	60
Clothing	30
Housing	05
Miscellaneous	05

(b) Calculate mean and median of the following:

Income (Rs)	*No. of workers*
10–20	5
20–30	15
30–40	20
40–50	20
50–60	10
60–70	5

8. (a) Calculate Karl Pearson's coefficient of correlation.

X	Y
1	3
2	5
3	6
4	8
5	10
6	11
7	13

(b) You are given the following data:

	X	Y
Mean	20	10
Standard Deviation	3	4

$r = +0.9$

Estimate the value of Y when $x = 25$.

9. (a) Following are the marks btained by two students A and B:

A	*B*
40	20
45	60
50	80
30	10
20	20
60	60

If the consistency of performance is the criterion for awarding a prize, who should get the prize?

(b) Find rank correlation coefficient and comment.

Marks in Mathematics	*Marks in Statistics*
60	64
70	58
50	72
40	44
80	86
90	80
85	95
96	81

10. (a) From the following details fit a straight line trend:

Year	*Production*
1989	160
1990	180
1991	184
1992	166
1993	188
1994	198
1995	184

(b) Calculate the premium at the age of 22 from the following data:

Age	20	25	30	35	40
Premium per 1000 Sum assured	30	40	55	75	100

First B. B. M. Examination, April 1997
(New Scheme)

BUSINESS MANAGEMENT (Paper – IV)
Business Mathematics and Statistics

Time: 3 Hours **Max. Marks: 90**

Instructions:

1. Neatness in work carries weightage.
2. Necessary calculations should be shown in the body of answers.

Part – I
(Business Mathematics)
SECTION – A

Answer any five of the following. Each question carries 3 marks. **(5 × 3 = 15)**

1. (a) Define linear combination of vectors.

(b) Find the value of (i) ${}^{9}P_4$ (ii) ${}^{21}C_{20}$.

(c) If $A = \begin{bmatrix} 1 & 3 & -2 & 5 \\ 3 & 1 & 2 & 6 \end{bmatrix}$ and $B = \begin{bmatrix} 4 & 7 & 1 & 3 \\ 2 & 5 & 2 & -3 \end{bmatrix}$ find $3A + 4B$.

(d) If $f(x) = x^{-1}$ find $f(x+1) + 2f(x) + \dfrac{1}{f(x)}$.

(e) Find $\dfrac{dy}{dx}$ if $y = \dfrac{x+1}{x}$.

(f) Evaluate $\int \left(4x^3 - \dfrac{3}{x} + \sqrt{x} \right) dx$.

(g) Evaluate $\lim\limits_{x \to \infty} \dfrac{(2n+1)^2}{n^2+1}$.

(h) Find the value of $|A| = \begin{vmatrix} 1 & 2 & 3 \\ 4 & 5 & 6 \\ 7 & 8 & 9 \end{vmatrix}$.

SECTION – B

Answer any two of the following. Each question carries 15 marks.

2. (a) If $A = \begin{bmatrix} -3 & -2 & -1 \\ 0 & -4 & 2 \\ -1 & -2 & -3 \end{bmatrix}$ and $B = \begin{bmatrix} 0 & -1 & -2 \\ -3 & -4 & -5 \\ -1 & 0 & -2 \end{bmatrix}$

Show that $(A+B)' = A' + B'$

(b) If $y = \dfrac{7x^5 + 6x^3 - 5x}{x^4}$ find $\dfrac{dy}{dx}$.

(c) If $\bar{a} = (2, 4, -1), \bar{b} = (3, -2, 0)$

find (i) $2\bar{a} + 3\bar{b} - \bar{c}$, (ii) $4\bar{a} - \bar{b} + 2\bar{c}$. **(5 + 5 + 5)**

3. (a) Solve by matrix method

$x + y = 3$

$y + z = 5$

$z + x = 4$

(b) If $A = \begin{bmatrix} 1 & 2 & 3 \\ 3 & -2 & 1 \\ 4 & 2 & 1 \end{bmatrix}$

Find $A^2 - 23A - 40I$.

(c) Evaluate

(i) $\lim\limits_{x \to 0} \dfrac{(x+1) - (x^2+1)}{x}$

(ii) $\lim\limits_{x \to 1} \dfrac{x^2 + x - 2}{x^2 - 3x + 2}$. **(5 + 5 + 5)**

4. (a) Find the number of permutations of the letters of the word "ASSIGNMENT". In how many of these will the 2 N's come together?

(b) Define average cost and average revenue. The average cost for a commodity is given by $AC = x + 4 + \dfrac{4}{x}$. Find the total cost and marginal cost.

(c) Evaluate $\int (x^3 + 4x - 5)\, dx$. **(5 + 5 + 5)**

5. (a) Find $\dfrac{dy}{dx}$ if $x = at^2, y = 2at$

(b) If $y = 3x^4 - 4x^2 + 18x$. Find $\dfrac{d^2y}{dx^2}$

(c) Evaluate (i) $\int (1 + e^{-x})\, dx$ (ii) $\int x \log x \, dx$. **(5 + 5 + 5)**

PART – II
(Statistics)
SECTION – A

Answer any five the following questions. Each carries 3 marks. **(5 × 3 = 15)**

6. (a) What are the sources of Primary data?

(b) Explain any three parts of a table?

(c) What are regression equations?

(d) What is a pie-chart?

(e) What do you mean by classification?

(f) What is the corresponding relative measure of standard deviation. Give the formula.

(g) What is Interpolation?

(h) Interpret the following

If $r = 0.925$

$r = -0.361$

$r =$ Zero

SECTION – B

Answer any two of the following. Each question carries 15 marks.

7. (a) In 1975 out of 1750 workers of a factory 1200 were members of a trade union. The number of women employed were 200 out of which 175 did not belong to any trade union. In 1980 the number of union workers increased to 1580 of which 1290 were men. On the other hand the number of non-union workers fell to 208 of which 180 were men.

In 1985, there were 1800 employees who belonged to the trade union and 50 who did not belong to any trade union of all the employees in 1985, 300 were women, of them only 8 did not belong to the trade union.

Tabulate the above information giving suitable title, table number etc...

(b) Represent the following data by a simple bar diagram.

Countries	*Production of sugar In quintals' 00000*
China	38
Japan	40
Indonesia	25
India	20
Pakistan	08
Sri Lanka	08

8. (a) Following data are given for marks in English (X) and marks in Maths (Y) at a certain examination:

	X	Y
Mean Marks	39.5	47.5
S.D. Marks	10.8	16.8

Co-efficient of correlation is $r = 0.42$. Find the regression equation of Y and X and estimate the marks in Maths when marks in English is 50.

(b) Find rank correlation coefficient for the following data and give your comments.

Marks in Accounts (X)	*Marks in Maths (Y)*
84	38
56	69

89	56
58	58
59	63
67	78
74	87
78	77

9. (a) Draw a Histogram from the following data regarding the monthly profits in rupees of 100 shops.

Profits (Rs)	*No. of Shops*
0–100	12
100–200	18
200–300	27
300–400	20
400–500	17
500–600	6

(b) From the above Histogram, locate the modal value graphically and verify this value with actual calculations.

10. (a) Estimate the value of sales for the year 1985 by using the method of least square for trend analysis.

Year	*Sales* ('000 *quintals*)
1979	100
1980	120
1981	140
1982	160
1983	180

(b) Form a difference table and find the value of $f(x)$ when $x = 9$.

X	$f(x)$
3	42
7	43
11	47
15	53
19	60

First B. B. M. Examination, November 1997
(New Scheme)

BUSINESS MANAGEMENT
Business Mathematics and Statistics (Paper – IV)

Time: 3 Hours **Max. Marks: 90**

Instructions:

1. Neatness in work carries weightage.
2. Necessary calculations should be shown in the body of answers.

Part – I
(Business Mathematics)
SECTION – A

Answer any five of the following. Each question carries 3 marks. **(5 × 3 = 15)**

1. (a) What do you mean by unit vector?

(b) Find the value of (i) ${}^{10}P_3$, (ii) ${}^{15}C_8$.

(c) Solve for A and B

$$2a + 3b = \begin{bmatrix} 1 & 2 \\ 2 & 0 \end{bmatrix} \text{ and } A + 2B = \begin{bmatrix} 2 & 1 \\ 1 & -1 \end{bmatrix}.$$

(d) If $f(x) = 2x^2 - 5x + 4$, for what values of x is $2f(x) = f(2x)$?

(e) Differentiate $\dfrac{3x - 1}{2x + 3}$ with respect to x.

(f) $\int \left(x^2 + 1\right)\left(2x^3 - 3\right) dx$.

(g) Evaluate $\begin{vmatrix} 2 & 0 & -1 \\ 0 & 1 & 5 \\ 1 & 3 & -4 \end{vmatrix}$.

(h) $\int_2^3 x^2 \, dx$.

SECTION – B

Answer any two of the following. Each question carries 15 marks. **(2 × 15 = 30)**

2. (a) If $A = \begin{bmatrix} 2 & 4 \\ 5 & 6 \end{bmatrix}$ and $B = \begin{bmatrix} 4 & 2 \\ 2 & 4 \end{bmatrix}$ find matrix such that $2A + B + 4X = 0$.

(b) Find the elasticity of demand for the demand function $x = 50 - 4p^2$ at $p = 1$ and at $p = 3$.

(c) A commodity was produced by using 3 units of labour and 2 units of capital. The total cost comes to Rs 62 If the commodity had been produced by using 4 units of labour and one unit capital, the cost would have come to Rs 56. What is the cost per unit of labour and capital? Use matrix method.

3. (a) $y = \dfrac{1}{\sqrt{4-x}}$ find $\dfrac{dy}{dx}$.

(b) If $\vec{a} = (3, 2, -1)$

$\vec{b} = (2, 0, -3)$

$\vec{c} = (1, -2, 3)$

What is the magnitude of the vector $2\vec{a} + \vec{b} + 3\vec{c}$?

(c) The total revenue R and the total cost C functions of a firm are given by $R = 30X - X^2$ and $C = 20 + 6X$.

(i) Find the marginal cost and marginal revenue.

(ii) Find the output at which $MR = MC$.

4. (a) In how many ways can the letters of the word STRANGE be arranged so that,

(i) The vowels are never separated.

(ii) The vowels never come together.

(b) Show that the matrix $P = \begin{bmatrix} 2 & 3 \\ 3 & 5 \end{bmatrix}$ satisfies the relation $p^2 + I = 7p$ where I is the unit matrix of the order 2 and hence find the inverse of P.

(c) Find $\dfrac{dy}{dx}$ if $y = (x^2 - 3)^4$.

5. (a) The marginal cost function for producing X units is $Y = 23 + 16X - 3X^2$ and the total cost for producing 1 unit is Rs 40. Find the total cost function and the average cost function.

(b) $\int (2x - 1)^{-1}\, dx$.

(c) How many different words can be made out of the letters of the word ALLAHABAD? In how many of these will the vowels occupy the even places?

PART – II

(Statistics)

SECTION – A

Answer any five of the following questions. Each question carries 3 marks. **(5 × 3 = 15)**

6. (a) What do you mean by frequency distribution?

(b) Define median.

(c) Mention any three merits of standard deviation.

(d) What do you mean by Analysis of Time series?

(e) Mention any three managerial applications of regression.

(f) What are cumulative frequency distributions?

(g) If $bxy = \frac{-8}{15}$ and $byx = \frac{-5}{6}$ find the value of r.

(h) Mention any three merits of the least squares method.

SECTION – B

Answer any two of the following. Each question carries 15 marks. **(2 × 15 = 30)**

7. (a) Prepare a blank table to show the exports of three companies *A*, *B* and *C* to the five countries the U.K. and U.S.A, Turkey, France and Germany in each of the years 1995 and 1996.

(b) Draw a suitable diagram to represent the following information.

	Factory X	*Factory Y*
Selling price per unit	Rs 400	Rs 600
Quantity sold	20 units	30 units
Total cost		
Wages (Rs)	3200	6000
Materials (Rs)	2400	6000
Miscellaneous Expenses (Rs)	1600	9000

8. (a) *X* Ltd is actively considering the following two mutually exclusive projects for adoption.

Year	*Project X Cash Project (Rs in lakhs)*	*Project Y Cash Profit (Rs in lakhs)*
1	10	5
2	5	25
3	20	45
4	40	30
5	60	30

Which is the most risky project (Use co-efficient of variation)

(b) Calculate Karl Pearson's Coefficient of correlation for the following data:

Price (in Rs)	*Demand (in 000 units)*
21	18
22	19
23	19
24	16
25	17
26	16
27	16

28	15
29	13
30	11

9. (a) Given below is the distribution of profits (in 000 rupees) earned by 94 book depots in Bangalore city.

Profit		*No. of Book depots*
Below	20	5
Below	30	14
Below	40	27
Below	50	48
Below	60	68
Below	70	83
Below	80	91
Below	90	94

Calculate median and mode.

(b) Fit a straight line trend by the method of least squares to the following data.

Year	*Earnings (Rs lakhs)*
1989	38
1990	40
1991	65
1992	72
1993	69
1994	60
1995	87
1996	95

Forecast the earnings for the year 1997.

10. (a) The information given below related to the advertisement and sales of a company.

	Advertisement expenditure (Rs lakhs)	*Sales (Rs lakhs)*
Arithmetic mean	20	100
Standard Deviation	3	12

Correlation coefficient between X and $Y = 0.8$.

(i) Find two regression equations

(ii) What should be the advertisement expenditure, if the company wants to attain a sales target of Rs 120 lakhs?

(b) Use Lagrange's formula and estimate from the following data the number of workers getting income not exceeding Rs 26 per month.

Income not exceeding (Rs)	*No. of workers*
15	36
25	40
30	45
35	48

First B. B. M Examination, April 1998
(New Scheme)

BUSINESS MATHEMATICS AND STATISTICS
Business Management (Paper – IV)

Time: 3 Hours **Max. Marks: 90**

Instructions:

1. Neatness in work carries weightage.
2. Necessary calculations should be shown in the body of answers.

Part – I
(Business Mathematics)
SECTION – A

Answer any five of the following questions. Each question carries 3 marks. **(5 × 3 = 15)**

1. (a) If $\vec{a} = (3, 1, 2)$ and $\vec{b} = (3, 4, 1)$ find $\vec{a} + \vec{b}$ and $\vec{a} - \vec{b}$.

(b) If $A = [1, 0, 9]$, $B = \begin{bmatrix} 3 \\ 4 \\ 5 \end{bmatrix}$ find AB and BA.

(c) Evaluate $\lim_{x \to 0} 7x^2 - 5x + 1$.

(d) Find $\frac{dy}{dx}$ if $y = \frac{3x - 1}{2x + 3}$.

(e) Find the value of (i) ${}^4P_4$ (ii) ${}^{100}C_{100}$.

(f) Evaluate $\int \left(2x^3 - \frac{4}{x} + \sqrt{x} \right) dx$.

(g) Find the value $|x| = \begin{vmatrix} 5 & 1 & 2 \\ 3 & 2 & 2 \\ 6 & 5 & 3 \end{vmatrix}$.

(h) If $f(x) = 3x^2 - 2x + 5$ find $f(-3)$, $f(2)$.

SECTION – B

Answer any two of the following. Each question carries 15 marks.

2. (a) If $A = \begin{bmatrix} 1 & 2 \\ 4 & 1 \end{bmatrix}$ and $B = \begin{bmatrix} 5 & 2 \\ -1 & 3 \end{bmatrix}$ find $A^2 - B^2$.

(b) Solve by using matrix method

$2x - y = 5$

$x - 4y = -1.$

(c) The total profit Y in rupees of furniture Co; from the manufacture and sale of 'X' chairs is given by $Y = \frac{-x^2}{1600} + 8x - 148$. How many chairs must the Co; sell to achieve the maximum profit? **(5 + 5 + 5)**

3. (a) If $y = \frac{2x-1}{3x+7}$ find $\frac{dy}{dx}$.

(b) If $\vec{a} = (3, 5, -4), \vec{b} = (1, 3, -2), \vec{c} = (1, 5, 3)$ find the magnitude of the vector $2\vec{a} + \vec{b} - 3\vec{c}$.

(c) In how many ways can the letters of the word COMMERCE be arranged? How many of these begin with 'M'? **(5 + 5 + 5)**

4. (a) Show that $\lim_{y \to 4} \frac{y^2 - 7y + 12}{y^3 - 64} = \frac{1}{48}$.

(b) Evaluate $\int_0^2 \left(4x^3 - 5x^2 + 6x + 9\right) dx$.

(c) Find the inverse of $A = \begin{bmatrix} 1 & 3 & 2 \\ 0 & 2 & 1 \\ 0 & 5 & 3 \end{bmatrix}$. **(5 + 5 + 5)**

5. (a) Show that $\begin{bmatrix} 3 & 4 & 7 \\ 2 & 1 & 3 \\ -5 & -1 & 2 \end{bmatrix} = -40$.

(b) If $A = \begin{bmatrix} 1 & 2 & 0 & 4 \\ 2 & 4 & -1 & 3 \end{bmatrix}$ and $B = \begin{bmatrix} 2 & 1 & 0 & 3 \\ 1 & -1 & 2 & 3 \end{bmatrix}$

Find Matrix 'X' such that $A - X = 3B$.

(c) If $f(x) = x^4 + \frac{1}{x^4}$ show that $f(-2) = f(2)$. **(5 + 5 + 5)**

PART – II

(Statistics)

SECTION – A

Answer any five the following questions. Each question carries 3 marks. **(5 × 3 = 15)**

6. (a) What are the limitations of statistics?

(b) Mention the important parts of the table.

(c) Distinguish between primary and secondary data.

(d) Why measures of dispersion are called secondary averages?

(e) What is linear correlation?

(f) If $\gamma = 0.78$ *S.D.* of *X* series = 15.96 *S.D.* of *Y* series = 2.56 find *bxy*.

(g) What is frequency polygon?

(h) what is interpolation?

SECTION – B

Answer any two of the following. Each question carries 15 marks.

7. (a) Draw a Histogram and locate mode from the following:

Profits (Rs)	*No. of shops*
0–100	13
100–200	18
200–300	27
300–400	20
400–500	17
500–600	6

(b) From the following data, find the missing item.

X	*Frequency*	
110	25	Mean wage = 115.86
112	17	
113	13	
117	15	
?	14	
125	8	
128	6	
130	2	

8. (a) In a sample study about the smoking habits in two towns following data were derived.

	Town A	*Town B*
Males in total population	52%	54%
Smokers	26%	28%
Male smokers	18%	20%

Tabulate the above data

(b) Below are given the wages earned by workers per week in a certain factory. Calculate number of workers earning Rs 35.

Monthly Income (Rs)	*No. of workers*
Up to Rs 10	50
20	150
30	300
40	500

50	700
60	800

9. (a) Calculate correlation between age and playing habits of students.

Age (years)	*No. of students*	*Regular player's*
15	250	200
16	200	150
17	150	90
18	120	48
19	100	30
20	80	12

(b) From the following data find the marks in English. If the marks in Hindi is 50.

	Hindi	*English*
S.D.	40	50
	10	16

$\gamma = +0.5$

10. (a) Fit an equation of the type $y = a + bx$ to the following data:

Year	*Production (000's tons)*
1991	70
1992	72
1993	88
1994	80
1995	90
1996	92

(b) Two workers on the same job show the following results over a long period of time.

	Worker A	*Worker B*
Mean	30minutes	25 minutes
S.D.	6 minutes	4 minutes

State which worker appears to be more consistent.

First B. B. M. Examination, Oct./Nov. 1998
(New Scheme)

BUSINESS MATHEMATICS AND STATISTICS (Paper – IV)
(Business Management)

Time: 3 Hours **Max. Marks: 90**

Instructions:

1. Neatness in work carries weightage.
2. Necessary calculations should be shown in the body of answers.

Part – I
(Business Mathematics)
SECTION – A

Answer any five of the following questions. Each carries 3 marks. **(5 × 3 = 15)**

1. (a) If $\vec{a} = (2, 0, 1)$ and $\vec{b} = (5, 1, 1)$ find $2\vec{a} + 3\vec{b}$ and $\vec{a} - 4\vec{b}$

(b) If $A = \begin{bmatrix} 1 & 3 \\ 4 & 5 \end{bmatrix}$ and $B = \begin{bmatrix} 4 & 5 \\ 7 & 6 \end{bmatrix}$ find $3A - 2B + I$ where I is a unit matrix.

(c) Evaluate $\lim_{x \to 1} \dfrac{4x^4 + 3x^2 - 1}{x^3 + 7}$.

(d) Differentiate $w.r.t\ x$ if $Y = ax^3 + bx + c$.

(e) Find the value of (i) ${}^5P_3$ (ii) ${}^{10}C_7$.

(f) Evaluate $\int (x^4 - 5x^3 + 16)\, dx$.

(g) Find the value of $|A| = \begin{vmatrix} 5 & -1 & 2 \\ -2 & -3 & -2 \\ 3 & 2 & 1 \end{vmatrix}$

(h) If $f(x) = \dfrac{x}{1 + x^2}$ find $f(3), f(-3)$.

SECTION – B

Answer any two of the following. Each question carries 15 marks.

2. (a) If $A = \begin{bmatrix} -1 & -2 \\ -3 & -4 \end{bmatrix}$ and $B = \begin{bmatrix} 3 & 4 \\ 5 & 2 \end{bmatrix}$ show that $AB \neq BA$.

(b) Solve by using Matrix Method.

$2x - 3y = 1$

$x + 5y = 7$

(c) Find the Average cost and Marginal cost if the total cost function of a commodity is given by $Y = 8X^2 + 9X + 3$ where Y is the total cost and X is the quantity produced.

(5 + 5 + 5)

3. (a) If $Y = (1 + 8x^2)(x^2 + 2)$ find $\frac{dy}{dx}$.

(b) If $\vec{a} = (2, 1, 3)$, $\vec{b} = (2, 4, 1)$, $\vec{c} = (0, 2, 1)$. Find (i) $\vec{a} - \vec{b}$ (ii) $3(\vec{c} - \vec{a})$.

(c) In how many ways can the letters of the word 'INTERMEDIATE' be arranged so that, the vowels always occupy even places? **(5 + 5 + 5)**

4. (a) Evaluate $\lim_{x \to 3} \frac{x^3 - 27}{x^2 - 9}$.

(b) Evaluate $\int_1^5 (3x^3 - 4x^2 + 5x + 3)\, dx$.

(c) Find the Adjoint of $A = \begin{bmatrix} -5 & 7 \\ -2 & 3 \end{bmatrix}$ and show that $A\,(\text{Adj } A) = |A| I$. **(5 + 5 + 5)**

5. (a) Evaluate $\begin{vmatrix} 7 & 6 & 1 \\ 5 & 3 & 8 \\ 3 & 2 & 4 \end{vmatrix}$.

(b) If $A = \begin{bmatrix} 0 & 2 & 3 \\ 2 & 1 & 9 \end{bmatrix}$ and $B = \begin{bmatrix} 7 & 6 & 3 \\ 1 & 4 & 5 \end{bmatrix}$ find the value of $2A - 9B$.

(c) If $f(x) = x^3 - 3x^2 - 4x - 1$ then find $f(x)$. **(5 + 5 + 5)**

PART – II
(Statistics)
SECTION – A

Answer any five of the following questions. Each question carries 3 marks. **(5 × 3 = 15)**

6. (a) Mention any three important functions of Statistics.
(b) Distinguish between classification and Tabulation.
(c) What are biased and unbiased errors?
(d) Mention any three properties of Arithmetic Mean.
(e) Give the Regression Equation of Y on X.
(f) Mention the types of correlation.
(g) What is an ogive?
(h) Define Mode.

SECTION – B

Answer any two of the following. Each question carries 15 marks.

7. (a) Draw a Pie Diagram to represent the following:

Males	Females	Girls	Boys	Total
2000	1800	4200	2000	10,000

(b) Mean weight of 150 students in a class is 60 Kg. The mean weight of Boys in the class is 70 Kg's and that of Girls is 55 Kg's. Find the number of Girls in the class.

8. (a). Represent the data by a simple bar diagram.

Countries	*Production of Rice (000's tons)*
A	38
B	42
C	29
D	28
E	18
F	11

(b) Find the value of Y when $X = 17$ by Newton's Advancing Difference Method.

X	Y
10	50
12	60
14	75
16	95
18	120
20	160

9. (a) Calculate Rank Correlation from the following:

Marks in Statistics	*Marks in Accountancy*
15	40
20	30
28	50
12	30
40	20
60	10
20	30
80	60

(b) Find the likely sales when Advertisement Expenditure is Rs 25 crores.

	Advt. Expenditure (crores)	*Sales (crores)*
Mean	20	120
S.D.	5	25

Correlation = +0.8

10. (a) Fit a straight line Trend by the method of least squares.

Year	*Production (000's tons)*
1990	70
1991	72

1992	88
1993	80
1994	90
1995	92

(b) State in which firm A and B has uniformity in Individual wages from the following:

	Firm A	*Firm B*
No. of Employees	50	100
Average wage per Employee	Rs 120	Rs 85
S.D.	3	4

First B. B. M. Examination, April/May 1999
(New Scheme)

BUSINESS MANAGEMENT (Paper – IV)
(Business Mathematics and Statistics)

Time: 3 Hours **Max. Marks: 90**

Instructions:

1. Neatness in work carries weightage.
2. Necessary calculations should be shown in the body of answers.

Part – I
(Business Mathematics)
SECTION – A

Answer any five of the following questions. Each carries 3 marks. **(5 × 3 = 15)**

1. (a) Given the vectors $u = (2, 3, 1)$, $v = (2, -1, 0)$ and $w = (0, -2, 2)$. Compute $(u - v + 2w)$ and $\left(\frac{5}{2}u - \frac{7}{2}v + \frac{9}{2}w\right)$.

   (b) If $A = \begin{bmatrix} 4 & -2 \\ 3 & -1 \end{bmatrix}$ and $B = \begin{bmatrix} 2 & 4 \\ 3 & 6 \end{bmatrix}$ find $AB - B$.

   (c) Evaluate $\lim_{x \to 3} \frac{x^2 - 9}{x - 3}$.

   (d) Differentiate *w.r.t.* x if $y = ax^2 - bx + c$.

   (e) Find the value of (i) ${}^8P_5$ (ii) ${}^9C_6$.

   (f) Evaluate $\int (x^2 - 1)^2 \, dx$.

   (g) Find 'y' if $\begin{vmatrix} -3 & -6 & 1 \\ 5 & y & -2 \\ 2 & -3 & 5 \end{vmatrix} = -7$.

   (h) If $f(x) = \sqrt{x^2 - 4}$ find $f(-4)$ and $f\left(\frac{9}{2}\right)$.

SECTION – B

Answer any two of the following. Each question carries 15 marks. **(2 × 15 = 30)**

2. (a) Find Matrix C if $A = \begin{bmatrix} 3 & 7 \\ 2 & 5 \end{bmatrix}$, $B = \begin{bmatrix} -3 & 2 \\ 4 & -1 \end{bmatrix}$ and $5C + 2B = A$.

(b) Solve by using matrix method

$$\frac{x}{7} + \frac{y}{5} = 1\frac{3}{7}$$

$$x + \frac{y}{4} = 4\frac{1}{4}.$$

(c) Given demand function $Q = \dfrac{20}{(P+1)}$ find the elasticity at point $P = 3$.

(5 + 5 + 5)

3. (a) If ${}^nP_4 : {}^nP_2 = 2 : 1$ find value of 'n'.

(b) In how many ways can the letters of the word STRANGE be arranged so that the vowels may appear in the odd places?

(c) A committee of 6 is to be formed from 7 Indians and 4 Pakistanis. In how many ways can this be done when the committee must contain at least 2 Pakistanis?

(5 + 5 + 5)

4. (a) If $3A + B = \begin{bmatrix} 3 & 2 \\ 8 & 3 \\ 10 & 4 \end{bmatrix}$ and $A - B = \begin{bmatrix} 1 & -2 \\ 0 & -3 \\ 2 & -4 \end{bmatrix}$ find A and B.

(b) Evaluate $\lim\limits_{x \to \infty} \dfrac{5 - 2x^2}{3x + 5x^2}$.

(c) Find the value of $\begin{vmatrix} 1 & 1 & 1 \\ 1 & (1+a) & 1 \\ 1 & 1 & (1+b) \end{vmatrix}$. **(5 + 5 + 5)**

5. (a) If $C = \begin{bmatrix} 1 & -3 \\ -6 & 20 \end{bmatrix}$ find C' and C^{-1}.

(b) Differentiate with respect to x

$$\frac{2x^2 + 3}{\sqrt{x}}.$$

(c) If $f(x) = 3x^2 + 4x - 9$ find $\dfrac{f(3) - f(-1)}{f(4)}$. **(5 + 5 + 5)**

PART – II

(Statistics)

SECTION – A

Answer any five of the following questions. Each question carries 3 marks. **(5 × 3 = 15)**

6. (a) Distinguish between primary data and secondary data.

(b) State any three objectives of classification.
(c) What are statistical errors?
(d) Explain briefly the measure of central tendency.
(e) Name the various components of time series.
(f) Distinguish between interpolation and extrapolation.
(g) What is a frequency polygon?
(h) What is a false-base-line?

SECTION – B

Answer any two of the following. Each question carries 15 marks. **(2 × 15 = 30)**

7. (a) Draw a rectangular diagram to represent the following information:

Particulars	*Factory A*	*Factory B*
Price per unit	Rs 15	Rs 12
Units sold	1,000	1,200
Cost of raw materials	Rs 5 per unit	Rs 5 per unit
Other costs	Rs 4 per unit	Rs 3 per unit

(b) The arithmetic mean of 50 items was 100. At the time of calculation two items 180 and 90 were wrongly taken as 100 and 10. What is the correct mean?

8. (a) A rupee spent on 'Khadi' is distributed as follows:

Farmer	–	20 Paise
Spinner	–	30 Paise
Weaver	–	25 Paise
Dyer	–	10 Paise
Agent	–	15 Paise
	Total	100 Paise

Present the data in the form of a pie diagram.

(b) Estimate the probable number of persons earning between 450 and 480 from the following table:

Income (Rs)	350–400	400–450	450–500	500–550	550–600
No. of persons	38	82	54	16	10

9. (a) Calculate Karl Pearson's coefficient of correlation between percentages of pass and failure from the following data:

No. of Students	800	600	900	700	500	400
No. passed	480	300	450	560	450	300

(b) From the following details estimate the value of A when $B = 55$:

Mean of $A = 39.5$

Mean of $B = 47.5$

Std. Deviation of $A = 10.8$

Std. Deviation of $B = 16.8$

Coefficient of correlation between A and $B = 0.42$.

10. (a) Fit a straight line trend to the data given below by the method of least squares:

Year	1989	90	91	92	93	94
Output (Rs crore)	672	824	968	1205	1464	1758

(b)

Particulars	*Factory A*	*Factory B*
No. of employees	50	100
Average wages per head per month	Rs 1,200	Rs 850
Variance of wages per head per month	81	64

In which factory is there greater variation in the distribution of wages per employee? Which factory pays more wages?

First Year B. B. M. Examination, Nov./Dec. 1999
(New Scheme)

BUSINESS MANAGEMENT
Business Mathematics and Statistics (Paper – IV)

Time: 3 Hours **Max. Marks: 90**

Instructions:
1. Neatness in work carries weightage.
2. Necessary calculations should be shown in the body of answers.

PART – I
(Business Mathematics)
SECTION – A

Answer any five of the following questions. Each carries 3 marks. **(5 × 3 = 15)**

1. (a) Given the vectors $p = (-1, 4, 3)$, $q = (1, -2, 0)$ and $r = (0, 4, -1)$ compute $(2p - q + r)$ and $\left(\frac{5}{2}p - \frac{3}{2}q + \frac{7}{2}r\right)$.

(b) If $A = \begin{bmatrix} 3 & -4 \\ 6 & -7 \end{bmatrix}$ and $B = \begin{bmatrix} 5 & 1 \\ -3 & -1 \end{bmatrix}$ find $A'B$.

(c) Evaluate $\lim_{x \to 3} \frac{x^2 + x - 12}{x - 13}$.

(d) Differentiate *w.r.t.* x if $y = 2x^4 - 3x^2 + 5$.

(e) Find the value of (i) ${}^9P_6$, (ii) ${}^8C_5$.

(f) Evaluate $\int (3 - 2x - x^4)\, dx$.

(g) Find 'x' if $\begin{vmatrix} 3 & -1 & 5 \\ x & 2 & 7 \\ -1 & 7 & 8 \end{vmatrix} = 4$.

(h) If $f(x) = x^2 - 3x + 2$. Find $f(0)$, $f(-2)$ and $f(3)$.

SECTION – B

Answer any two of the following. Each question carries 15 marks.

2. (a) Find the matrix B if $A = \begin{bmatrix} 4 & 1 \\ 2 & 3 \end{bmatrix}$ and $A + 2B = A^2$.

(b) Solve by using Matrix method $\frac{x}{5}+\frac{y}{6}+1=\frac{x}{6}+\frac{y}{5}=28$.

(c) If the demand law is $p=\frac{10}{(x+1)^2}$, find the elasticity of demand in terms of x.

(5 + 5 + 5)

3. (a) If $^nP_5 = 30\ ^nP_4$ find 'n'.

(b) In how many different ways can 10 examination papers be arranged in a row, so that the best and the worst papers may never come together?

(c) A man has 6 friends. In how many ways may he invite one or more of them to a dinner? **(5 + 5 + 5)**

4. (a) If $A + B = \begin{bmatrix} 4 & 6 \\ 3 & -5 \end{bmatrix}$ and $A - B = \begin{bmatrix} 2 & 4 \\ 1 & -1 \end{bmatrix}$ find A and B.

(b) Evaluate $\lim_{x \to \infty} \frac{6-5x^2}{4x+15x^2}$.

(c) Find the value of $\begin{vmatrix} 91 & 94 & 97 \\ 92 & 95 & 98 \\ 93 & 96 & 99 \end{vmatrix}$. **(5 + 5 + 5)**

5. (a) If $A = \begin{bmatrix} 5 & -2 \\ 7 & -3 \end{bmatrix}$ find A' and A^{-1}.

(b) If $y = (1-5x)^6$ find $\frac{dy}{dx}$.

(c) If $f(x) = 2x^2 - 9x + 7$ find $\frac{f(2) - f(-4)}{f(3)}$ **(5 + 5 + 5)**

PART – II
(Statistics)
SECTION – A

Answer any five of the following questions. Each question carries 3 marks. **(5 × 3 = 15)**

6. (a) What is primary data?

(b) Give the meaning of class frequency and class magnitude.

(c) State any four purposes of tabulation.

(d) Bring out any three characteristics of an ideal average.

(e) What are cyclical fluctuations?

(f) Explain the term extrapolation.

(g) What is probable error?

(h) Explain the term regression.

SECTION – B

Answer any two of the following. Each question carries 15 marks.

7. (a) From the family budgets of two middle class families given below, draw a suitable diagram.

Family	*Total Income*	*Expenditure items*					
		Food	*Clothing*	*Rent*	*Fuel*	*Others*	*Total*
A	Rs 4,000	155	800	400	600	1000	4300
B	Rs 6,000	2000	1000	600	800	1200	5600

(b) The Arithmetic mean of 98 items is 50. Two items 60 and 70 were left out at the time of calculation. What is the correct mean of all the items?

8. (a) Draw a Pie chart to represent the following data of the proposed expenditure by the Karnataka Governor for 1998–99:

Items	*Expenditure (in Rs lakhs)*
Agriculture	600
Industry	400
Irrigation	450
Education	300
Miscellaneous	250

(b) Given below are the marks obtained by 200 students of University in Statistics paper:

Marks	40–50	50–60	60–70	70–80	80–90
No. of Students	32	42	55	40	31

Interpolate the number of students who have obtained less than 65 marks.

9. (a) Given: No. of pairs = 8

	X	*Y*
A. Mean	74.50	125.50
Std. Deviation	13.07	15.85

Summation of products of corresponding deviations from respective means = 2176.

Calculate the coefficient of correlation between *x* and *y*.

(b) For two variables *X* and *Y* the regression equation of *X* on *Y* is $X = 5Y - 7$ and the regression equation of *Y* on *X* is $Y = 0.1X + 1.7$.

Find the means of *X* and *Y* and the coefficient of correlation between *X* and *Y*.

10. (a) The number of units of a product exported during 1990–97 is given below. Fit a straight line trend and estimate the export for the year 1998.

Year	1990	91	92	93	94	95	96	97
No. of units exported (in '000s)	12	13	13	16	19	23	21	23

(b) An analysis of the monthly wages paid to the workers in firms *A* and *B* belonging to the same industry gives the following results:

	Firm A	*Firm B*
No. of workers	500	600
Average monthly wages	Rs 186	Rs 175
Variance of distribution of wages	Rs 81	Rs 100

Which firm has larger wage bill? In which firm is there greater variability in individual wages?

First Year B. B. M. Examination, April/May 2000
(New Scheme)

BUSINESS MANAGEMENT

Business Mathematics and Statistics

Time: 3 Hours **Max. Marks: 90**

Instructions:

1. Neatness in work carries weightage.
2. Necessary calculations should be shown in the body of answers.

PART – I
(Business Mathematics)
SECTION – A

Answer any five of the following questions. Each carries 3 marks. **(5 × 3 = 15)**

1. (a) What do you mean by a diagonal matrix? Give an example.

(b) If $f(x) = \dfrac{x^2 - 4}{x + 2}$ find $f(-3)$ and $f(-2)$.

(c) If $A = \begin{bmatrix} 5 & 3 \\ 7 & 2 \end{bmatrix}$ and $B = \begin{bmatrix} 6 & -2 \\ -8 & 1 \end{bmatrix}$ find $B'\,A'$.

(d) If $B = \begin{bmatrix} 2 & 4 \\ 1 & 6 \end{bmatrix}$ find $\frac{1}{2}B^2$.

(e) Evaluate $\lim\limits_{x \to a} \dfrac{\sqrt{x} - \sqrt{a}}{x - a}$.

(f) Differentiate *w.r.t.* x: If $y = ax^2 + bx + c$.

(g) Evaluate $\int \left(8x^3 - 3x^2 + 2x + 1\right) dx$.

(h) Find the number of permutations of the letters of the word STATISTICS.

SECTION – B

Answer any two of the following. Each question carries 15 marks. **(15 × 2 = 30)**

2. (a) If $A = \begin{bmatrix} 2 & 4 & 7 \\ 9 & 0 & -3 \\ -6 & -8 & -5 \end{bmatrix}$ and $B = \begin{bmatrix} 1 & 2 & 5 \\ 3 & 7 & 9 \\ 4 & -2 & -7 \end{bmatrix}$

show that $(AB)' = B'\,A'$.

(b) Evaluate $\lim_{x \to \infty} \frac{(x+5)(2x+1)}{(3x+1)(x-2)}$

(c) If $y = \frac{x^2 - 3x + 5}{x^2 + 3x + 5}$ find $\frac{dy}{dx}$. **(5 + 5 + 5)**

3. (a) If $\begin{vmatrix} 3 & 4 & x \\ 2 & 1 & 3 \\ -5 & -1 & 2 \end{vmatrix} = -40$ Find x.

(b) Solve by matrix method

$x + y + z = 6$

$x + 2y + 3z = 15$

$-x + y - z = -2$

(c) Evaluate $\int \left(\frac{1}{\sqrt{4 - 7x}} \right) dx$. **(5 + 5 + 5)**

4. (a) If $\vec{a} = (4, 3, 1)$, $\vec{b} = (-2, 6, 1)$, $\vec{c} = (3, -4, 1)$ find

(i) $2\vec{a} + \vec{b} + 3\vec{c}$

(ii) $\vec{a} + 2\vec{b} - \vec{c}$.

(b) A student is allowed to choose 5 questions out of 7 in Geometry and 3 questions out of 5 in Trigonometry in an examination. In how many ways can he choose them?

(c) Evaluate $\int_1^5 \left(3x^3 - 4x^2 + 5x + 3\right) dx$. **(5 + 5 + 5)**

5. (a) Evaluate $\lim_{x \to -2} \frac{x^2 + 3x + 2}{x^3 + 8}$.

(b) Differentiate $w.r.t.\ x$

$y = \frac{3x + 2}{e^x + 4}$.

(c) If $A = \begin{bmatrix} 1 & 2 & -1 & 4 \\ 2 & 4 & 3 & -1 \end{bmatrix}$, $B = \begin{bmatrix} 2 & 1 & 3 & 1 \\ 1 & -1 & 2 & 1 \end{bmatrix}$

find a matrix Y such that $A + 2Y = 4B$. **(5 + 5 + 5)**

PART – II
(Statistics)
SECTION – A

Answer any five of the following questions. Each question carries 3 marks. **(5 × 3 = 15)**

6. (a) Give three methods of collecting primary data.
   (b) Distinguish between classification and tabulation.
   (c) What is a Histogram?
   (d) Define Median. State the merits of median.
   (e) State demerits of Mean deviation.
   (f) What is meant by positive and negative correlation?
   (g) What are the Regression lines?
   (h) What are the uses of Analysis of Time Series?

SECTION – B

Answer any two of the following. Each question carries 15 marks. **(15 × 2 = 30)**

7. (a) Calculate mean deviation from median from the following data:

Marks less than	*No. of students*
80	100
70	90
60	80
50	60
40	32
30	20
20	13
10	5

   (b) The following table shows the results of B. B. M. students of Bangalore University for the last three years. Represent the data by sub-divided Bar diagram:

Year	*I Class*	*II Class*	*III Class*	*Failed*	*Total*
1997	600	1600	2600	1600	6400
1998	700	2100	3100	1500	7400
1999	600	2600	3600	1600	8400

8. (a) Find out the number of students securing more than 50 marks but less than 60 marks in the University Examination from the following:

Marks	0–20	20–40	40–60	60–80	80–100
No. of Students	5	26	85	54	30

   (b) Below are given the annual production (in thousand tonnes) of a fertiliser factory:

Year	*Production*
1992	70
1993	75
1994	90
1995	91

1996	95
1997	98
1998	100

(i) Fit a straight line trend by the method of least squares and tabulate the trend values.

(ii) Estimate the likely production for the year 2000.

9. (a) Calculate the two regression equations from the following data and estimate X when $Y = 26$.

X	10	12	13	17	18	20	24	30
Y	5	6	7	9	13	15	20	21

Calculate co-efficient of correlation also.

(b) Calculate the co-efficient of rank correlation for the following data of marks obtained by 9 students in Accountancy and Statistics at an Examination.

Marks in Accountancy	*Marks in Statistics*
80	30
55	8
60	17
36	13
60	25
25	12
90	13
15	2
60	5

10. (a) Draw an histogram and Frequency Polygon for the following data:

Age in years	*No. of persons*
10–20	3
20–30	16
30–40	22
40–50	35
50–60	24
60–70	15
70–80	2

(b) In the house of Lok Sabha there were 600 members present. During the discussion on a motion put to vote, 400 voted in favour of the resolution.

The Government members in the house were 380. The members belonging to the opposition voted for the resolution were 65. All the members are belonging to either of the two groups and there were no absentees.

Tabulate the above information.

First Year B. B. M. Examination, Oct./Nov. 2000
(New Scheme)

BUSINESS MANAGEMENT
Business Mathematics and Statistics (Paper – IV)

Time: 3 Hours **Max. Marks: 90**

Instructions:

1. Neatness in work carries weightage.
2. Necessary calculations should be shown in the body of answers.

PART – I
(Business Mathematics)
SECTION – A

Answer any five of the following questions. Each carries 3 marks. **(5 × 3 = 15)**

1. (a) What is a "Triangular Matrix"? Give an example.

(b) If $\begin{bmatrix} 2 & -3 \\ 4 & 1 \end{bmatrix} + 2A = \begin{bmatrix} 6 & -3 \\ 2 & 5 \end{bmatrix}$ find A.

(c) Evaluate $\lim_{x \to \infty} \dfrac{3x + 2}{4x - 7}$.

(d) Differentiate *w.r.t. x*: if $y = x^2 . e^x$

(e) Evaluate $\int (2x - 3)^5 \, dx$.

(f) If $\vec{a} = (3, 2, 1), \vec{b} = (1, 2, 3)$

find (i) $\vec{a} + 3\vec{b}$, (ii) $2\vec{a} - \vec{b}$.

(g) Find the value of ${}^{10}P_2$, ${}^{14}C_{10}$ and ${}^{9}C_2$.

(h) If the total cost $C = 200 + 6x$ and total revenue $R = 10 + x - x^2$. Find marginal cost and marginal revenue.

SECTION – B

Answer any two of the following. Each question carries 15 marks. **(2 × 15 = 30)**

2. (a) If $3A + 2B = \begin{bmatrix} 21 & 16 & 1 \\ 21 & 2 & 12 \end{bmatrix}$ and $2A - 3B = \begin{bmatrix} -12 & -11 & 5 \\ 1 & -16 & 8 \end{bmatrix}$

Find A and B.

(b) Evaluate $\lim_{x \to 1} \dfrac{x^2 - 1}{x^2 - 3x + 2}$

(c) Find the number of permutation of the letters of the word ASSESSMENT. In how may of these do the S's come together? **(5 + 5 + 5)**

3. (a) Compute the inverse of the matrix

$$A = \begin{bmatrix} 3 & 5 & 7 \\ 2 & -3 & 1 \\ 1 & 1 & 2 \end{bmatrix}.$$

(b) If the total revenue and the total cost of a firm are given by $R(x) = 30x - x^2$ and $C(x) = 20 + 6x$ find the output at which the profit is maximum.

(c) If $\vec{a} = (3, 2, -1)$, $\vec{b} = (5, -1, 2)$, $\vec{c} = (-2, 3, 1)$ find **(5 + 5 + 5 = 15)**

(i) $\vec{a} + \vec{b} + \vec{c}$

(ii) $3\vec{a} + 2\vec{b} + \vec{c}$

(iii) $\vec{a} + 3\vec{b} - 2\vec{c}$.

4. (a) Solve by using matrix method.

$4x - 3y = 5$

$x + 2y = 4$

(b) Evaluate $\lim_{x \to 2} \left[\dfrac{1}{x-2} - \dfrac{2}{x^2 - 2x}\right]$.

(c) If $y = x^3(2x^2 - 3x - 3)$ find $\dfrac{dy}{dx}$. **(5 + 5 + 5)**

5. (a) Solve for x: If $\begin{vmatrix} 6 & -2 & -4 \\ x & 2 & -1 \\ -5 & 3 & x \end{vmatrix} = 0$.

(b) Evaluate $\int_1^2 \left(3x^3 + 4x^2 + 2x + 5\right) dx$

(c) Differentiate $w.r.t.\ x$ $\dfrac{1 - e^x}{1 + e^x}$. **(5 + 5 + 5)**

PART – II

(Statistics)

SECTION – A

Answer any five of the following questions. Each question carries 3 marks. **(5 × 3 = 15)**

6. (a) What is secondary data? Give two examples.

(b) What is Pie diagram?

(c) Define standard Deviation.

(d) What are the assumptions of Interpolation?

(e) What are the components of time series?

(f) For a distribution *C.V.* is 22.5% and the value of arithmetic mean is 7.5. Find the value of *S.D.*

(g) In a Distribution $N = 10$ and $r = 0.6$ find probable error.

(h) Distinguish between coordination and Regression.

SECTION – B

Answer any two of the following. Each question carries 15 marks. **(2 × 15 = 30)**

7. (a) In a sample study about coffee habit in two towns, the following information was received:

Town A: Females were 40% Total coffee drinkers were 46% and male non-coffee drinkers were 20%.

Town B: Males were 55% Male non-coffee drinkers were 30% and female coffee drinkers were 15%. Represent the above data in a tabular form.

(b) Draw an Ogive curve from the data given below and locate the median.

Profit [Rs. in lakhs]	*No. of shops*
10–20	6
20–30	8
30–40	12
40–50	18
50–60	25
60–70	16
70–80	8
80–90	5
90–100	2

8. (a) From the following table of yearly premium for policies maturing at different ages, Estimate the premium for policies maturing at the age of 47 years.

Age (in years)	45	50	55	60	65
Premium (in Rs.)	287	240	208	186	171

(b) Calculate the value of Co-efficient of Mean deviation [from median] of the following data.

Marks	*No. of Students*
10–20	2
20–30	6
30–40	12
40–50	18
50–60	25
60–70	20
70–80	10
80–90	7

9. (a) Find the following table:

(i) Fit a straight line trend by the method of least squares.

(ii) Calculate the Trend values.

Year	*Production [in '000 tons]*
1990	12
1991	10
1992	14
1993	11
1994	13
1995	15
1996	16

(b) You are given below the following information about advertisement expenditure and sales:

	Advertisement (x) Expenses (Rs. crores)	*Sales (y) (Rs. crores)*
Mean	20	120
S.D.	5	25

Correlation co-efficient 0.8

(i) Calculate the two regression equations.

(ii) Find the likely sales when advertisement is 25 crores.

10. (a) Find Karl Pearson's Co-efficient of correlation of the following data.

Age of Husband	20	22	23	25	25	28	29	30	30	34
Age of Wife	18	20	22	24	21	26	26	25	27	29

(b) An Employee offers a cash gift of 5% of the average monthly wages in his factory to every employee.

Monthly wages	*No. of employees*
20–30	28
30–40	32
40–50	45
50–60	60
60–70	56
70–80	40
80–90	20

Calculate cash gift per employee.

First Year B. B. M. Examination, April/May 2001
(New Scheme)

BUSINESS MANAGEMENT (Paper – IV)
Business Mathematics and Statistics

Time: 3 Hours **Max. Marks: 90**

Instructions:

1. Neatness in work carries weightage.
2. Necessary calculations should be shown.

PART – I
(Business Mathematics)
SECTION – A

Answer any five of the following questions. Each carries 3 marks. **(5 × 3 = 15)**

1. (a) What is localised vector?

(b) If $u = x^2 + y^2$, find $\frac{\partial u}{\partial x}, \frac{\partial u}{\partial y}$.

(c) Integrate $2x^2 - 7x + 15$.

(d) Find the value of ${}^6P_4, {}^{16}P_5$.

(e) If $\vec{a} = (6, 4, -3), \vec{b} = (5, -1, 0)$ find $3\vec{a} + 2\vec{b} - \vec{c}$.

(f) If $X = \begin{bmatrix} 3 & 6 \\ 1 & 4 \end{bmatrix}, Y = \begin{bmatrix} 6 \\ 5 \end{bmatrix}$ find XY.

(g) Define Matrix.

(h) If the total cost $= C = 100x + 5x^2$ find Marginal cost when $x = 4$.

SECTION – B

Answer any two of the following. Each question carries 15 marks. **(2 × 15 = 30)**

2. (a) If $A = \begin{bmatrix} 2 & 1 & 2 \\ 3 & 1 & 1 \\ 1 & 3 & 2 \end{bmatrix}$ $B = \begin{bmatrix} -1 & 1 & 2 \\ 2 & -1 & 1 \\ 1 & 2 & 1 \end{bmatrix}$

Prove that $(AB)' = B'A'$.

(b) How many diagonals are there in a polygon of 15 sides?

(c) If $\vec{a} = (4, 6, -2), \vec{b} = (-3, -2, 0), \vec{c} = (0, -3, -6)$ find

(i) $3\vec{a} - 4\vec{b} + \vec{c}$

(ii) $6\vec{a} - 2\vec{b} - 4\vec{c}$.

3. (a) Evaluate $\lim_{x \to \infty} \dfrac{3x^2 + 5x - 2}{5x^2 + 4x + 1}$.

(b) Evaluate $\int \dfrac{\log x}{x^2}$.

(c) For the demand function $2x - 5y = 7$, where "x" is the number of units and "y" is the price per unit, find the Total Revenue, Marginal Revenue and Average Revenue.

4. (a) Solve using Matrix method

$$3x - 2y + 4z = 8$$
$$2x + y - z = 1$$
$$4x - 3y + 2z = 4.$$

(b) If $y = 9x^5 - 6x^4 + \frac{1}{4}x^3 - 8x^2 + 10$ find $\dfrac{d^2y}{dx^2}$.

(c) Integrate $\dfrac{3x + 4x^2 + 5x^3}{x^6}$.

5. (a) Find the Average Cost and Marginal cost if the total cost function of a commodity is given by $y = 9x^2 + 8x + 4$, where "y" is the total and "x" is the quantity produced.

(b) If $A = \begin{bmatrix} 2 & 1 & 1 \\ -1 & 0 & 1 \\ 1 & 3 & -1 \end{bmatrix}$ calculate $A^2 - 5A + 9I$.

(c) Evaluate $\lim_{x \to 1} \dfrac{x^2 + x + 1}{x + 1}$.

PART – II
(Statistics)
SECTION – A

Answer any five of the following questions. Each question carries 3 marks. **(5 × 3 = 15)**

6. (a) What are the functions of Statistics?

(b) Distinguish between Classification and Tabulation.

(c) Find *A.M.* if 12.4, 12.6, 12.9, 12.1 and 12.3 cms.

(d) If $C.V. = 3.99\%$, $\sigma = 2.68$ find Mean.

(e) What are the components of time series?

(f) What is F.B.L?

(g) What are the desired qualities of a good measure of dispersion?

(h) What is a questionnaire?

SECTION – B

Answer any two of the following. Each question carries 15 marks. **(2 × 15 = 30)**

7. (a) From the following data find the missing frequency when mean is 15.38:

Central value:	10	12	14	16	18	20
Frequency:	3	7	?	20	8	5

(b) The monthly profits in rupees of 1000 shops are distributed as follows:

Profits per Shop	*No. of Shops*
0–100	12
100–200	18
200–300	27
300–400	20
400–500	17
500–600	06

Determine mode.

8. (a) Twenty percent of the workers in a firm employing a total of 2,000 earn less than Rs 2 per hour, 440 earn from Rs 2 to 2.24 per hour, 24% earn from Rs 2.25 to 2.49 per hour, 370 earn from Rs 2.50 to 2.74 per hour, 12% earn from Rs 2.75 to 2.99 per hour and the rest earn Rs 3 or more per hour. Set up a frequency table and calculate the median wage.

(b) You are given the following data relating to aptitude scores and productivity index.

Aptitude scores (X):	9	18	18	20	20	23
Productivity index (Y):	33	23	33	42	29	32

Find the coefficient of correlation between aptitude scores and productivity index.

9. (a) The following are some of the particulars of the distribution of weights of boys and girls in a class.

	Boys	*Girls*
Number	100	50
Mean weight	60 Kgs	45 Kgs
Variance	9	4

(i) Find the Standard Deviation.

(ii) Which of two distributions is more variable?

(b) Using interpolation estimate the business done in 1985 from the following data:

Year	*Business done (in lakh Rs)*
1982	150
1983	235
1984	365
1986	525
1987	780

10. (a) Obtain the regression of Y on X and X on Y from the following table and estimate the blood pressure when age is 50.

Age (X)	*Blood Pressure (Y)*
56	147
42	125
72	160
36	118
63	149
47	128
55	150
49	145
38	115
42	140
68	152
60	155

(b) Fit a straight line trend by the method of least squares to the following data. Assuming that the same rate of change continues, what would be the predicted earnings for the year 1997?

Year	*Earnings (Rs lakhs)*
1988	38
1989	40
1990	65
1991	72
1992	79
1993	60
1994	87
1995	95

First Year B. B. M Examination, Oct./Nov. 2001
(New Scheme)

BUSINESS MANAGEMENT (Paper – IV)
Business Mathematics and Statistics

Time: 3 Hours **Max. Marks: 90**

Instructions:

1. Neatness in work carries weightage.
2. Necessary calculations should be shown.

PART – I
(Business Mathematics)
SECTION – A

Answer any five questions. Each question carries 3 marks. **(5 × 3 = 15)**

1. (a) What is Linear Combinations?

(b) If $\vec{a} = (3, -4), \vec{b} = (-2, 0), \vec{c} = (4, 4)$.

(i) $\vec{a} + \vec{b}$

(ii) $6\vec{a}$.

(c) Differentiate Diagonal Matrix and Scalar Matrix.

(d) Find 'n' if ${}^nC_{n-2} = 28$.

(e) Differentiate Scalar and Vector.

(f) Find ${}^nC_{n-4}$.

(g) Evaluate $\lim_{x \to 1} \frac{x^2 - 3x + 2}{x^2 - 4x + 3}$.

(h) Evaluate $\int (2x + 1)^2 (x + 1)\, dx$.

SECTION – B

Answer any two of the following. Each question carries 15 marks. **(2 × 15 = 30)**

2. (a) Solve for X and Y.

$$\begin{bmatrix} X \\ Y \end{bmatrix} = \begin{bmatrix} 1 & 3 \\ 4 & 7 \end{bmatrix} \begin{bmatrix} -1 \\ 3 \end{bmatrix}.$$

(b) Find the number of Permutations of the word "ACCOUNTANTS".

(c) If $\vec{a} = (1, -1, 2), \vec{b} = (0, 2, -3), \vec{c} = (2, 0, -3), \vec{d} = (-1, 2, 3)$ find

$3[2\vec{a} + \vec{b}] + 4[\vec{c} - 2\vec{d}]$ and its magnitude.

3. (a) If $u = x^4 + y^4 + z^4 + 4x^3 y + 4y^3 z + 4z^3 x$

find $\frac{\partial u}{\partial x}, \frac{\partial u}{\partial y}$ and $\frac{\partial u}{\partial z}$

(b) Evaluate $\int_0^1 \left(x^3 - 3x^2 + 2x + 1\right) dx$.

(c) Find the value of $\begin{vmatrix} 1 & 2 & 3 \\ 2 & 4 & 6 \\ 1 & -1 & 0 \end{vmatrix}$.

4. (a) The total cost C, of making "x" units of a product is $C = 0.00003x^3 - 0.045x^2 + 8x + 25{,}000$. Find the marginal cost at 1000 units output.

(b) Evaluate $\lim_{x \to -\frac{1}{3}} \frac{9x^2 - 1}{3x + 1}$.

(c) Evaluate $\int_1^2 \frac{x^2}{x + 2} dx$.

5. (a) In how many ways 5 passengers can sit in a compartment having 16 vacant seats?

(b) Differentiate *w.r.t* 'x' $7 + \frac{4}{x} - \frac{2}{x^2}$.

(c) Integrate $\int x^n \log x \, dx$.

PART – II

(Statistics)

SECTION – A

Answer any five of the following questions. Each question carries 3 marks. **(5 × 3 = 15)**

6. (a) Define Statistics.

(b) Write a note on misuse and distrust of statistics.

(c) What are the merits and demerits of Quartile Deviation?

(d) How do you calculate mode when it is ill defined?

(e) Explain with examples (i) Secular Trend (ii) Seasonal Variation (iii) Irregular Variation.

(f) What is time series analysis?

(g) If $C - V = 17.75\%$, $\overline{X} = 58.375$ Find standard deviation.

(h) How do you Interpret 'r' when it is (a) + 1, (b) –1, (c) 0?

SECTION – B

Answer any two of the following. Each question carries 15 marks. **(2 × 15 = 30)**

7. (a) Five percent of the workers in a firm employing a total of 2000 men earn less than Rs 2.00 per hour, 480 earn from Rs 2.00 to Rs 2.24 per hour, 35 percent earn from Rs 2.25 to Rs 2.49 per hour, 370 earn from Rs 2.50 to Rs 2.74 per hour, 12 percent earn from Rs 2.75 to Rs 2.99 per hour and the rest earn Rs 3.00 or more per hour. What is median wage?

   (b) Ten competitors in a beauty contest are ranked by three judges in the following order:

First Judge	:	1	5	4	8	9	6	10	7	3	2
Second Judge	:	4	8	7	6	5	9	10	3	2	1
Third Judge	:	6	7	8	1	5	10	9	2	3	4

   Use the rank correlation coefficient to discuss which pair of judges have the nearest approach to common tastes in beauty.

8. (a) You are given below the following information about advertisement and sales:

	Adv. Exp (X) *(Rs. crores)*	*Sales (Y)* *(Rs. Crores)*
Mean	20	120
S.D.	5	25

   Correlation coefficient = 0.8

   (i) Calculate the two regression equations.

   (ii) Find the likely sales when advertisement expenditure is Rs 25 crores.

   (b) Estimate the number of students who get more than 48 but not more than 50 marks from the following data:

Marks upto	: 45	50	55	60	65
No. of Students	: 447	484	505	511	514

9. (a) Draw the Pie diagram for the following data:

Food	12.9%
Clothing	12.5%
Oil	27.2%
Ghee	15.4%
Transport	15.9%
Social Service	16.1%

   (b) Given below are the figures of production of a factory:

Year	*Production (000' tonnes)*
1992	80
1993	90
1994	92
1995	83
1996	94
1997	99
1998	92

   (i) Fit a straight line trend.

10. (a) The arithmetic mean of the following series is 30.5 marks. Find the missing figure:

Marks	*Frequency*
10	8
20	10
?	20
40	15
50	7

(b) Lives of two models of refrigerators in a recent survey are:

Life (No. of years)	*Number of Refrigerators*	
	Model A	*Model B*
0–2	5	2
2–4	16	7
4–6	13	12
6–8	7	19
8–10	5	9
10–12	4	1

What is the average life of each of these models of refrigerators? Which model has greater uniformity?

First Year B. B. M. Examination, April/May 2002
(New Scheme)

BUSINESS MANAGEMENT (Paper – IV)
Business Mathematics and Statistics

Time: 3 Hours **Max. Marks: 90**

Instructions:

1. Neatness in work carries weightage.
2. Necessary calculations should be shown.

PART – I
(Business Mathematics)
SECTION – A

Answer any five questions. Each question carries 3 marks. **(5 × 3 = 15)**

1. (a) What is symmetric matrix? Give an example.

(b) Find the Adjoint of $\begin{bmatrix} -5 & 7 \\ -2 & 3 \end{bmatrix}$.

(c) If $u = (x + y)(x - y)$ find $\frac{du}{dx}$ *and* $\frac{du}{dy}$.

(d) Evaluate $\int \left(3x^4 - \frac{5}{x} + \sqrt[3]{x} \right) dx$.

(e) Evaluate $\lim_{x \to \infty} \frac{(x + 5)(2x + 1)}{(3x + 1)(x - 2)}$.

(f) Find the value of ${}^{8}P_{3}$, ${}^{10}C_{2}$ and ${}^{12}C_{8}$.

(g) The total cost C of making x litres of a product is $C = 10 + 30\sqrt{x}$ Find the marginal cost of 225 litres of output.

(h) If $\vec{a} = (-3, -1, 2)$, $\vec{b} = (2, 3, -4)$ find

(i) $\vec{a} + \vec{b}$

(ii) $3\vec{a} - 2\vec{b}$.

SECTION – B

Answer any two of the following. Each question carries 15 marks. **(2 × 15 = 30)**

2. (a) If $A = \begin{bmatrix} -2 & 1 \\ 3 & 2 \end{bmatrix}$ $B = \begin{bmatrix} -3 & 2 \\ -1 & 5 \end{bmatrix}$ find $(A^2 B^2)^2$.

(b) $\lim_{x \to 0} \dfrac{x}{\sqrt{1+x} - \sqrt{1-x}}$

(c) Differentiate *w.r.t. x*. $(x^2+3)(x^4-5)$.

3. (a) Prove that $A = \begin{bmatrix} 1 & 2 & 2 \\ 2 & 1 & 2 \\ 2 & 2 & 1 \end{bmatrix}$ satisfies the equation $A^2 - 4A - 5I = 0$ where I is the identity matrix and 0 is the zero matrix.

(b) Evaluate $\lim_{x \to 1} \dfrac{\sqrt[4]{x} - 1}{\sqrt[6]{x} - 1}$.

(c) If $y = \dfrac{ax^2 + bx}{cx + d}$ find $\dfrac{dy}{dx}$.

4. (a) Solve for x : $\begin{vmatrix} 2 & 4 & 10 \\ 4 & 2x & 20 \\ 6 & 2 & -4 \end{vmatrix} = 0$.

(b) Evaluate $\int_0^2 \left(5x^4 + 2x^3 + 4x^2 + 5x + 10\right) dx$.

(c) If $\vec{a} = (2, -1, 1)$, $\vec{b} = (1, 2, -4)$, $\vec{c} = (2, 1, -3)$, $\vec{d} = (-2, 1, 4)$ find $3(2\vec{a} + \vec{b}) + 4(\vec{c} - 2\vec{d})$ and the magnitude.

5. (a) The price of 4 Accounting books, 2 Commerce books, 3 Economics book is Rs 134, The cost of one Accountancy book, 3 Commerce books and 2 Economics book is Rs 81. The cost of 2 Accounting books, one Commerce book and 5 Economics book is Rs 130. Find the rate per book of each using matrix method.

(b) How many different arrangements can made by using all the letters in the word. "KARNATAKA"?

(c) The total cost of x units is y rupees, where, $y = \dfrac{x^3}{300} - 0.2x^2 + 27.5x$ and all units made are sold at Rs 10 per unit. At what two points does marginal cost equal marginal revenue?

PART – II

(Statistics)

SECTION – A

Answer any five of the following. Each question carries 3 marks. **(5 × 3 = 15)**

6. (a) Mention the functions of statistics.

(b) Distinguish between classification and tabulation.

(c) What is a Pie chart?

(d) Indicate three limitations of Arithmetic Mean.

(e) Mean and variance of 100 items and found to be 40 and 169. What is its co-efficient of variation?

(f) Interpret correlation in the following cases:

(i) When $r = +1$, (ii) When $r = -1$.

(g) Calculate two regression coefficients. When $r = -0.99$, $\sigma x = 10$ and $\sigma y = 12$.

(h) Mention any three methods of measuring secular trend.

SECTION – B

Answer any two of the following. Each question carries 15 marks. **(2 × 15 = 30)**

7. (a) Form a continuous frequency table from the following data, using class intervals such as 40–50, 50–60 etc, and determine the modal value.

90,	78,	86,	51,	96,	104,	51,	78,	50,	72
76,	49,	77,	90,	84,	76,	42,	74,	69,	70
68,	69,	104,	80,	79,	54,	79,	73,	58,	91
65,	60,	77,	78,	67,	50,	84,	76,	110,	53
74,	40,	60,	42,	82,	41,	61,	75,	115,	81

(b) From the following calculate mean deviation (From median)

Marks	*No. of Students*
10–14	12
15–19	29
20–24	59
25–29	75
30–34	46
35–39	19

8. (a) An analysis of the monthly wages paid to the workers in two firms A and B belonging to the same industry gives the following results.

	Firm A	*Firm B*
Number of workers	250	300
Average monthly wages (in Rs)	90	85
Variance of distribution of wages	121	196

(i) Which of the firm *A* or *B* has a larger wage bill?

(ii) In which of the forms *A* or *B* is there greater variability in individual wages?

(b) Below are given the wages earned by workers per week in a certain factory. Interpolate the number of workers earning more than Rs 150 per week.

Weekly wages	*Number of workers*
Below 100	10
Below 120	40
Below 140	60
Below 160	100
Below 180	150
Below 200	170

9. (a) The following are the annual profits in thousands of rupees in a certain business.
   (i) Fit a straight line trend to these figures by the method of least squares.
   (ii) Plot the actual and trend values on a graph.

Year	*Profit (in Rs)*
1991	60
1992	72
1993	65
1994	75
1995	85
1996	80
1997	95

(b) From the following data
   (i) Determine the value of correlation coefficient through regression co-efficients.
   (ii) Calculate the two regression equations.

X :	6	8	5	9	12
Y :	12	10	13	16	18

10. (a) Calculate Karl Pearson's co-efficient of correlation from the following data:

X	7	9	13	16	19	21	25	34
Y	11	13	16	16	19	26	23	36

(b) Draw an ogive curve from the following data and find out the medium and quartiles.

Classes	*Frequency*
50–60	15
60–70	23
70–80	45
80–90	60
90–100	50
100–110	23
110–120	11